35

Topics in Heterocyclic Chemistry

For further volumes:
http://www.springer.com/series/7081

Aims and Scope

The series Topics in Heterocyclic Chemistry presents critical reviews on present and future trends in the research of heterocyclic compounds. Overall the scope is to cover topics dealing with all areas within heterocyclic chemistry, both experimental and theoretical, of interest to the general heterocyclic chemistry community.

The series consists of topic related volumes edited by renowned editors with contributions of experts in the field.

Janine Cossy
Editor

Synthesis of Saturated Oxygenated Heterocycles I

5- and 6-Membered Rings

With contributions by
M.A. Brimble • K.P. Kaliappan • A.J. Moreno-Vargas •
K. Palanichamy • M.A. Perry • J.D. Rainier • S.D. Rychnovsky •
N. Sizemore • L.A. Stubbing • P. Vogel

Springer

Editor
Janine Cossy
ESPCI
Laboratory of Organic Chemistry
Paris Cedex 05
France

ISSN 1861-9282 ISSN 1861-9290 (electronic)
ISBN 978-3-662-50909-8 ISBN 978-3-642-41473-2 (eBook)
DOI 10.1007/978-3-642-41473-2
Springer Heidelberg New York Dordrecht London

Printed on acid-free paper

Springer is part of Springer Science+Business Media (www.springer.com)

Preface

Two volumes are dedicated to the synthesis of saturated oxygenated heterocycles and consist of eight chapters covering the synthesis of 5- to 16-membered ring cyclic ethers and lactones. Rather than offer an exhaustive description of the synthesis of cyclic ethers and lactones, these volumes present methods and strategies to synthesize heterocycles and thus helping the reader to find suitable methods for obtaining a desired saturated oxygenated heterocycle. The first volume comprises five chapters and the second volume three chapters.

In chapter entitled "Synthesis of Substituted Tetrahydrofurans," J. D. Rainier outlines the advances that have been made during the last 10 years in the synthesis of tetrahydrofurans such as nucleophilic additions to acetals and hemiacetals, cycloadditions, oxidative cyclizations, furan reductions, Prins-pinacol cascades, ring-opening of bicyclic substrates, and nucleophilic substitutions.

In chapter "Synthesis of Saturated Tetrahydropyrans," S.D. Rychnovsky, M. A. Perry, and N. Sizemore review the common strategies to access tetrahydropyrans such as the formation of O1–C2, C2–C3, C3–C4, O1–C6, and C2–C3 bonds, as well as C2 functionalization of lactols and lactones.

The chapter "Synthesis of Saturated Six-Membered Ring Lactones" by K. P. Kaliappan and K. Palanichamy describes various selected methods such as lactonization of δ-hydroxy acid derivatives, oxidation, electrophilic cyclization, intramolecular nucleophilic displacement, radical and reductive cyclizations, palladium-catalyzed lactonization, as well as carbonylation and carboxylation.

The synthesis of 7-oxabicyclo[2.2.1]heptanes and derivatives is reported in the chapter "Synthesis of 7-Oxabicyclo[2.2.1]heptane and Derivatives" has been written by P. Vogel and A. J. Moreno-Vargas. Most of the methods, reported to access 7-oxabicyclo[2.2.1]heptanes, are Diels–Alder reactions, but non-Diels–Alder reactions such as electrophilic cyclizations have also been included in this chapter. As 7-oxabicyclo[2.2.1]heptane derivatives can be good precursors of other oxygenated heterocycles, their ring cleavage either by cleavage of a C-O or a C-C bond have been reported. In addition, as 7-oxabicyclo[2.2.1]heptane derivatives are extremely useful synthons, few syntheses of natural products and bioactive compounds, using these synthons, have been described.

In chapter "Synthesis of 5,6- and 6,6-Spirocyclic Compounds," M. A. Brimble and L. A. Stubbing describe a number of recently reported and useful methods to synthesize 5,6- and 6,6-spirocyclic compounds, including their applications to the synthesis of natural products and bioactive compounds containing spiroacetal scaffolds. One can find dehydrative spirocyclization of dihydroxyketones, metal-catalyzed addition/elimination of allylic alcohols, acid-catalyzed spirocyclization of hemiacetals, spirocyclization of exo- and endocyclic enol ethers, transition-metal-catalyzed hydroalkoxylation of alkynes, electrophilic cyclization and oxa-Michael cyclization, intramolecular hetero-Michael addition, ring-opening of epoxides and cyclopropanes, cycloadditions, furan oxidation, intramolecular hydrogen abstraction, reductive cyclizations, ring-closing metathesis, and rearrangements.

In chapter "Synthesis of Seven-Membered-Ring Ethers and Lactones," O. Piva describes the access to saturated oxygenated 7-membered cyclic ethers, by ring expansion of oxygenated structures, by formation of C–O and C–C bonds using different methods. For 7-membered cyclic lactones, oxidative processes, halolactonization, lactonization of ω-hydroxyacids, tandem Suzuki coupling and lactonization, and ring enlargement are reported.

For the synthesis of 8- to 10-membered ring ethers, in chapter "Synthesis of Eight- to Ten-Membered-Ring Ethers," J. M. Contelles and E. Soriano focus on the formation of carbon–carbon double bonds by metathesis, as well as on the formation of carbon–carbon single bonds. The authors also report on the cyclization to form C–O bonds, ring expansion, ring-opening, and rearrangement.

Chapter "Synthesis of 12- to 16-Membered-Ring Lactones" is dedicated to the synthesis of 12- to 16-membered ring lactones. In this chapter, M. Kalesse and M. Cordes present an overview of the macrocyclization of seco-acids as well as new effective procedures to access 12- to 16-membered ring lactones such as ring-closing metatheses of alkynes and olefins. The authors also report the use of ketene sources and benzodioxinones to produce macrocyclic lactones. Nitrile oxide-olefin cycloaddition, intramolecular C–H oxidative macrolactonization, and Yamaguchi and Mukaiyama macrocyclization as well as macrolactonization via thioester or using phosphorus reagents are described.

I would like to express my sincere gratitude to all the authors of these chapters for their efforts and outstanding contributions. I would also like to thank B. Maes for giving me the opportunity to edit these two volumes, to Elizabeth Hawkins and Tanja Jaeger from Springer for coordinating the project, and to Fairin Miriam John Bennet for the editing process, for her help and patience.

Finally, I hope that this book will be a good source of inspiration for those planning the synthesis of saturated oxygenated heterocycles, for solving specific synthetic problems, or for elaborating on new synthetic tools.

Paris, France Janine Cossy

Contents for Volume I

Contents for Volume II

Synthesis of Substituted Tetrahydrofurans

Jon D. Rainier

Contents

Abstract This chapter outlines the advances that have been made in the synthesis of substituted tetrahydrofurans over the period of time from 2005 to 2012. Included are nucleophilic substitutions, [3+2]-cycloadditions, oxidative cyclizations, and the

J.D. Rainier (✉)
Department of Chemistry, University of Utah, 315 South 1400 West, Salt Lake City, UT 84112, USA
e-mail: rainier@chem.utah.edu

J. Cossy (ed.), *Synthesis of Saturated Oxygenated Heterocycles I*, Topics in Heterocyclic Chemistry 35, DOI 10.1007/978-3-642-41473-2_1,

ring opening of bicyclic oxanorbornenes as well as the application of these methods to the generation of tetrahydrofuran-containing natural products.

Keywords [3+2]-Cycloaddition • Acetal • Allylsilane • Cobalt • Cyclization • Furan • Grubbs catalyst • Hemiacetal • Iodoether • Manganese • Michael addition • Osmium • Osmium tetroxide • Oxanorbornene • Oxidative cyclization • Palladium • Permanganate • Prins • Prins-pinacol • Ring-closing • Ring-opening • ROM/CM • ROM/RCM • Ruthenium • Ruthenium tetroxide • Tetrahydrofuran

1 Introduction

Substituted tetrahydrofurans have been popular targets for synthetic chemists. This is due to both the challenges associated with the synthesis of substituted heterocycles and the presence of tetrahydrofuran rings in natural and non natural targets including lignans [1], nucleosides [2], annonaceous acetogenins, and macrolides [3, 4]. This chapter covers the important advances that have been made in this area from the period 2005 to 2012 with a focus on some of the newer technologies including oxidative cyclizations, [3+2]-cycloadditions, and the fragmentation of bicyclic ring systems (for a review that covers the period up to 2005, see [5]).

2 Nucleophilic Additions to Acetals and Hemiacetals

The addition of nucleophiles to cyclic acetals and hemiacetals is an effective method of building tetrahydrofurans. These reactions give tetrahydrofurans from readily available starting materials and generally proceed with high diastereoselectivity that is generally predictable using Woerpel's model where the stereochemical outcome is explained using stereoelectronic effects [6]. In this model the nucleophile adds from the "inside" of the favored envelope conformer due to developing eclipsed interactions with the corresponding "outside" approach (see **3**) (Scheme 1).

Hong and coworkers have taken advantage of hemiacetal reductions to synthesize lignan natural products. In the course of their studies, they discovered that the relative stereochemistry of the products that come from the activation and reduction of hemiketal **4** was dependent upon the conditions used to carry out the reaction (Scheme 2) [7]. Namely, they found that 2,5-*cis*-tetrahydrofuran isomer **5**, as needed for the synthesis of (–)-futokadsurin A and (–)-veraguensin, came from the rapid $NaBH_3CN$ reduction of the intermediate oxocarbenium ion that comes

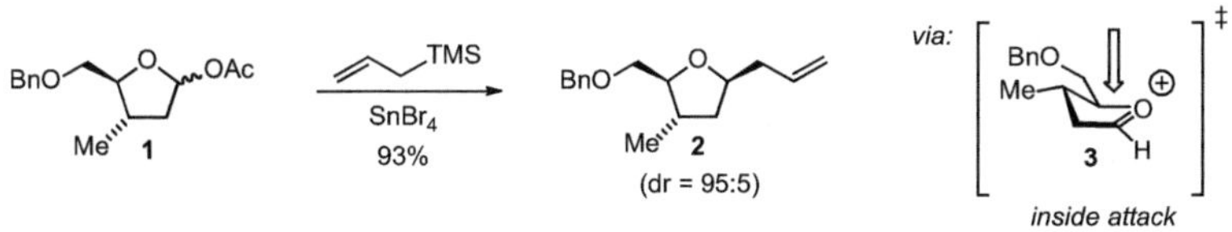

Scheme 1 Stereoelectronic model for tetrahydrofuran synthesis from oxocarbenium ion addition by Woerpel et al. [6]

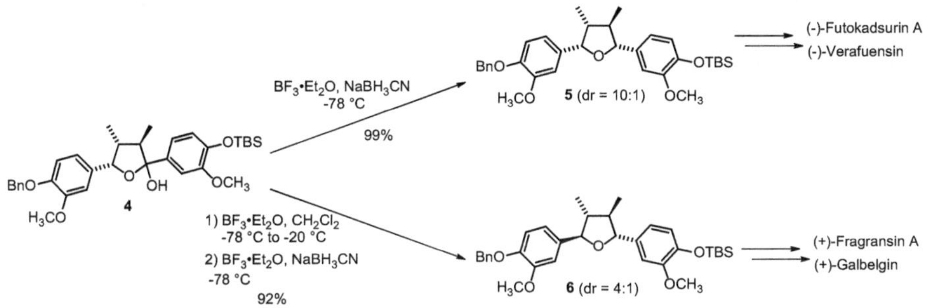

Scheme 2 Synthesis of lignan natural products from lactol reductions by Hong et al. [7]

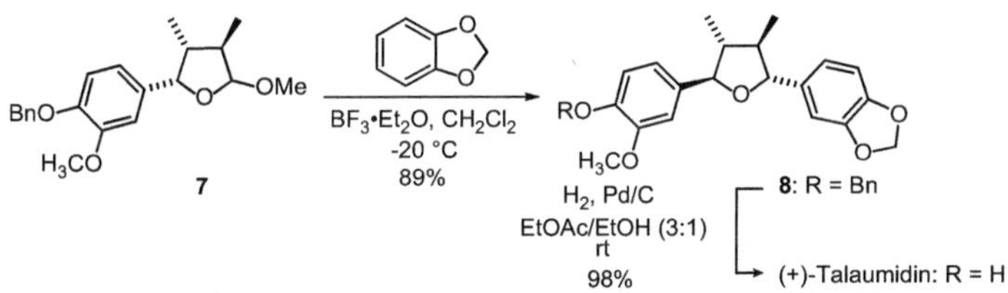

Scheme 3 Synthesis of (+)-talaumidin by Hong et al. [7]

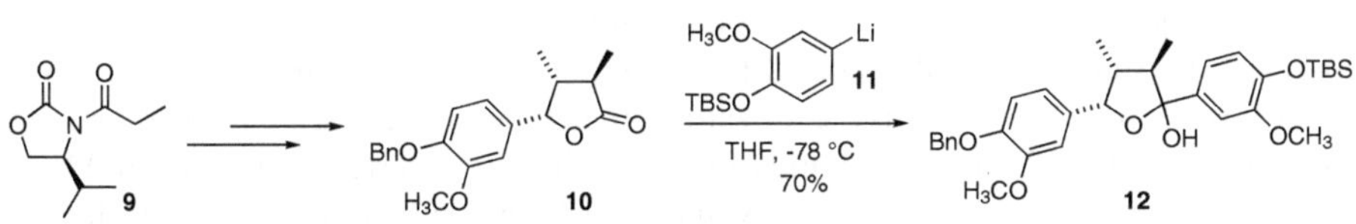

Scheme 4 Synthesis of lignan natural product precursors by Hong et al. [7]

from **4**. They also found that the generation of the oxocarbenium ion in the absence of reducing agent led to equilibration at C-1 and to the all *trans*-diastereomer **6** after a subsequent reduction step. Furan **6** was taken on to the lignan natural products (+)-fragransin A and (+)-galbelgin.

Hong et al. were able to couple the epimerization sequence outlined in Scheme 2 with a Friedel–Crafts arylation to generate (+)-talaumidin precursor **8** (Scheme 3).

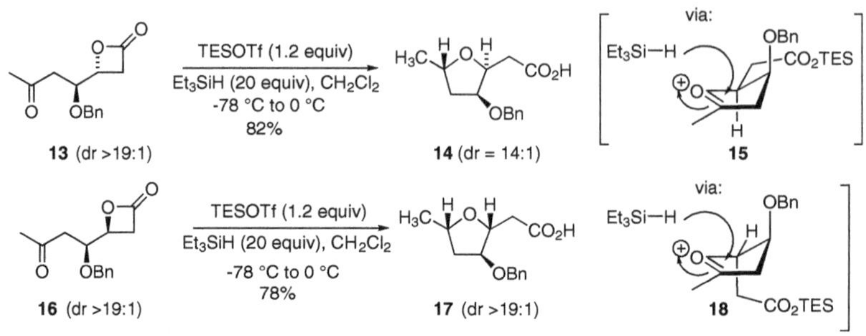

Scheme 5 Keto-β-lactone cyclization to tetrahydrofurans by Romo and Mitchell [8]

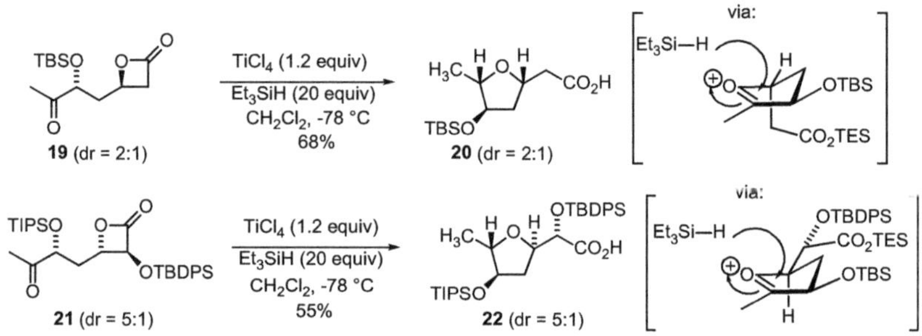

Scheme 6 Keto-β-lactone cyclizations to tetrahydrofurans by Romo and Mitchell [8]

Hong et al. took a relatively straightforward approach to the lignan precursors (Scheme 4). The absolute stereochemistry was established using oxazolidinone **9** and an Evans aldol reaction. Conversion to γ-lactone **10** was followed by the addition of the appropriate aryl lithium **11** to give the requisite hemiketal **12**. Talaumidin precursor **7** came from the reduction of **10**.

In a unique twist on the use of oxocarbenium reduction to generate tetrahydrofurans, Romo and Mitchell reported the generation of 2,5-*cis*- and 2,5-*trans*-tetrahydrofurans from the reductive cyclization of ketones having a pendant β-lactone (Scheme 5) [8]. This reaction builds upon previously reported results from Mead et al. and leads to the stereospecific synthesis of highly substituted tetrahydrofurans [9]. For example, when *anti*-β-lactone **13** was subjected to TESOTf and excess Et_3SiH, they isolated 2,5-*trans*-tetrahydrofuran **14** in 82 % yield as a 14:1 mixture of diastereomers, while the corresponding *syn*-β-lactone **16** gave 2,5-*cis*-tetrahydrofuran **17** in 78 % yield as a >19:1 mixture of diastereomers. The stereochemical outcome of these reactions is consistent with the Woerpel et al. model as depicted for the reduction of **15** and **18**.

In addition to the stereochemistry, Romo and coworkers found that the position of the ether and the reaction conditions were important. When $TiCl_4$ or $BF_3{\cdot}Et_2O$ were used instead of TESOTf in the reactions of **13** and **16**, the corresponding furan

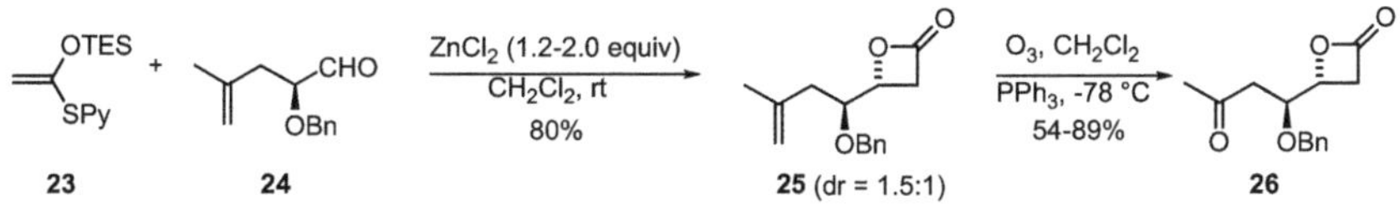

Scheme 7 TMAL process to β-lactones by Romo et al. [10]

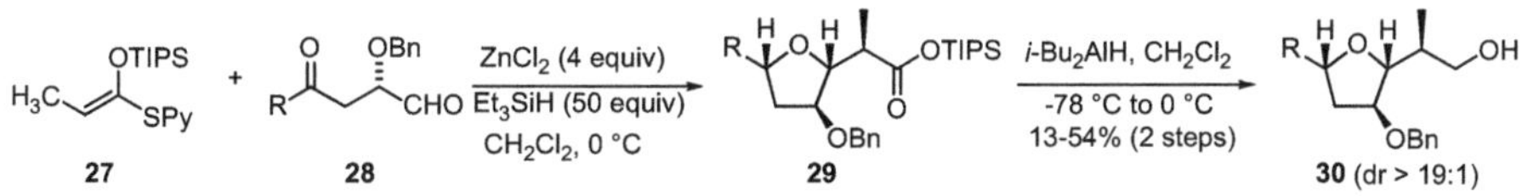

Scheme 8 TMAL, reductive cyclization to tetrahydrofurans by Romo et al. [10]

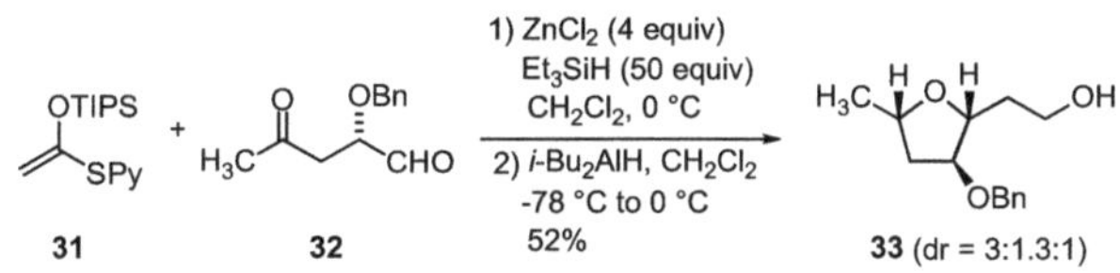

Scheme 9 TMAL, reductive cyclization with unsubstituted ketene by Romo et al. [10]

was formed as the major product. When the ether was moved so that it was adjacent to the ketone, furan formation was no longer a problem (Scheme 6). For example, tetrahydrofuran **20** was formed when *trans*-δ-silyloxy-keto-β-lactones **19** was subjected to $TiCl_4$, while **22** came from subjecting the corresponding *cis*-isomer **21** to the same conditions.

The β-lactone cyclization precursors come from Romo and Yang's tandem Mukaiyama aldol lactonization (TMAL) process (Scheme 7) [10]. As an example of this, lactone **25** was the product of the $ZnCl_2$-mediated cyclization of α-benzyloxy aldehyde **24** with thio-ketene acetal **23**. Ozonolytic cleavage of the alkene gave **26**.

Romo and coworkers were able to expand on their tetrahydrofuran synthesis by finding conditions for a three-component cascade sequence that resulted in 2,5-*cis*-tetrahydrofurans from keto-aldehydes, ketene acetals, and Et_3SiH (Scheme 8) [11]. Optimal conditions required an excess of $ZnCl_2$ and Et_3SiH to give tetrahydrofurans in high diastereoselectivity. The two-step yields, with alkyl ketones, were in the 42–54 % range with the lowest yield of 13 % coming from the use of a benzylic ketone (R = Ph). The authors found it easier to isolate the products after reduction of the crude reaction mixture.

Ketene substitution also impacted the outcome of the cascade sequence (Scheme 9). Unsubstituted ketene acetal **31** gave reasonable yields of tetrahydrofuran **33** but with diminished diastereoselectivity.

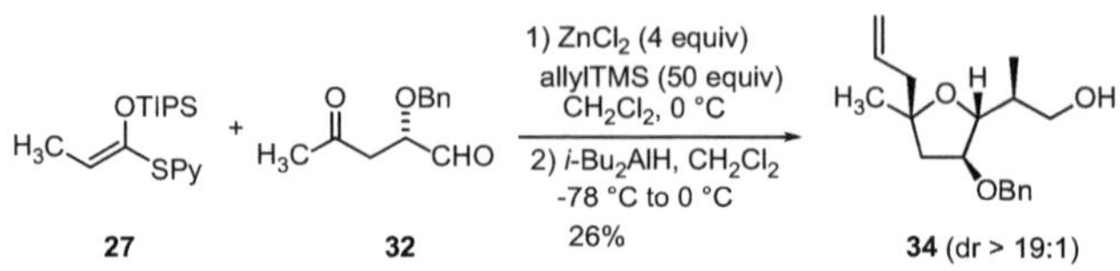

Scheme 10 TMAL, allylsilane additions to give tetrahydrofurans by Romo et al. [10]

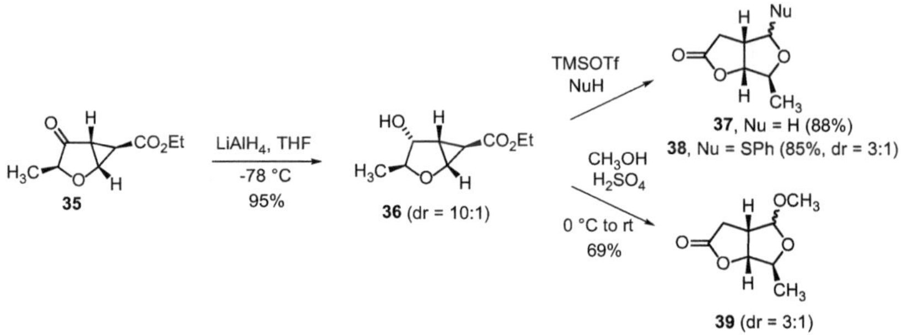

Scheme 11 Heterolytic cyclopropane ring openings by Gharpure et al. [12]

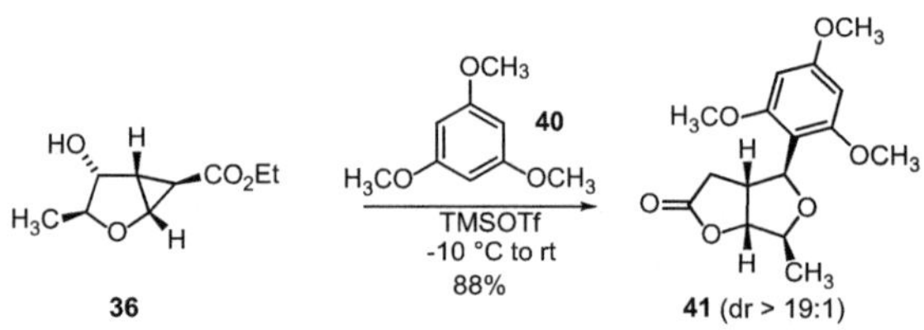

Scheme 12 Generation of aryl *C*-glycosides by Gharpure et al. [12]

Finally, the intermediate oxocarbenium ion could be trapped with nucleophiles other than hydride. The use of allylsilane as the nucleophile in the reaction of **27** with **32** gave **34** with excellent diastereoselectivity (Scheme 10). Because of the poorer nucleophilicity of allylsilane when compared to triethylsilane, greater amounts of the furan by-product were isolated here.

Gharpure and coworkers have found furanocyclopropanes to be rich precursors to highly substituted tetrahydrofurans [12]. For example, by subjecting alcohol **36** to TMSOTf and Et_3SiH or TMSOTf and PhSH, they were able to both regioselectively open the cyclopropane adjacent to the furan oxygen and generate bicyclic substrates **37** and **38**, respectively (Scheme 11). Methanol adduct **39** was obtained when the reaction was promoted by H_2SO_4. Interestingly, when ketone **35** was subjected to similar conditions, exclusive generation of pyran products was observed.

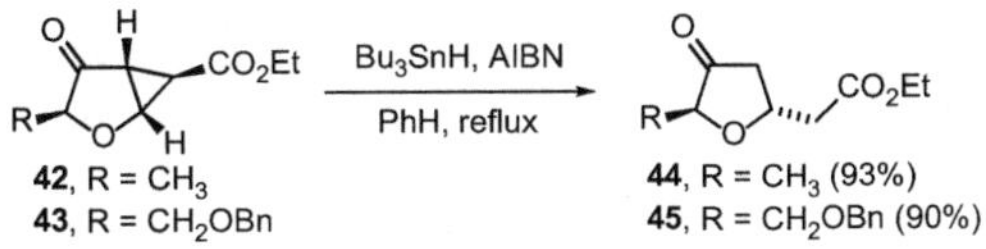

Scheme 13 Homolytic cyclopropane ring-opening strategy by Gharpure et al. [12]

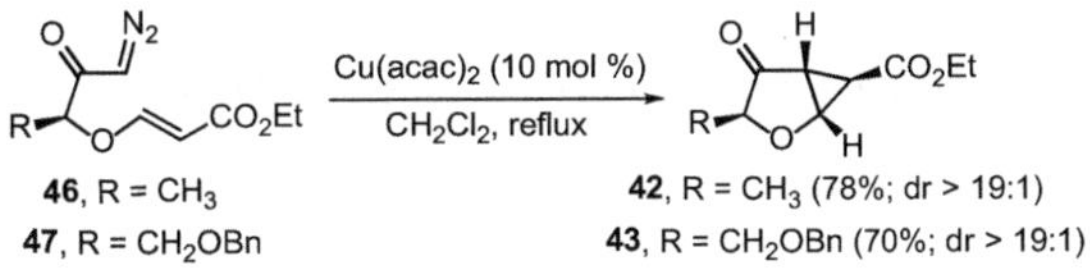

Scheme 14 Synthesis of cyclopropyl furanones by Gharpure et al. [12]

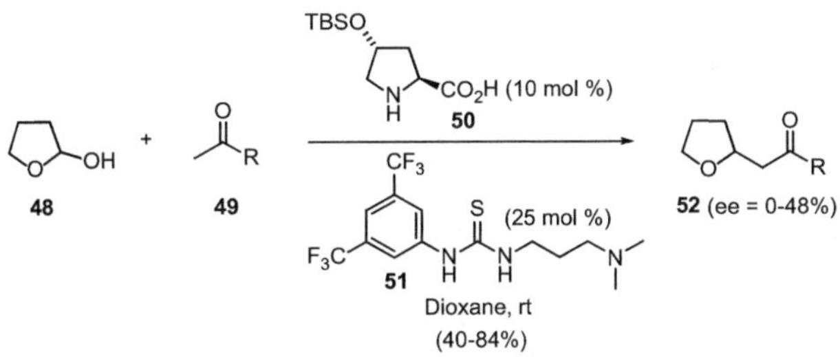

Scheme 15 Use of organocatalytic aldol-Michael reactions to tetrahydrofurans by McQuade et al. [13]

Gharpure and coworkers also found that the oxocarbenium ion from the cyclopropane opening reacts with carbon nucleophiles. *C*-Aryl furan **41** resulted from the use of 1,3,5-trimethoxybenzene (**40**) as the nucleophile in the ring opening (Scheme 12).

Cyclopropyl furans **42** and **43** can also be induced to undergo ring opening distal to the furan oxygen. For example, by employing free radical conditions, the reduction of **42** or **43** resulted in 2,5-*trans*-furanones **44** and **45**, respectively, via cyclopropylcarbinyl radical ring opening and reduction (Scheme 13). Cyclopropylfuranones **42** and **43** came from the diazo decomposition of the corresponding vinylogous carbonates **46** and **47**, respectively (Scheme 14).

In a sequence that is the equivalent of an oxocarbenium-induced C–C bond-forming reaction, McQuade and coworkers have used organic catalysts to carry out the coupling of 2-hydroxy tetrahydrofuran **48** with methyl ketones **49** to give 2-alkyl furans (Scheme 15) [13]. Mechanistically the reaction is proposed to proceed through the addition of a thiourea-stabilized enolate from the reaction of **49** with **51** to an iminium intermediate that comes from the condensation of **48** with proline catalyst **50**. Hydrolysis and Michael cyclization or displacement of the ammonium ion subsequent to the Mannich reaction gives the observed product **52**.

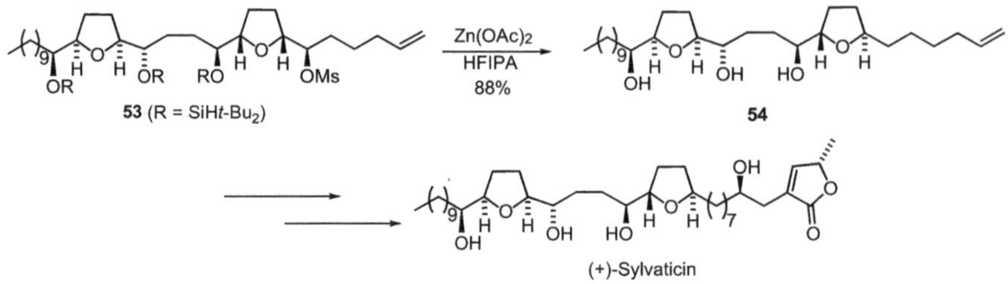

Scheme 16 Oxidative cyclization strategy to (+)-sylvaticin by Donohoe et al. [14]

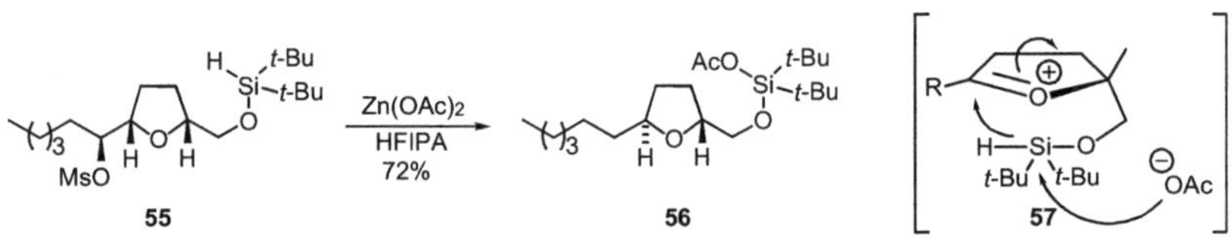

Scheme 17 Conversion of *cis*-tetrahydrofurans into *trans*-tetrahydrofurans by Donohoe et al. [15]

The Donohoe group has developed a creative solution to converting 2,5-*cis*-tetrahydrofurans that come from their oxidative cyclization (Sect. 4.5) into the corresponding 2,5-*trans*-isomers [14]. During their synthesis of (+)-sylvaticin, they found that by treating **53** with Zn(II) in hexafluoroisopropanol (HFIPA), they could induce its conversion to 2,5-*trans*-tetrahydrofuran **54** in 88 % yield (Scheme 16).

Donohoe et al. has carried out deuterium-labeling experiments that are consistent with a mechanism that involves an initial [1,2]-hydride shift and a subsequent intramolecular oxocarbenium ion reduction as depicted by **57** (Scheme 17) [15].

3 Cycloadditions

Cycloadditions have become a valuable means of generating tetrahydrofurans. They typically involve the condensation of an aldehyde with a 1,3-dipole in an overall [3+2]-cycloaddition process. With one exception, the focus of this section is on tetrahydrofuran formation from dipolar cycloadditions of either push–pull cyclopropanes or allylsilanes with aldehydes.

3.1 Cyclopropane–Aldehyde [3+2]-Cycloadditions

The recent renewed interest in the use of push–pull cyclopropanes to generate tetrahydrofurans has been at least partly driven by results from the Johnson

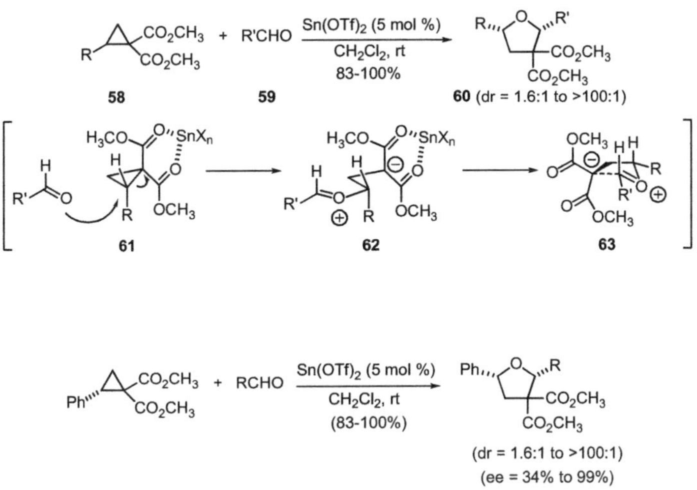

Scheme 18 [3+2]-Cycloadditions of push–pull cyclopropanes by Johnson et al. [16, 17, 86]

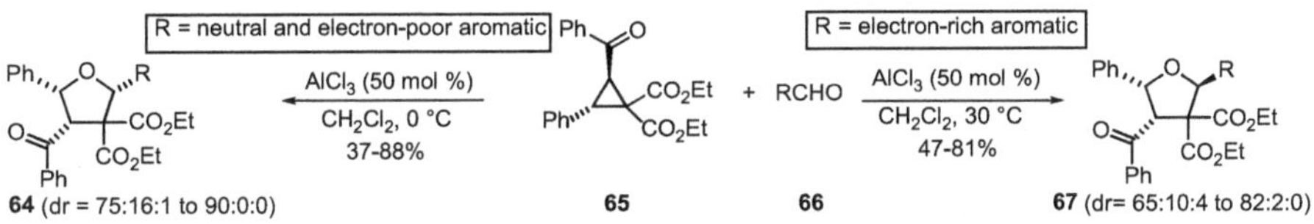

Scheme 19 Synthesis of tetrahydrofurans from *trans*-cyclopropanes by Yang et al. [19]

laboratories. For example, they demonstrated the ability to synthesize a wide range of 2,5-*cis*-tetrahydrofurans in high diastereoselectivity from the reaction of donor–acceptor cyclopropanes with aldehydes (Scheme 18) [16, 86]. The reaction has been successful even with tetrasubstituted cyclopropanes resulting in the generation of quaternary substituted tetrahydrofurans [17]. Through an analysis of the scope of the reaction along with competition experiments, the authors proposed an asynchronous mechanism whereby the reaction is initiated by attack of the aldehyde on the cyclopropane as depicted by the **61–62** transformation [18].

Yang and coworkers have also examined cyclopropane [3+2]-cycloadditions to give highly substituted tetrahydrofurans (Scheme 19) [19]. Interestingly, the $AlCl_3$-mediated cycloaddition of *trans*-cyclopropane **65** with electronically neutral or electron-deficient aryl aldehydes led to a predominance of *cis*-tetrahydrofuran **64**, while the use of electron-rich aryl aldehydes resulted in the generation of the corresponding *trans*-tetrahydrofurans **67**. The authors demonstrated that the reaction was reversible and that the *trans* isomer resulted from the equilibration of the *cis* isomer.

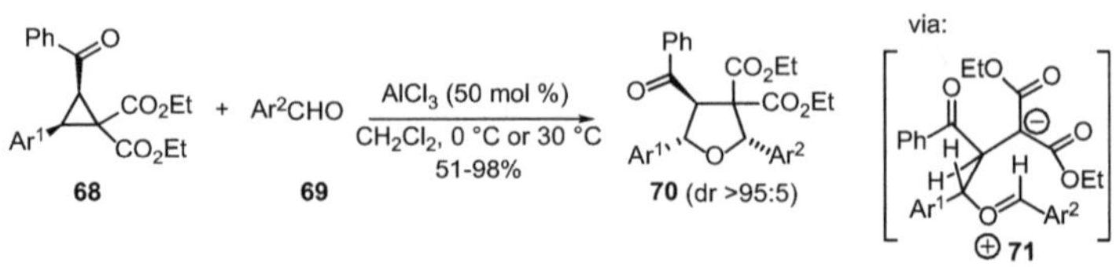

Scheme 20 Synthesis of tetrahydrofurans from *cis*-cyclopropanes by Yang et al. [20]

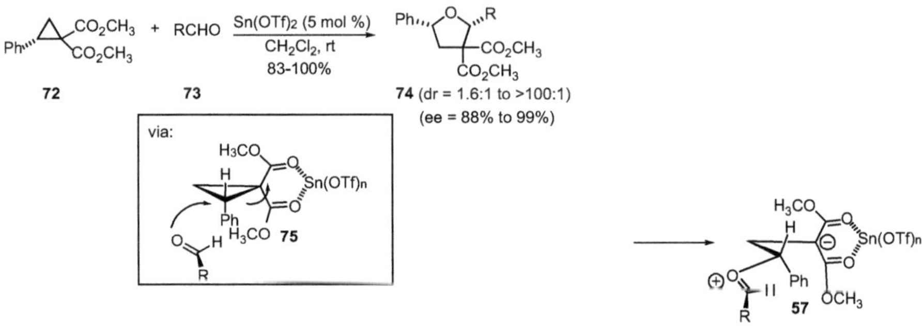

Scheme 21 Diastereoselective [3+2]-cycloadditions by Johnson et al. [21]

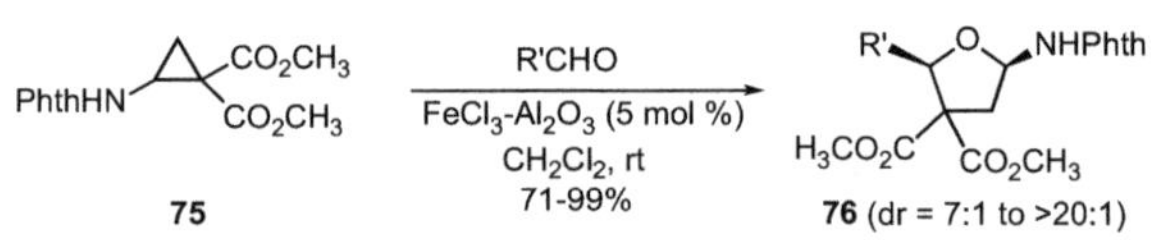

Scheme 22 Synthesis of aminotetrahydrofurans from [3+2]-cycloadditions by Waser et al. [22]

The corresponding *cis*-cyclopropane **68** gave 2,4-*trans*-, 2,5-*cis*-tetrahydrofuran **70** regardless of whether electron-rich or electron-poor aldehydes were used (Scheme 20) [20]. Yang et al. believes that the cyclization of **68** proceeds in a stepwise manner with all substituents occupying *pseudo*-equatorial positions as indicated by **71**.

Critical for the formulation of their proposed stepwise S_N2 addition mechanism, Johnson et al. found that excellent chirality transfer occurred when enantiomerically pure cyclopropane **72** was employed in the [3+2]-cycloaddition (Scheme 21) [21].

Waser and coworkers have reported the first use of amino cyclopropanes in [3+2]-cycloadditions with aldehydes to give amino tetrahydrofurans with high levels of diastereoselectivity (Scheme 22) [22]. A wide range of activators could be used in these reactions including normally inert Lewis acids like $FeCl_3$ on Al_2O_3. In contrast to the Johnson et al. work mentioned above, racemic tetrahydrofurans resulted from the use of enantiomerically enriched aminocyclopropanes implying that the reaction proceeds through a zwitterionic intermediate.

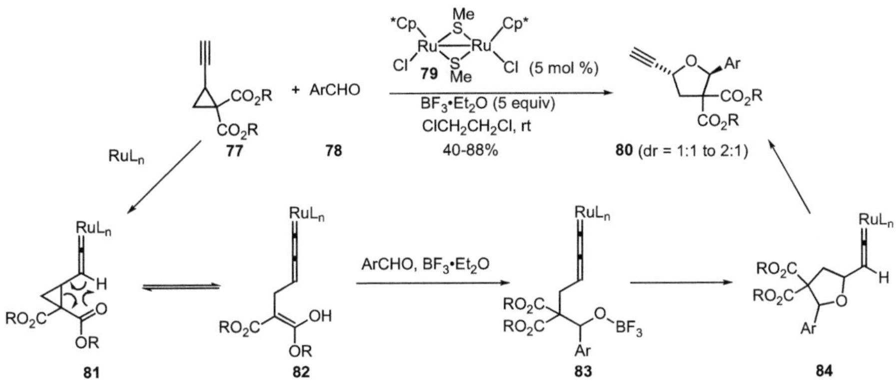

Scheme 23 Ru-allenylidene [3+2]-cycloadditions to tetrahydrofurans by Sakata, Nishibayashi et al. [23]

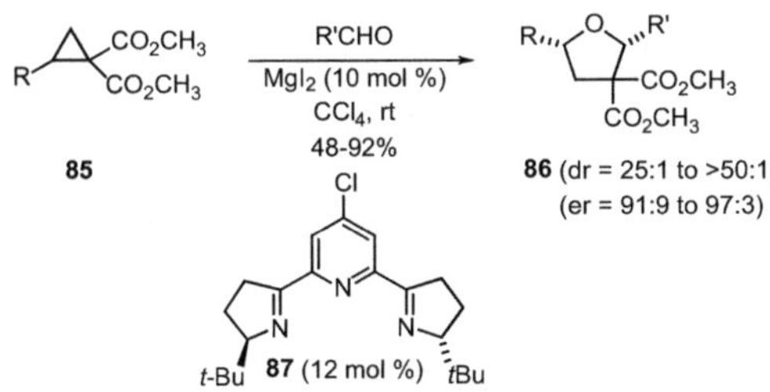

Scheme 24 Enantioselective [3+2]-cycloadditions to tetrahydrofurans by Johnson et al. [24]

Sakata, Nishibayashi, and coworkers have uncovered a unique metal allenylidene-induced [3+2]-cycloaddition reaction between ethynylcyclopropanes and aldehydes to give alkynyl-substituted tetrahydrofurans **80** (Scheme 23) [23]. The reaction requires a mixture of ruthenium dimer **79** and excess $BF_3{\cdot}Et_2O$ and was proposed to proceed via cyclopropyl ring-opening and [1,5]-hydride shift to give allenyl-Ru intermediate **82**. Condensation with the $BF_3{\cdot}Et_2O$-activated aldehyde and cyclization gives the tetrahydrofuran **84**.

In addition to their use of optically active cyclopropanes (Scheme 21), Johnson and coworkers have also developed an enantioselective variant of their [3+2]-cycloaddition where racemic cyclopropane **85** was converted into enantiomerically enriched tetrahydrofurans **86** using chiral PyBox ligand **87** and MgI_2 (Scheme 24) [24].

Johnson and Campbell have applied their [3+2]-cycloaddition chemistry to an efficient synthesis of (+)-polyanthellin A [25]. As illustrated in Scheme 25, they generated the core polyanthellin architecture over a three-step sequence. From acyclic precursor **87**, intramolecular cyclopropanation gave [3+2]-cycloaddition precursor **88**. By subjecting **88** to aldehyde **89**, they were able to both generate the polyanthellin tetrahydrofuran ring and establish a bis-olefin which was used in a subsequent ring-closing metathesis to give the polyanthellin architecture. In order

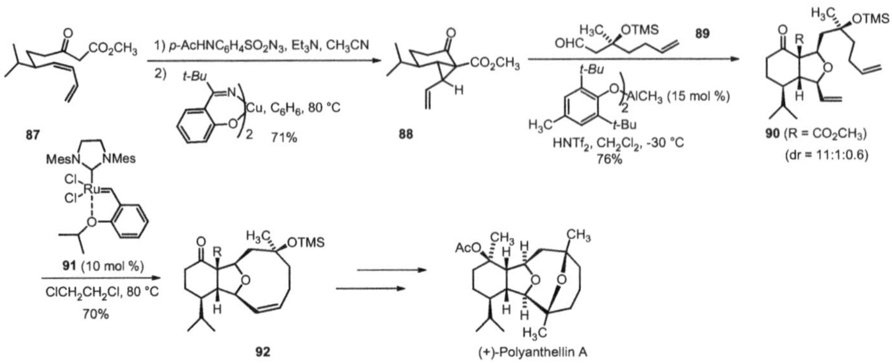

Scheme 25 Use of [3+2]-cycloadditions in the total synthesis of (+)-Polyanthellin A by Johnson and Campbell [25]

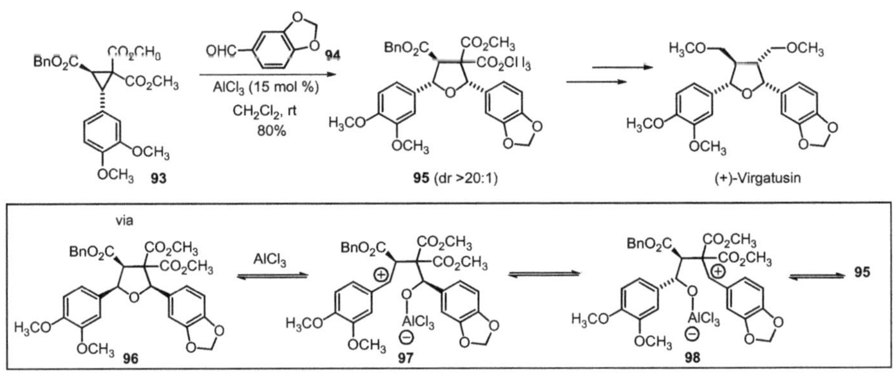

Scheme 26 Use of [3+2]-cycloadditions to (+)-Virgatusin by Johnson et al. [26]

to avoid decomposition of the aldehyde during the [3+2]-cycloaddition, the authors screened a number of Lewis acids and ultimately found that a mixture of Yamamoto's MAD catalyst and $HNTf_2$ was optimal.

Johnson and coworkers have also applied their cycloaddition chemistry to the synthesis of (+)-virgatusin (Scheme 26) [26]. From enantiomerically pure *trans*-cyclopropane **93**, treatment with piperonal **94** in the presence of $AlCl_3$ gave 2,5-*cis*-tetrahydrofuran **95**. Interesting was that **95** resulted from retention of the stereochemistry on the cyclopropane which, in a similar fashion to the Yang et al. observation (Scheme 19), is a consequence of reversible tetrahydrofuran formation. As a result of this, the benzyl ester stereochemistry on the cyclopropane was what ultimately dictated the C2 and C5 stereocenters.

In their total synthesis of (+)-isatisine A, Kerr and Karadeolian utilized a [3+2]-cycloaddition strategy to the tetrahydrofuran (Scheme 27) [27]. The cycloaddition between indole-2-carboxaldehyde **99** and vinylcyclopropane **100** was carried out

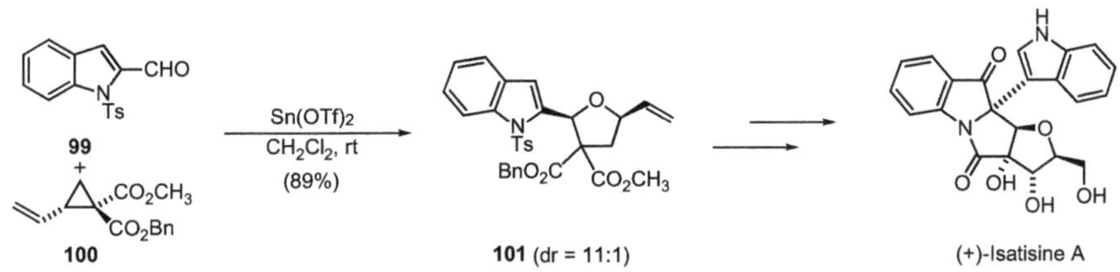

Scheme 27 [3+2]-Cycloaddition to the tetrahydrofuran moiety in (+)-isatisine A by Kerr et al. [27]

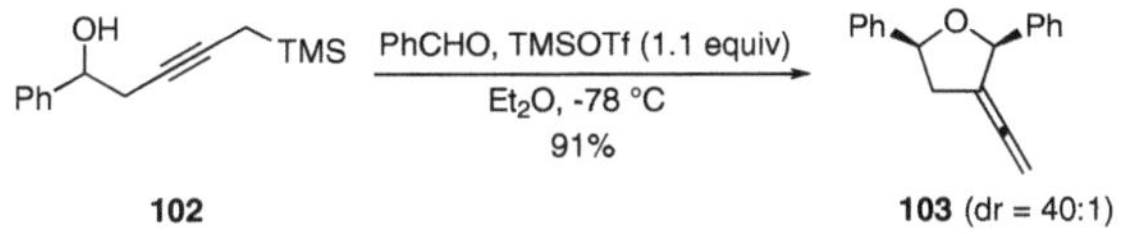

Scheme 28 Propargylsilane Prins cyclization to substituted tetrahydrofurans by Cho et al. [28]

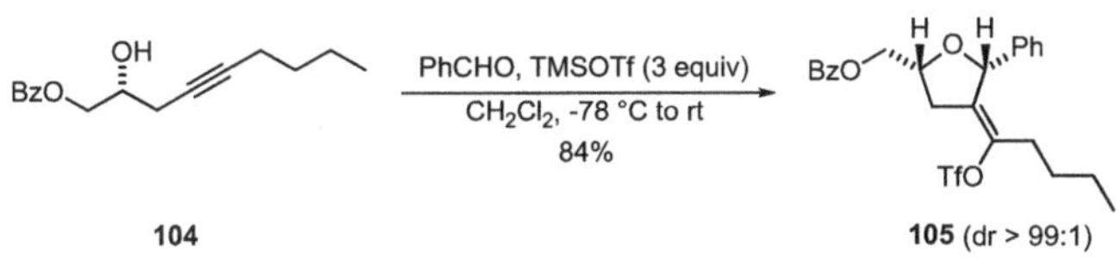

Scheme 29 Alkyne-Prins cyclization to tetrahydrofurans by Cho et al. [29]

using $Sn(OTf)_2$ and resulted in the generation of tetrahydrofuran **101** in 89 % yield as an 11:1 mixture of 2,5-*cis*- and 2,5-*trans*-isomers.

3.2 *Prins Cyclizations*

Although Prins cyclization normally gives tetrahydropyrans, Cho et al. found that the reaction can give tetrahydrofurans by utilizing propargylsilanes as substrates [28]. For example, the condensation of propargyl alcohol **102** with benzaldehyde in the presence of TMSOTf gave allenyl furan **103** in 91 % yield (Scheme 28). The reaction was amenable to the use of aldehydes, ketones, and acetals as coupling partners.

Building on their propargylsilane results, Cho and coworkers found internal alkynes to also give 2,5-*cis*-tetrahydrofurans but having a vinyl triflate at the three-position when TMSOTf was used to promote the reaction (Scheme 29) [29]. By subjecting optically active propargyl alcohol **104** to benzaldehyde and TMSOTf, they generated tetrahydrofuran **105** in 84 % yield as a single

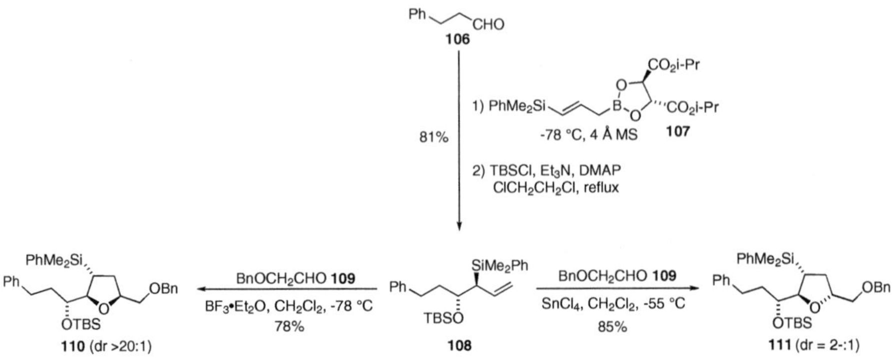

Scheme 30 Use of [3+2]-cycloadditions to tetrahydrofurans by Roush and Micalizio [31]

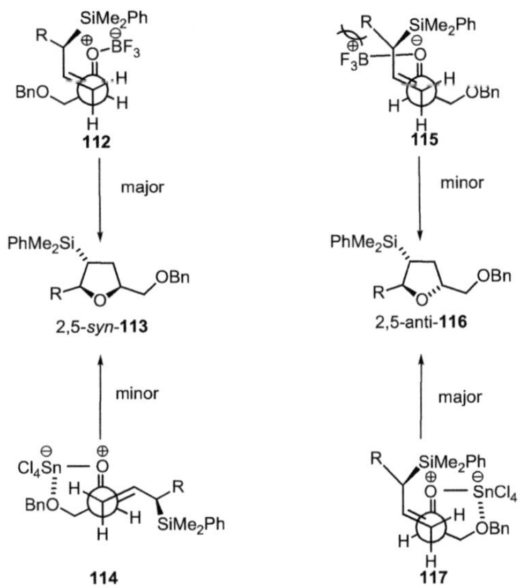

Scheme 31 Mechanistic hypothesis for *cis*- and *trans*-tetrahydrofuran formation by Roush and Micalizio [31]

stereoisomer. Interestingly, ketones rather than vinyl triflates were isolated when Et_2O was used as the solvent instead of CH_2Cl_2.

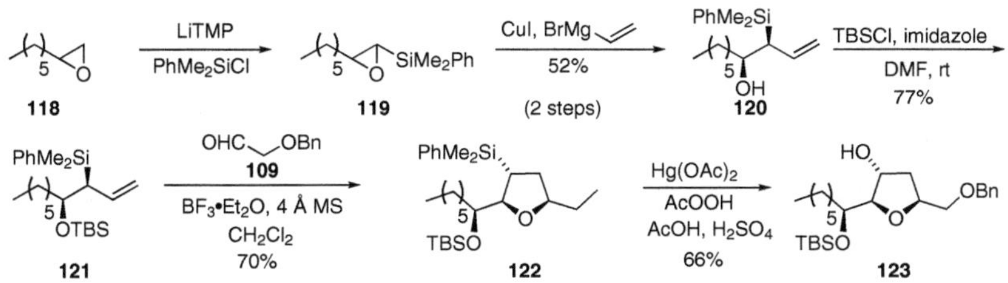

Scheme 32 [3+2]-Cycloadditions to tetrahydrofurans by Hodgson and Salik [32]

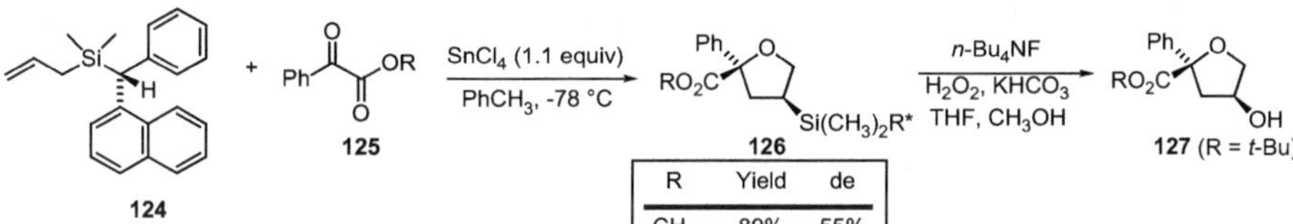

R	Yield	de
CH_3	89%	55%
i-Pr	92%	71%
C_6H_{11}	88%	80%
t-Bu	94%	85%

Scheme 33 Chiral allylsilane [3+2]-cycloadditions by Akiyama et al. [33]

3.3 Allylsilane–Aldehyde Cycloadditions

Allylsilane–aldehyde [3+2]-cycloadditions are a very powerful means of generating tetrahydrofurans as they can be used to give either 2,5-*cis*- or 2,5-*trans*-isomers depending upon the Lewis acid used to promote the reaction [30]. As an example of the power of these reactions, Roush and Micalizio described the reaction of allylsilane **108** (available from the condensation of allylborane **107** with aldehyde **106**) with aldehyde **109** in the presence of $BF_3{\cdot}Et_2O$ to give *cis*-tetrahydrofuran **110** in 78 % yield and in >20:1 dr (Scheme 30) [31]. In a complementary fashion, the use of $SnCl_4$ as the Lewis acid gave an 85 % yield (dr = 20:1) of *trans*-tetrahydrofuran **111**.

Roush and Micalizio proposed that the stereochemical outcome of the reactions was governed by the synclinal transition states illustrated in Scheme 31. The use of $BF_3{\cdot}Et_2O$ leads to transition state **112** having the Me_2PhSi substituent synclinal to the carbonyl oxygen atom. Transition state **115** is disfavored because of steric interactions between BF_3 and the allylsilane substituent. $SnCl_4$ interacts in a bidentate fashion and proceeds via *syn*clinal transition state **117**.

Hodgson and Salik have also examined allylsilane–aldehyde [3+2]-cycloadditions to tetrahydrofurans. What distinguishes their work is the manner in which the precursor to the cycloaddition is generated (Scheme 32) [32]. From epoxide **118**, deprotonation and trapping of the resulting anion with $PhMe_2SiCl$ gave epoxysilane **119**. Addition of vinyl magnesium bromide in the presence

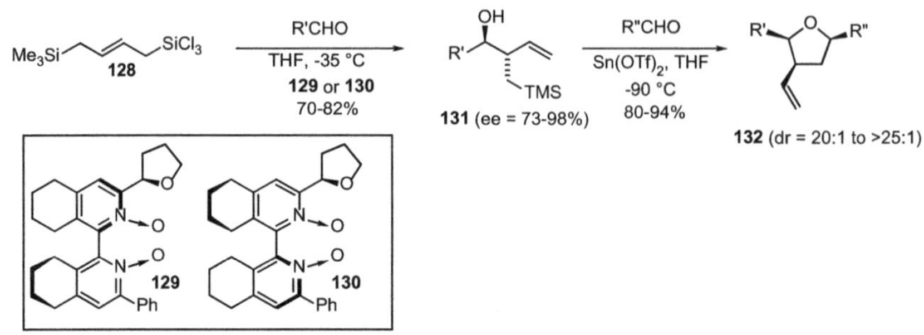

Scheme 34 Homoallylsilane [3+2]-cycloadditions to tetrahydrofurans by Kocovsky et al. [34]

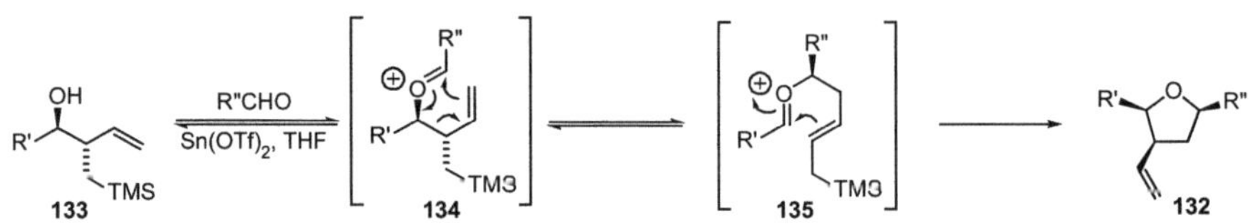

Scheme 35 Mechanistic rationale for the generation of tetrahydrofurans by Kocovsky et al. [34]

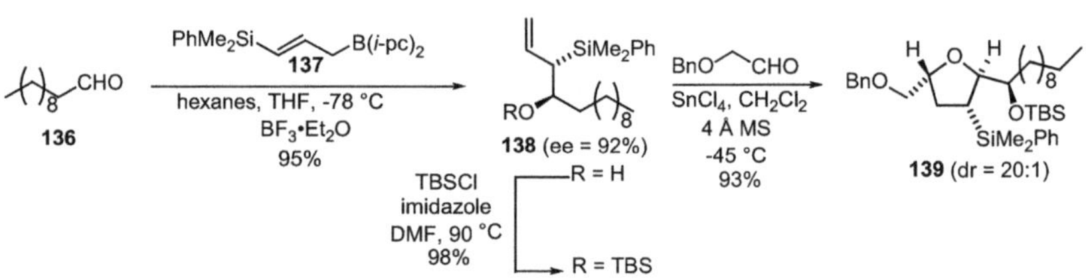

Scheme 36 [3+2]-Approach to asimicin by Roush et al. [36]

of CuI gave *syn*-hydroxysilane cyclization precursor **120** in 52 % yield for the two steps. TBS protection was followed by [3+2]-cycloaddition with α-benzyloxyacetaldehyde in the presence of $BF_3{\cdot}Et_2O$ to give tetrahydrofuran **122** in 70 % yield with complete diastereocontrol. The utility of the silyl group was demonstrated through its conversion to alcohol **123** using Fleming's oxidative protocol.

Akiyama and coworkers have examined the use of chiral allylsilane **124** in [3+2]-cycloadditions with α-ketoesters **125** (Scheme 33) [33]. When run in the presence of $SnCl_4$ (1.1 equivalent), the cycloaddition of **124** gave tetrahydrofuran **126** with 85 % diastereomeric excess for the *t*-butyl ester.

Kocovsky and coworkers have recently described a triple allylation that leads to tetrahydrofurans in high diastereo- and enantioselectivity (Scheme 34) [34]. From bis-allylsilane **128**, a chiral base-catalyzed aldehyde condensation gives homoallylic alcohol **131**. The best catalysts for the reaction were dioxides **129**

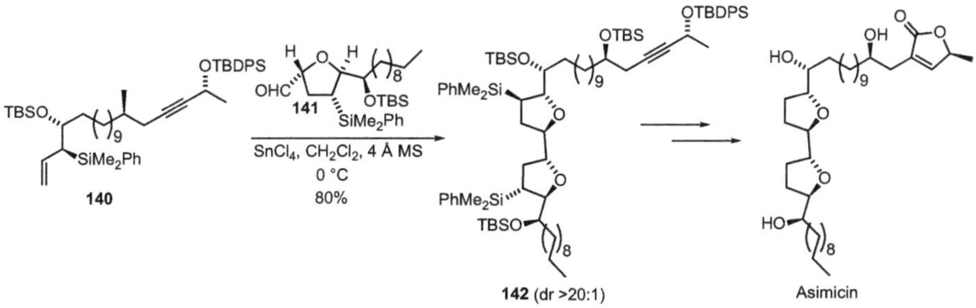

Scheme 37 [3+2]-Approach to Asimicin. Coupling and completion by Roush and Tinsley [36]

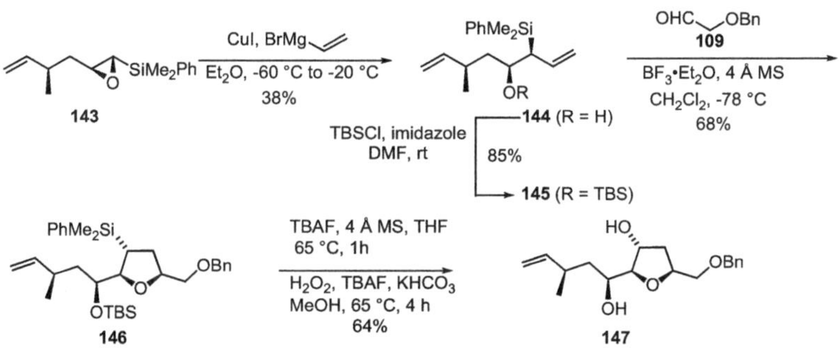

Scheme 38 Approach to the lytophilippines by Hodgson and Salik [32]

and **130** giving alcohols in high enantioselectivity with aromatic and α,β-unsaturated aldehydes and with moderate enantioselectivity with aliphatic aldehydes. The Kočovský team utilized homoallylsilane **131** in condensation reactions with aldehydes to give substituted tetrahydrofurans **132** in both high yields and diastereoselectivity.

The authors believe that the transformation outlined above proceeds via the mechanism depicted in Scheme 35 where an oxonia-Cope rearrangement applied to **134** is followed by an intramolecular allylsilane addition to activated oxonium ion **135**. The stereoselectivity in the generation of **132** is believed to be a result of chair transition states in both the oxonia-Cope reaction and the allylsilane cyclization.

The Roush group has applied their [3+2]-cycloaddition to a number of natural product syntheses (see [31] and [35, 87, 88]). One example is outlined in Schemes 36 and 37 where they applied their chemistry to the total synthesis of the annonaceous acetogenin natural product asimicin utilizing two [3+2]-annulations to build the bis-tetrahydrofuran ring system [36]. To the first tetrahydrofuran ring, aldehyde **136** was subjected to a Brown allylboration using diisopinocampheylborane **137** to efficiently give optically active allylsilane **138**. The generation of 2,5-*trans*-tetrahydrofuran **139** followed their previous results using $SnCl_4$ as the Lewis acid.

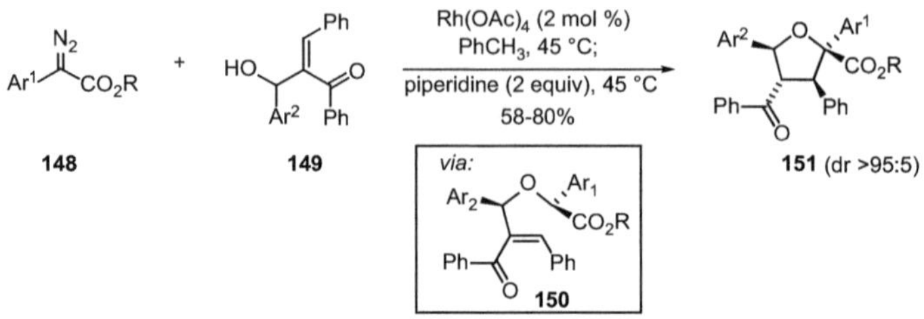

Scheme 39 Oxonium ylide approach to tetrahydrofurans by Hu et al. [37]

In the key step to completing asimicin, Roush and Tinsley used their [3+2]-cycloaddition to combine allylsilane **140** and tetrahydrofuran **141** (Scheme 37). The reaction was promoted by $SnCl_4$ and resulted in bis-tetrahydrofuran **142** in 80 % yield as essentially a single diastereomer. Following protecting group and oxidation state modification, the synthesis of asimicin was completed using Trost's two-step approach to butenolides.

Hodgson and Salik were able to use their [3+2]-cycloaddition chemistry to generate the tetrahydrofuran portion of the lytophilippines (Scheme 38). The substituted epoxide **143** was employed to give tetrahydrofuran **147** after a Tamao–Fleming oxidation of the silyl-tetrahydrofuran product.

3.4 Other Cycloadditions

Hu and coworkers have identified a one-flask approach to tetrahydrofurans that combined an Rh-carbene O–H insertion reaction with a C–C forming Michael reaction (Scheme 39) [37]. This sequence enabled them to access subsituted tetrahydrofuran derivatives **151** with high levels of diastereoselection. They propose that the observed selectivity was dictated by the O–H insertion reaction (see **150**).

4 Oxidative Cyclizations

Inspired by the synthesis of tetrahydrofurans that are imbedded in natural products, a number of groups have examined the metal-catalyzed oxidative cyclization of 1,5-dienes and/or hydroxyolefins where the alkenes in both instances are inactivated. Metals that have been shown to be capable of carrying out these transformations include Mn, Ru, Pd, Os, and Co. As is outlined in this section, these reactions have several significant advantages over other tetrahydrofuran-

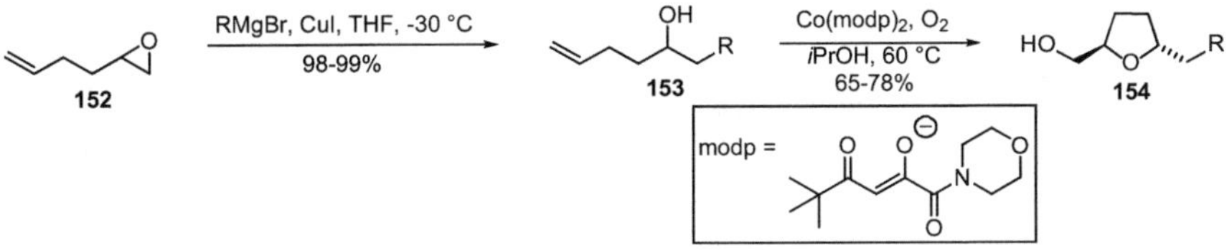

Scheme 40 Co(II)-catalyzed cyclizations to tetrahydrofurans by Sinha et al. [39]

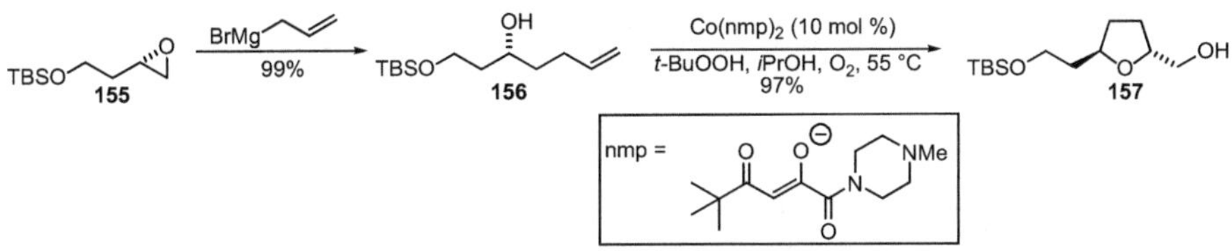

Scheme 41 Co(II)-catalyzed cyclization to tetrahydrofurans by Pagenkopf and Mora [40]

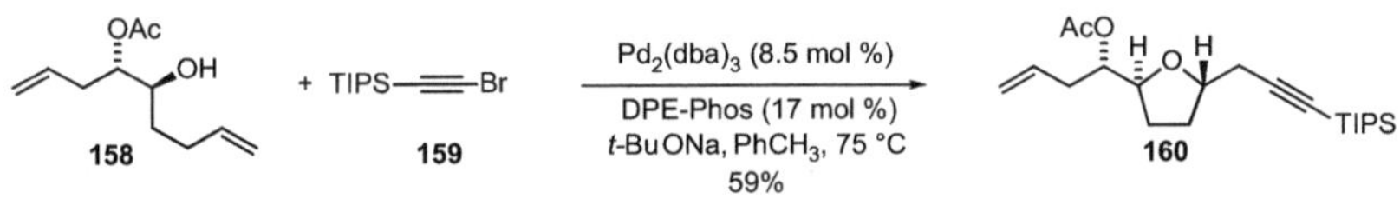

Scheme 42 Oxyalkynylations to tetrahydrofurans by Waser and Nicolai [42]

forming sequence including the ready availability of the starting substrates and the high levels of diastereocontrol that are often observed.

4.1 Cobalt-Catalyzed Cyclizations

Co(II)-catalyzed cyclizations were among the first metal-catalyzed oxidative cyclizations to be studied [38]. Sinha and coworkers have recently reported the synthesis of a library of mono- and bis-tetrahydrofuran substrates using a Co(II)-catalyzed cyclization (Scheme 40) [39]. It is worth noting that the readily available cyclization precursors came from the addition of alkyl or aryl cuprates to 1,2-epoxyhexene **152**.

Morra and Pagenkopf have utilized Co(II) catalyst that have been developed in their laboratory to generate tetrahydrofurans [40]. During their synthesis of the C18–C34 subunit of amphidinolide C, they used 10 mol% of water soluble $Co(nmp)_2$ to give 2,5-*trans*-tetrahydrofuran **157** in 97 % yield on a multi-gram scale (Scheme 41) (Forsyth has also used this chemistry see: [41]).

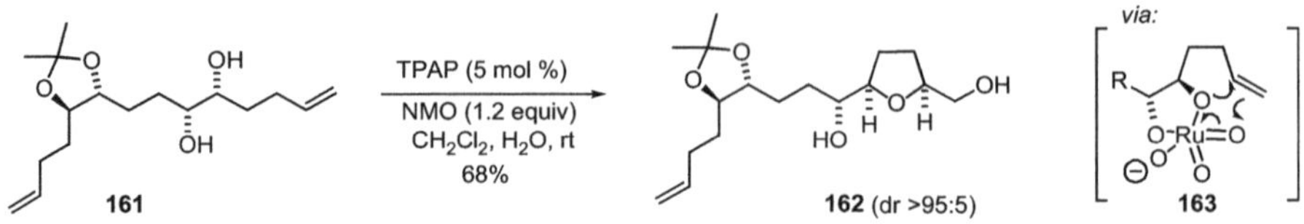

Scheme 43 Dihydroxy olefin cyclizations by Cheng and Stark [45]

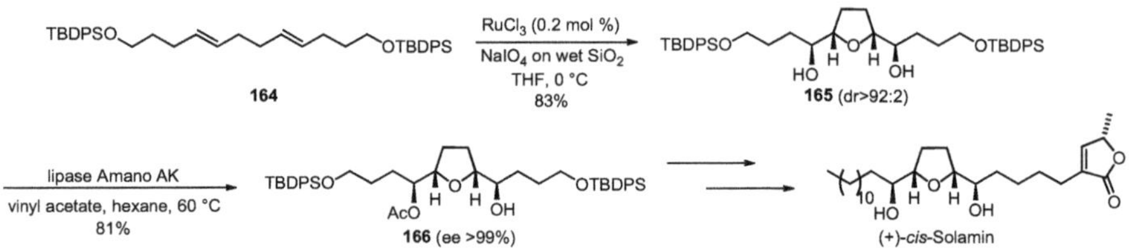

Scheme 44 Oxidative cyclization approach to *cis*-solamin by Göksel and Stark [46]

4.2 Palladium-Catalyzed Cyclizations

Palladium has also been used to carry out oxidative cyclizations of hydroxyolefins. One advantage to the use of Pd is that it is amenable to tandem cyclization–coupling sequences. To this end, Nicolai and Waser have examined oxyalkynylation reactions between hydroxyolefins and bromoacetylenes where 2,5-*trans*-tetrahydrofurans were generated from simple starting materials [42]. Illustrated in Scheme 42 is their use of this chemistry to generate tetrahydrofuran **160** in 59 % yield as a single stereoisomer. Mechanistically, they proposed a Pd(0)/Pd(II) catalytic cycle which contrasts their previously proposed Pd(II)/Pd(IV) cycle that was invoked for related lactone cyclizations [43].

4.3 Ruthenium-Catalyzed Cyclizations

RuO_4 was first shown to be effective at carrying out the oxidative cyclization of 1,5-dienes to give 2,5-*cis*-tetrahydrofurans by Sharpless and coworkers [44]. The Stark group has recently been able to optimize these reactions. For example, Stark and Cheng have described the oxidative conversion of 5,6-dihydroxy olefins into enantiomerically pure tetrahydrofurans using RuO_4 (Scheme 43) [45]. One of their more impressive examples involved the use of diene **161** to generate tetrahydrofuran **162** in 68 % yield as a single diastereomer. Over the course of their work, the authors demonstrated that the formation of Ru diester **163** was critical for successful cyclizations.

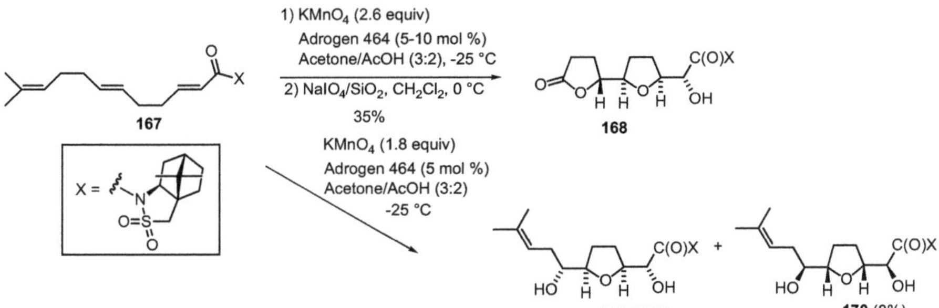

Scheme 45 Generation of tetrahydrofurans using $KMnO_4$ by Brown et al. [48]

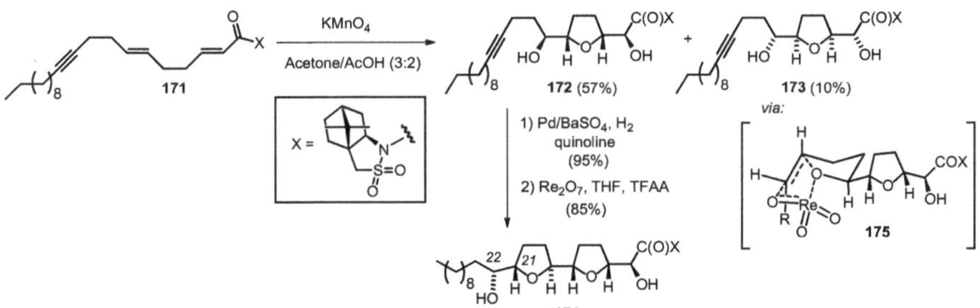

Scheme 46 Chemoselective dienyne cyclizations by Hu and Brown [50]

Illustrated in Scheme 44 is Stark and coworkers' use of RuO_4 to synthesize the tetrahydrofuran ring of the annonaceous acetogenin *cis*-solamin [46]. From diene **164** (available in four steps from cyclododecatriene), oxidative cyclization using RuO_4 formed in situ led to an 83 % yield of *meso*-tetrahydrofuran **165** as the major diastereomer. Lipase desymmetrization of **165** gave **166** in 81 % yield as a single enantiomer and diastereomer. Subsequent to the generation of **166**, differentiation of the two primary hydroxyl groups and functionalization gave *cis*-solamin.

4.4 Manganese-Mediated Cyclizations

Permanganate was the first reagent that was demonstrated to have the ability to carry out the oxidative cyclization of 1,5-dienes [47]. Brown and coworkers have continued the development of this area by discovering a diastereoselective $KMnO_4$-mediated 1,5-diene cyclization that leads to 2,5-*cis*-tetrahydrofuran derivatives. They have used a camphorsultam chiral auxiliary to control the facial selectivity and have found that polyunsaturated substrates can be chemoselectively manipulated. As an

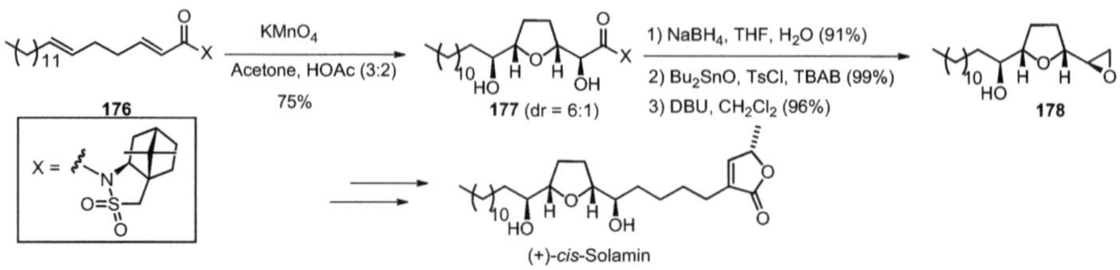

Scheme 47 $KMnO_4$ approach to (+)-*cis*-solamin by Brown et al. [51]

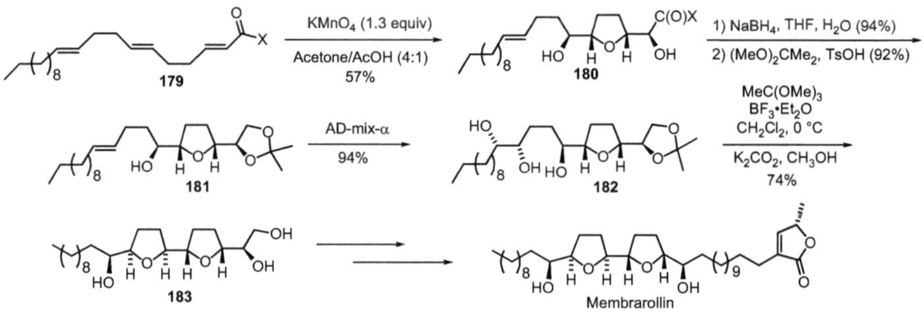

Scheme 48 Synthesis of membrarollin by Brown et al. [48]

example of both of these concepts, Brown and coworkers discovered that polyenes can undergo chemo- and stereoselective mono-tetrahydrofuran formation when the amount of $KMnO_4$ is carefully controlled (Scheme 45) [48]. Thus, while the oxidation of triene **167** using 2.6 equivalents of $KMnO_4$ gave bicycle **168** in 35 % yield after treatment with $NaIO_4$, when **167** was treated with 1.8 equivalents of $KMnO_4$, mono-tetrahydrofurans **169** and **170** were isolated in 52 % and 8 % yields, respectively.

Further building on the chemoselectivity of the $KMnO_4$ process, Brown and coworkers found that dienynes such as **171** undergo chemoselective and stereoselective oxidative cyclizations to give tetrahydrofuran derivatives, for example, **172** and **173** in good yields (Scheme 46). In addition to this, in a subsequent transformation they found that the *Z*-alkene that results from the Lindlar reduction of **173** gives tetrahydrofuran derivative **174** after a second oxidative cyclization using Re_2O_7 [49]. They propose that **174** comes from substrate-based stereocontrol, e.g., **175**, making the stereochemical outcome of the second tetrahydrofuran ring dependent upon the outcome of the initial $KMnO_4$ reaction [50].

Brown and coworkers have applied their $KMnO_4$ oxidative cyclization technology to the synthesis of several tetrahydrofuran-containing natural products. For example, the treatment of sultam cyclization precursor **176** with $KMnO_4$ in acetone and

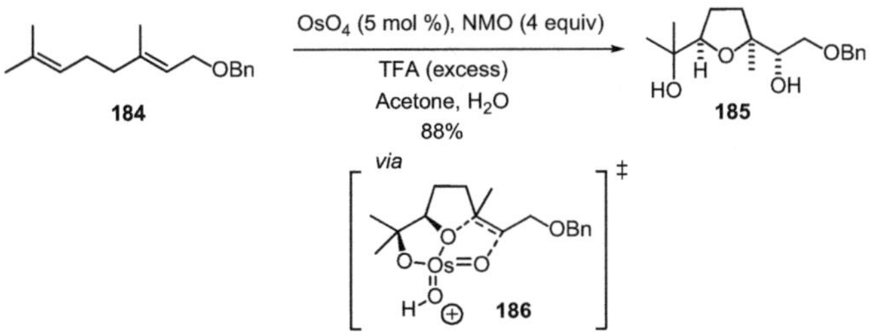

Scheme 49 Use of OsO_4 to generate tetrahydrofurans by Donohoe et al. [53, 54]

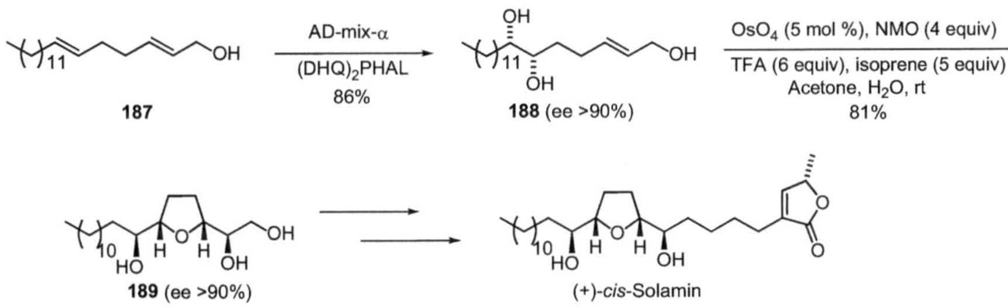

Scheme 50 Asymmetric furan cyclization and the formal total synthesis of (+)-*cis*-solamin by Donohoe et al. [55]

acetic acid gave tetrahydrofuran **177** as the major product (Scheme 47). Epoxide formation and side chain manipulation gave *cis*-solamin [51].

Brown et al. also carried out the synthesis of membrarollin (Scheme 48) [48]. Following the oxidative cyclization of triene **179** to give **180**, dihydroxylation of the remaining olefin and a subsequent orthoformate-initiated cyclization of **182** gave the second tetrahydrofuran ring of membrarollin as **183**.

4.5 Osmium-Catalyzed Cyclizations

Piccialli and coworkers first described the use of OsO_4 to induce the oxidative cyclization of 1,5-dienes in 1998 [52]. Subsequently, the Donohoe group has been at the forefront of the development and use of OsO_4 to generate tetrahydrofurans [53, 54]. Critical to their success has been the discovery that acids (CSA, TFA, citric acid, $Cu(OTf)_2$) enable them to lower their OsO_4 loadings to 5 mol% (Scheme 49). Presumably, the role of the acid in the reaction is to activate the intermediate osmate ester which induces cyclization onto the pendant olefin via transition state **186**. The only downside to their approach has been that the use of acid negates the use of chiral amine ligands. Regardless, if one considers that the

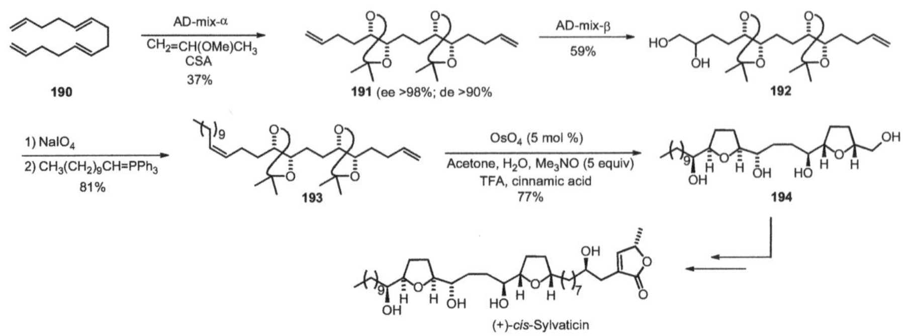

Scheme 51 Donohoe et al. Use of oxidation cyclizations to (+)-*cis*-Sylvaticin [56]

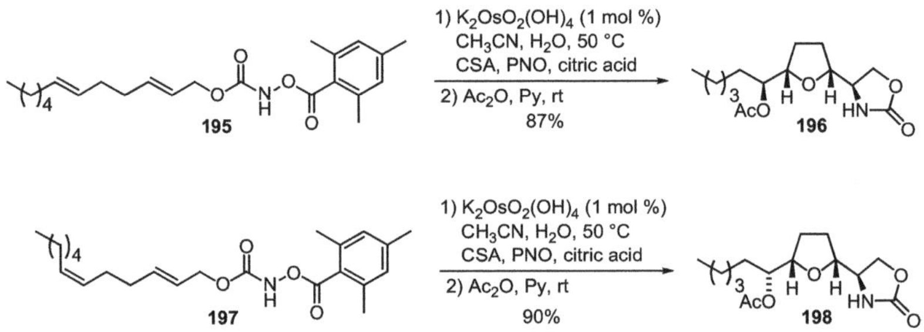

Scheme 52 Aminohydroxylation strategy to tetrahydrofurans by Donohoe et al. [58]

products of this reaction come from readily available starting materials, the reaction is impressive. For example, the oxidative cyclization of geraniol derivative **184** using a mixture of OsO_4, NMO, and TFA gave tetrahydrofuran **185** as a single diastereomer in 88 % yield.

Donohoe et al. has also developed a two-step cyclization sequence to generate optically active tetrahydrofurans [55]. As illustrated for the their work targeting the formal total synthesis of (+)-*cis*-solarin, Donohoe et al. were able to generate optically active tetrahydrofuran **189** by using an asymmetric Sharpless dihydroxylation to initially build diol **188** and by then subjecting **188** to their oxidative cyclization conditions (Scheme 50). Interestingly, the oxidative cyclization reaction of **188** requires the use of Os(VI) rather than Os(VIII); Donohoe et al. were able to employ catalytic OsO_4 by including a sacrificial olefin (isoprene) in the reaction.

In addition to the (+)-*cis*-solamin example outlined above, the Donohoe group has also applied their method in the synthesis of the annonaceous acetogenin (+)-sylvaticin (Scheme 51) [56], (for the application of the oxidative cyclization chemistry to bicyclic ring systems see: [57]). Exposure of a mixture of tetradecatetraene isomers (commercially available as a mixture of three stereoisomers) or the pure *E*,*E*-isomer to AD-mix-α gave **191** in an 18 % yield after bis-acetonide

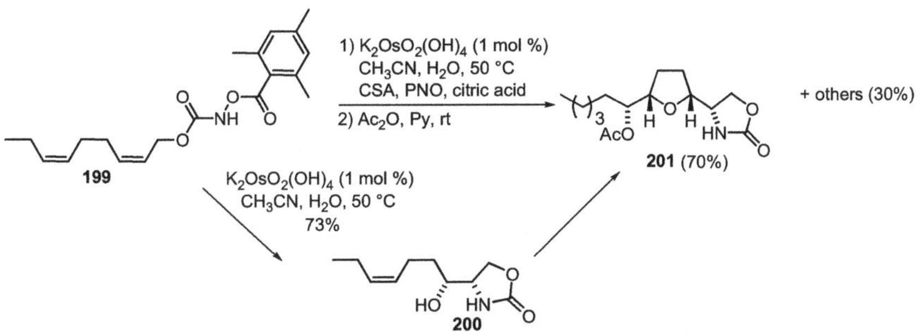

Scheme 53 (*Z*,*Z*)-Diene cyclizations to tetrahydrofurans by Donohoe et al. [58]

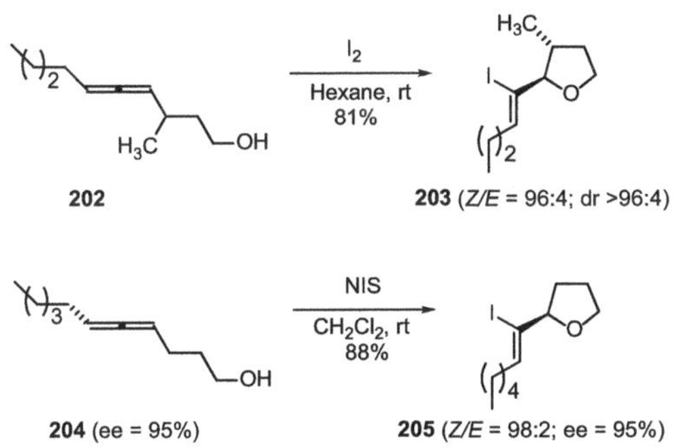

Scheme 54 Iodoetherification strategy to tetrahydrofurans by Fu, Ma et al. [60]

generation (37 % yield from the pure isomer). A second asymmetric dihydroxylation effectively differentiated the remaining terminal olefins to give diol **192** in 59 % yield along with 31 % of the corresponding tetraol. As an aside, the authors showed that the tetraol could be recycled to generate more **192**. Periodate-mediated oxidative cleavage of the diol and Wittig olefination gave bis-cyclization precursor **193**. Impressively, when **193** was exposed to the oxidative cyclization conditions under acidic conditions, both hydrolysis of the bis-acetonide and oxidative cyclization occurred to provide bis-tetrahydrofuran **194** in 77 % yield. Attachment of the requisite butenolide gave (+)-*cis*-sylvaticin.

In addition to dihydroxylations, Donohoe and coworkers have also used aminohydroxylation of 1,5-dienes to synthesize tetrahydrofuran substrates [58]. As in the dihydroxylation/cyclization sequence, they either employed a one-pot or a two-pot protocol. An example of the one-pot reaction is illustrated in Scheme 52. Exposure of (*E*,*E*)-diene **195** to aminohydroxylation conditions efficiently gave 2,5-*cis*-tetrahydrofuran **196**. Based on the proposed mechanism for the reaction, it is not surprising that the reaction of (*E*,*Z*)-diene **197** gave diastereomeric 2,5-*cis*-tetrahydrofuran **198** in 90 % yield. The aminohydroxylation

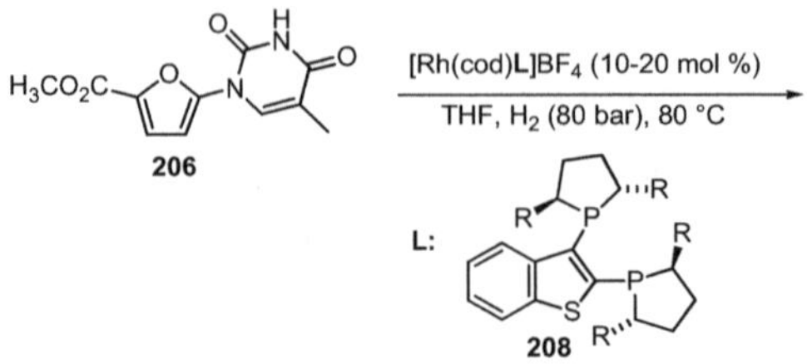

R	Loading	Yield	ee
CH_3	20 mol %	10%	0%
CH_2CH_3	10 mol %	60%	32%
$CH(CH_3)_2$	10 mol %	98%	72%

Scheme 55 Furan reductions to deoxynucleosides by Albert et al. [62]

sequence differed from Donohoe et al. previous oxidative cyclizations in that stoichiometric pyridine *N*-oxide was used to oxidize Os(IV) to Os(VI) and that *N,O*-carbamates were used as internal oxidants to bring the oxidation state of osmium to Os(VIII).

The application of the one-pot conditions to (*Z*,*Z*)-substrate **199** was not as clean as the reactions of either **195** or **197** (Scheme 53). By subjecting **199** to a stepwise sequence where aminohydroxylation was followed by cyclization, the authors were able to conclude that the problematic step for the use of **199** was the oxidative cyclization to **201** rather than the aminohydroxylation to **200**. Similar problems were found for the (*Z*,*E*)-diene that corresponds to **199**.

4.6 *Intramolecular Iodoetherification*

In 1981, the work of Rychnovsky and Bartlett demonstrated the utility of iodoetherification to build tetrahydrofurans [59]. Others have recently built upon these studies. For example, Fu, Ma, and coworkers have carried out iodoetherification of allenyl alcohols [60]. Representative examples are given in Scheme 54 where **204** (available from a Claisen rearrangement of the corresponding enantiomerically enriched propargyl alcohol) underwent cyclization to give **205** in high yield, with high diastereoselectivity and with complete preservation of the optical purity of the starting allene. When substitution was present in the linker between the alcohol and the allene, 2,3-*trans*-tetrahydrofuran **203** (R $=$ CH_3) was isolated.

5 Furan Reductions

While not as widely used as other methods of generating tetrahydrofurans, furan reductions can be an effective means of generating saturated rings. Outlined here is a single example which represents the state of the art in 2012 [61]. In an effort to

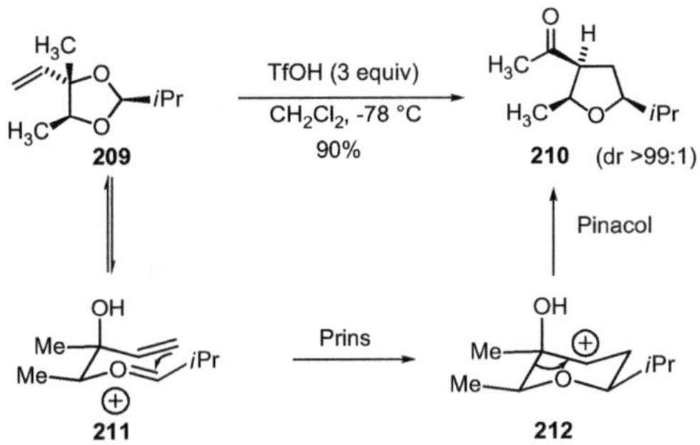

Scheme 56 Prins-Pinacol cascade to substituted tetrahydrofurans by Overman et al. [64]

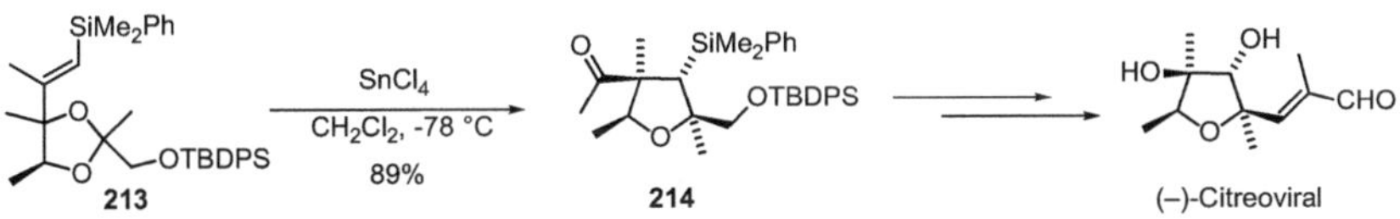

Scheme 57 Prins-pinacol cascade to (–)-citreoviral by Overman et al. [65]

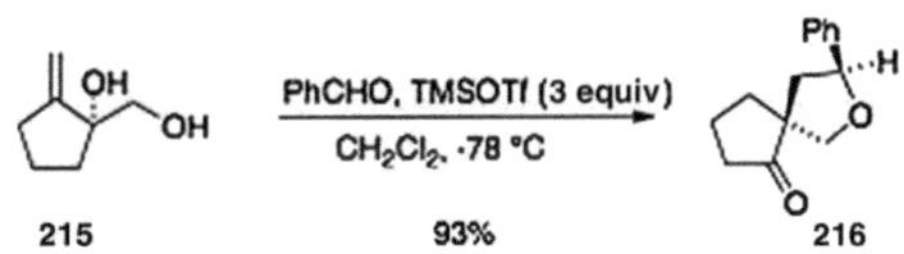

Scheme 58 Synthesis of spiro-fused tetrahydrofurans from Prins-pinacol cascades by Cho et al. [66]

generate enantiomerically pure 2,3-dideoxynucleoside analogs, Albert and coworkers examined the asymmetric hydrogenation of 2,5-disubstituted furans (Scheme 55) [62]. A survey of several different homogenous catalysts led them to conclude that the steric bulk of the ligand was important both for activity and selectivity. Their best result came from the use of the butiphane-type Rh-catalyst **208** which gave 3,4-dideoxynucleoside **207** in 98 % yield and in 72 % ee from the reduction of furan **206**.

6 Prins-Pinacol Cascades

The Overman group has studied and utilized Prins-pinacol cascades to generate highly substituted tetrahydrofuran rings over the past 20 years [63]. An example from the Overman group is given in Scheme 56 where acetal **209** (from the

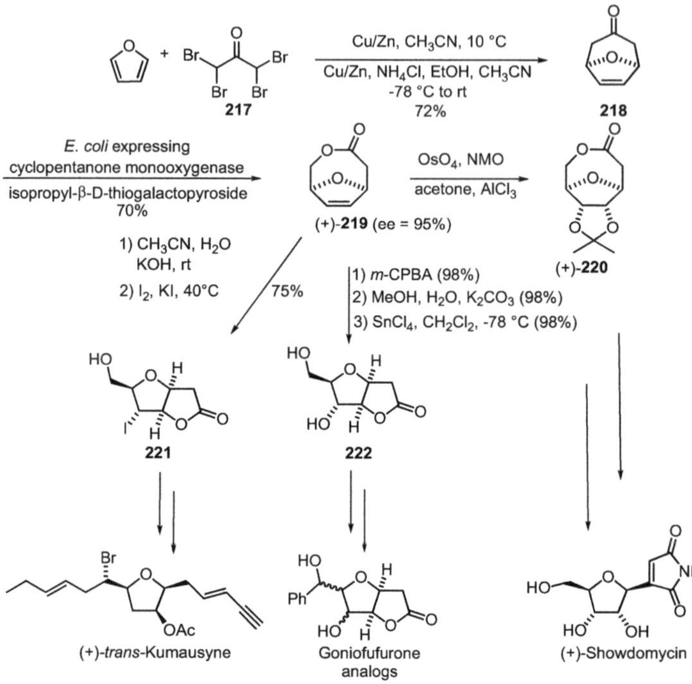

Scheme 59 Microbial approach to tetrahydrofurans by Mihovilovic et al. [67]

corresponding *syn*-diol) was used to give 2,3,5-*cis*-tetrahydrofuran **210** in 90 % yield as a single diastereomer [64]. The reaction of **209** has been proposed to proceed via chair TS **211** to initially give **212** where the methyl group that is attached to the oxonium ion lies in the equatorial position in the transition state. A subsequent pinacol rearrangement gives **210**.

A particularly impressive example of the Prins-pinacol chemistry is illustrated in Scheme 57 where tetrahydrofuran **214** having five contiguous stereocenters was generated in 89 % yield from the treatment of acetal **213** with $SnCl_4$. As illustrated, **214** served as a precursor to the natural product (–)-citreoviral [65].

Not surprisingly, other groups have also examined Prins-pinacol reactions. Cho et al. have used Prins-pinacol cascades to build spiro-fused tetrahydrofurans (Scheme 58) [66].

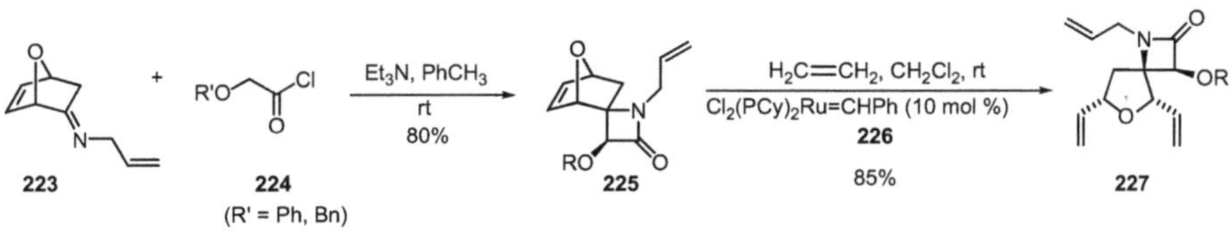

Scheme 60 ROM/RCM approach to spiro-fused tetrahydrofurans by Plumet et al. [68]

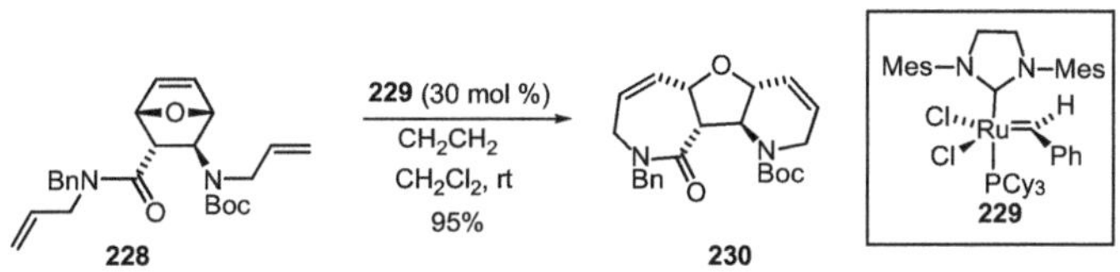

Scheme 61 ROM/RCM to fused tetrahydrofurans by Winkler et al. [69]

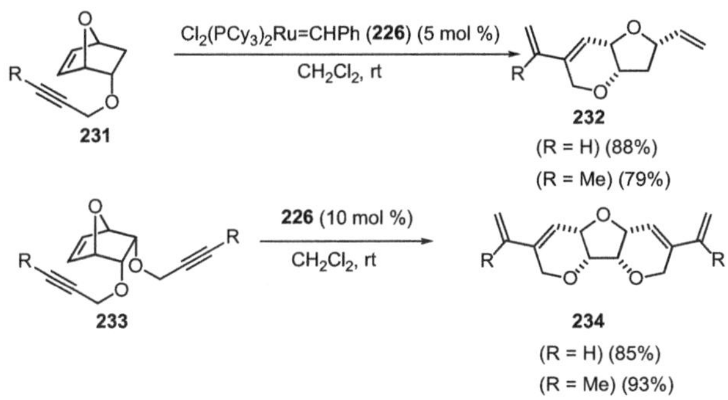

Scheme 62 Enyne metathesis route to polycyclic tetrahydrofurans by Plumet et al. [70]

7 Ring-Opening of Bicyclic Substrates

Oxabicyclic substrates are attractive precursors to tetrahydrofurans as a consequence of the ease with which they are synthesized and that they lead to the exclusive generation of 2,5-*cis*-tetrahydrofurans. Included in this section is one example of an enzymatic approach to tetrahydrofurans from an oxabicyclic ring system and several examples of ring-opening metathesis approaches to tetrahydrofurans.

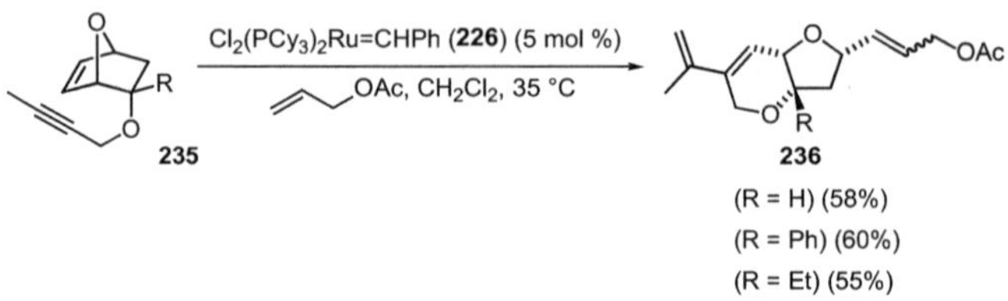

Scheme 63 Enyne cross-metathesis to tetrahydrofurans by Plumet et al. [70]

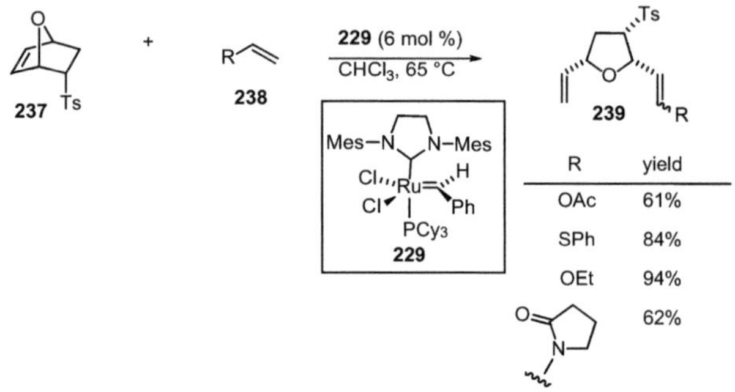

Scheme 64 Regioselective ROCM reactions by Rainier and Liu [71]

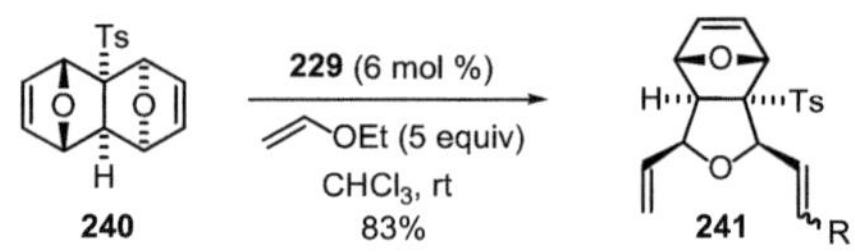

Scheme 65 Regioselective generation of tricyclic tetrahydrofurans by Rainier and Liu [71]

7.1 Microbial Oxidation of Oxabicyclic Substrates

Mihovilovic and coworkers have utilized a microbial Baeyer–Villiger oxidation to desymmetrize readily available 8-oxabicyclo[3.2.1]octene **218**. Their key reaction used *E. coli* in a whole-cell oxidation to give multi-gram quantities of **219** in high optical purity (Scheme 59) [67]. To demonstrate its utility they transformed **219** into **220**, **221**, and **222** thus intercepting intermediates that had previously been converted into the natural products (+)-*trans*-kumausyne, goniofufurone, and (+)-showdomycin.

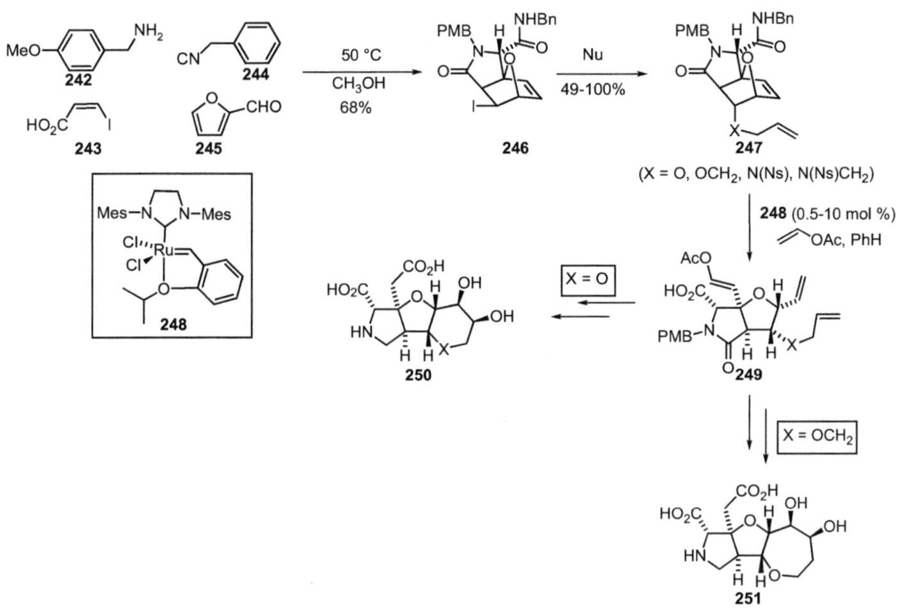

Scheme 66 ROM/CM to generate furan-containing glutamate ligands by Oikawa et al. [72]

7.2 Ring-Opening Metathesis

A number of groups have examined the generation of substituted tetrahydrofurans from ring-opening metathesis/cross-metathesis (ROM/CM) reactions of 7-oxabicyclo[2.2.1]heptenes. The Plumet group has had a tremendous impact on this area. One example of their work is highlighted in Scheme 60 where they demonstrated the use of ROM/CM reactions and ring-opening metathesis/ring-closing metathesis (ROM/RCM) reactions to generate 2,5-*cis*-tetrahydrofurans having spiro-fused β-lactams at C3 [68]. The ring-opening precursors **225** were generated from allylamine **223** using a Staudinger reaction. ROM/RCM of **225** using the first generation Grubbs catalyst **226** and ethylene enabled them to efficiently generate the spiro-fused furan **227**. The corresponding ROM/RCM was limited to ethers as the use of an ester or thioether only afforded ROM/CM products.

Winkler and coworkers have developed tandem ROM/RCM that leads to furan-containing polycyclic substrates [69]. As an example of their work, the ROM/RCM cascade of readily available oxanorbornene **228** using the second generation Grubbs ruthenium catalyst **229** gave tricyclic substrate **230** in 95 % yield (Scheme 61). The use of this strategy to generate even more complex substrates than **230** was somewhat problematic.

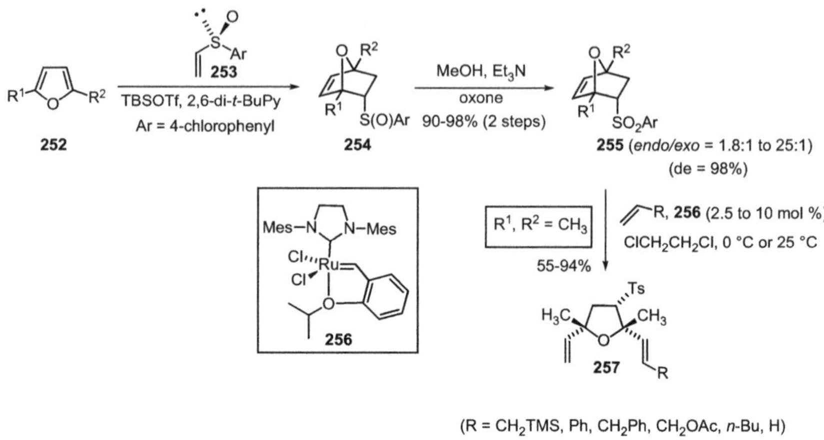

Scheme 67 Optically active 7-oxanorbornenes in ROCM reactions by Martin and Benjamin [73]

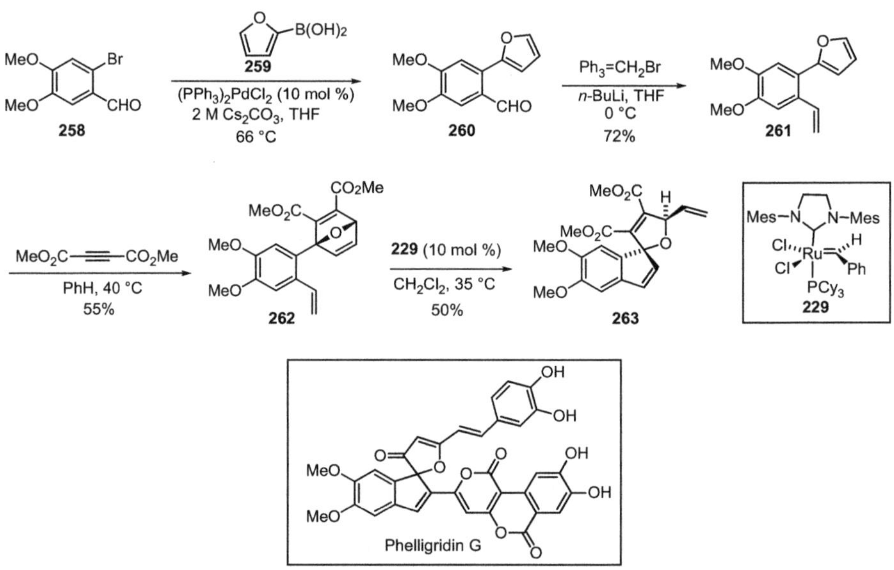

Scheme 68 ROM/RCM approach to the Phelligridin G scaffold by Cooper and Wright [74]

Plumet and coworkers successfully carried out ring-opening, ring-closing enyne metathesis reactions to give bicyclic and tricyclic tetrahydrofuran derivatives [70]. For example, the ring-opening and ring-closing enyne metathesis of **231** resulted in the generation of bicyclic derivative **232**. In a similar fashion, the reaction of oxanorbornene **233** gave tricyclic tetrahydrofuran **234** (Scheme 62).

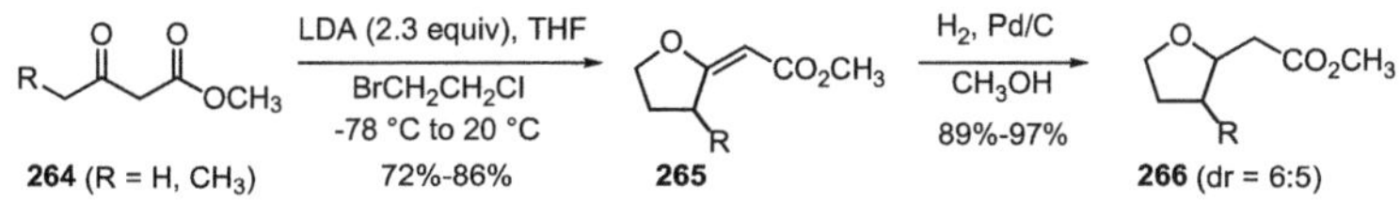

Scheme 69 Nucleophilic substitution/cyclization strategy to tetrahydrofurans by Langer et al. [76]

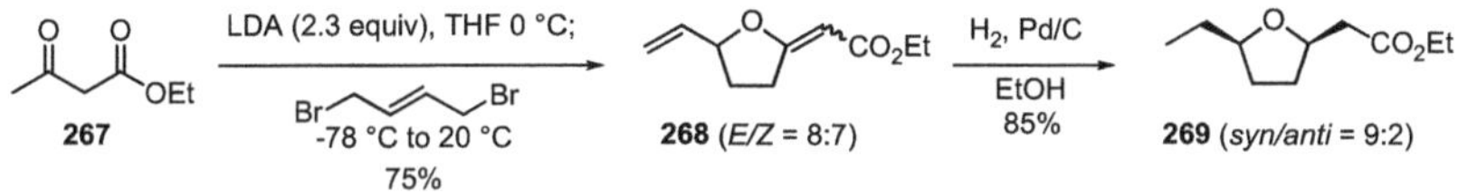

Scheme 70 Nucleophilic substitution/cyclization strategy to tetrahydrofurans by Langer et al. [76]

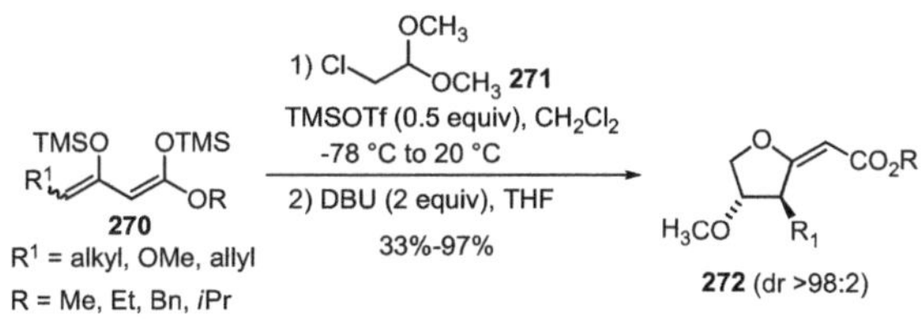

Scheme 71 Silyl enol ethers as precursors of 2-alkylidene tetrahydrofurans by Langer et al. [77]

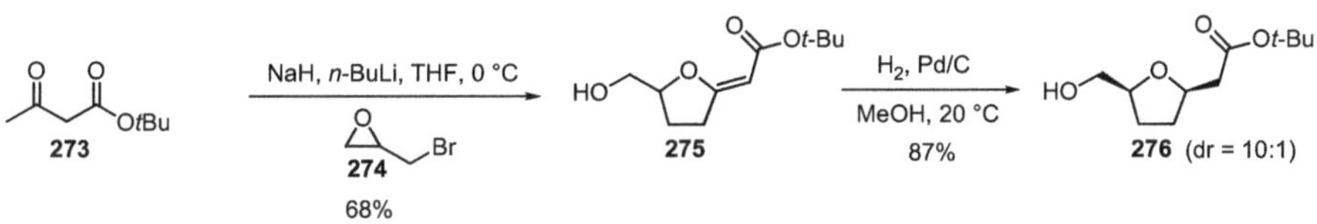

Scheme 72 Epibromohydrin approach to tetrahydrofurans by Langer et al. [78]

The enyne metathesis sequence could also be coupled with a subsequent cross-metathesis. As a demonstration of this, Plumet and coworkers exposed **235** to the first generation Grubbs catalyst and allyl acetate to give bicyclic substrate **236** in good yield (Scheme 63).

Liu and Rainier have reported ROM/CM reactions of 7-oxanorbornenes with both neutral and electron-rich olefins to give the corresponding furans (Scheme 64) [71]. Of particular interest in these studies was that the reaction was highly regioselective giving 2,3,5-trisubstituted furans having the substituted alkene at C2 as illustrated by **239**. While the majority of the substrates utilized a tosyl group on the oxanorbornene, other substituents (ester, benzyloxymethyl) showed similar selectivity.

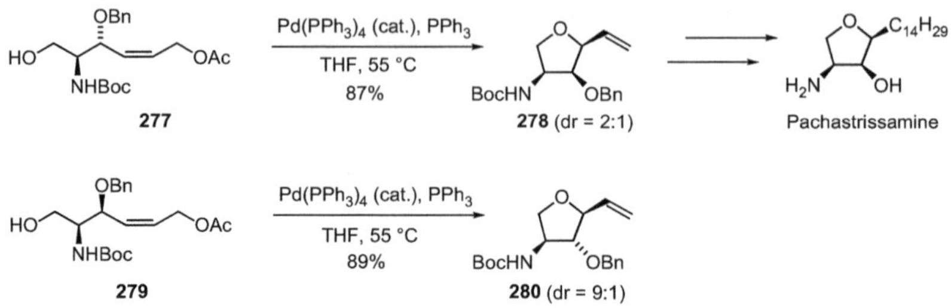

Scheme 73 π-Allyl cyclizations to the pachastrissamine tetrahydrofuran by Passiniemi and Koskinen [79]

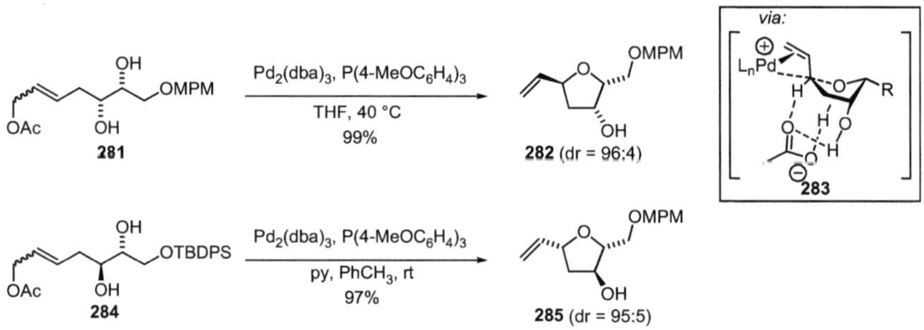

Scheme 74 Use of Pd-cyclizations to generate tetrahydrofurans by Gandon, Roulland et al. [80]

In addition to the reactions outlined above, regioselective ROM/CM was used to generate tricyclic tetrahydrofuran **241** (Scheme 65).

Oikawa and coworkers have employed unsymmetrical 7-oxanorbornenes in regioselective ROM/CM reactions and have used the products from these reactions to generate glutamate analogs **250** and **251** [72]. As illustrated in Scheme 66, by combining amine **242** with furfural **245**, benzylisonitrile **244**, and β-iodoacrylic acid **243**, they were able to generate oxanorbornene **246** in 68 % yield from a tandem Ugi/Diels-Alder cascade. The iodide in **246** was amenable to substitution with oxygen and nitrogen nucleophiles to give tricyclic metathesis precursor **247**. The authors subsequently used the Hoveyda–Grubbs second-generation catalyst **248** to carry out an ROM/CM reaction with vinyl acetate to give tetrahydrofuran **249** which was taken on to glutamate analogs.

Benjamin and Martin have employed optically active oxanorbornenes in regioselective ROM/CM reactions (Scheme 67) [73].

From an interest in the phelligridin G scaffold Cooper and Wright have reported a tandem ROM/RCM approach to oxaspirocycles (Scheme 68) [74]. The ROM/RCM reaction of readily available oxabicycle **262** using the Grubbs second-generation catalyst **229** resulted in the generation of spirocycle **263** in 50 % yield.

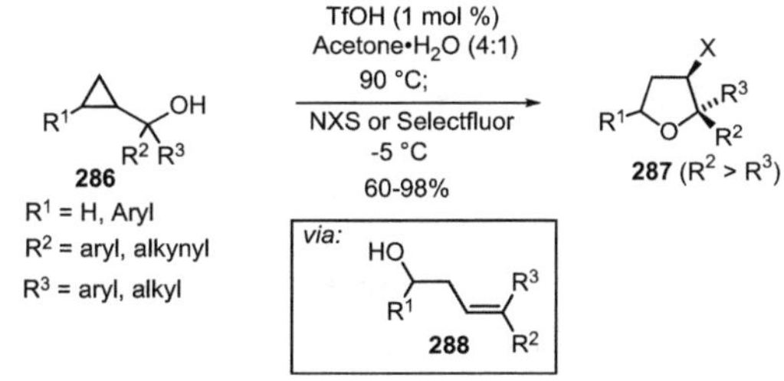

Scheme 75 Halocyclization to Tetrahydrofurans by Chan et al. [81]

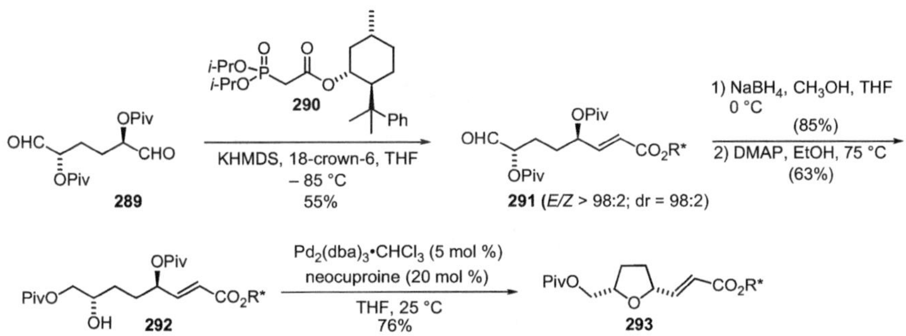

Scheme 76 Pd-Catalyzed cyclizations to *cis*-tetrahydrofurans by Rein and Vares [82]

8 Nucleophilic Substitutions

Another approach to substituted tetrahydrofurans has involved the intramolecular addition of oxygen nucleophiles to electrophilic carbon atoms [75]. Outlined here are recent efforts in this area.

Langer and coworkers have synthesized a number of substituted tetrahydrofurans from alkylidene tetrahydrofurans which in turn were generated from an alkylation/cyclization sequence of β-ketoesters [76]. For example, alkylation of the dianion of β-ketoester **264** with 1-bromo-2-chloroethane resulted in the synthesis of 2-alkylidene tetrahydrofuran **265** (Scheme 69).

In an analogous fashion to the results with 1-bromo-2-chloroethane, the treatment of the dianion from ethyl acetoacetate with 1,4-dibromo-2-butene resulted in the generation of 2-alkylidenetetrahydrofuran **268** as a 1:1 mixture of olefin isomers in 75 % yield (Scheme 70). Hydrogenation of **268** gave 2,5-*cis*-tetrahydrofuran **269**.

Langer et al. have also carried out alkylation and cyclization reactions from bis-silyl enol ether **270** using 1-chloro-2,2-dimethoxyethane **271** as the electrophile to give 2-alkylidiene tetrahydrofuran **272** (Scheme 71) [77].

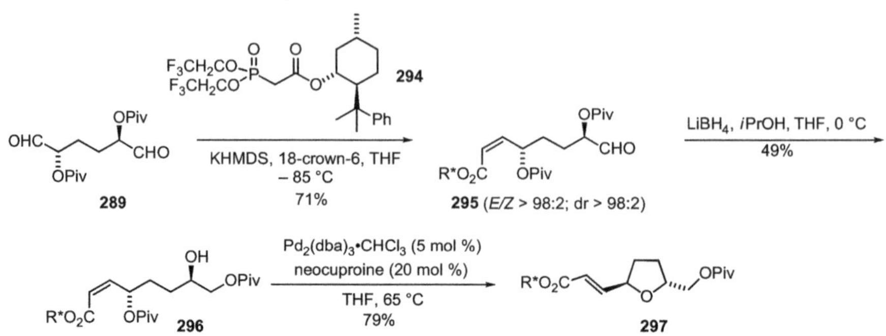

Scheme 77 Pd-Catalyzed cyclizations to *trans*-tetrahydrofurans by Rein and Vares [82]

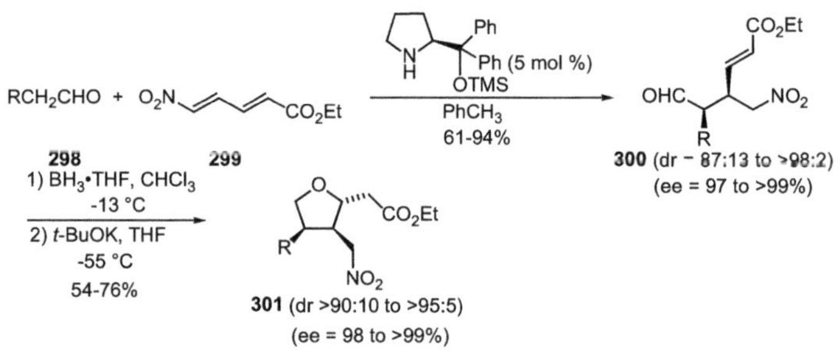

Scheme 78 Tandem Michael additions to tetrahydrofurans by Ender et al. [83]

As illustrated in Scheme 72, Langer et al. were also able to generate 2-alkylidene tetrahydrofuran substrates from the alkylation of β-ketoesters with epibromohydrins (Scheme 72) [78]. Hydrogenation gave 2,5-*cis*-tetrahydrofuran **270** with relatively high diastereocontrol.

Koskinen et al. have employed Pd-catalyzed π-allyl cyclizations to generate tetrahydrofurans including the tetrahydrofuran present in the cytotoxic natural product pachastrissamine (Scheme 73) [79]. From *Z*-allyl acetate **277** they were able to generate tetrahydrofuran **278** as the major product in 59 % yield. Interestingly, they found the stereochemistry adjacent to the alkene to be an important factor in the selectivity. The use of the *syn*-isomer **279** resulted in the generation of tetrahydrofuran **280** in 81 % yield with 9:1 diastereoselectivity. Tetrahydrofuran **278** was subsequently transformed into pachastrissamine.

Gandon, Roulland, and coworkers have also used Pd-catalyzed π-allyl cyclizations to generate tetrahydrofurans (Scheme 74) [80]. In a related fashion to Koskinen et al. results, they found that product stereochemistry was dependent upon the relative stereochemistry of the stereocenters on the allylic acetate. With *cis*-diol **281** they isolated *trans*-tetrahydrofuran **282** as the major product, while the use of the corresponding *trans*-diol **284** resulted in the generation of *cis*-tetrahydrofuran **285**.

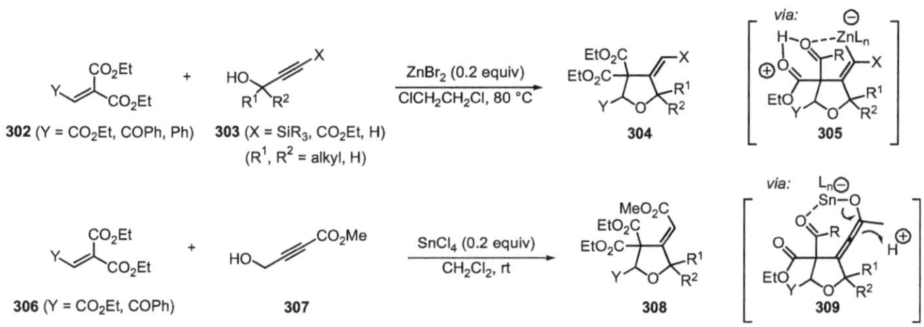

Scheme 79 [3+2]-Cycloaddition approach to tetrahydrofurans by Yamazaki et al. [84]

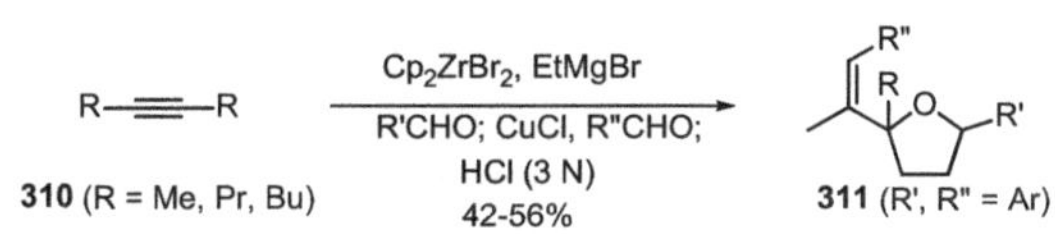

Scheme 80 Zirconacyclopentene approach to tetrahydrofurans by Xi et al. [85]

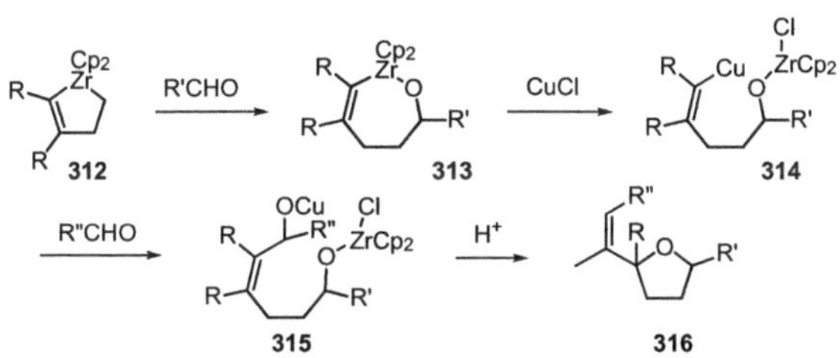

Scheme 81 Proposed mechanism for tetrahydrofuran formation from zirconacyclopentene by Xi et al. [85]

The diastereoselectivity was significantly diminished when the hydroxyl was replaced with an ether or by hydrogen. Based on this and DFT calculations, the authors propose a counterion-directed reaction where acetate both coordinates to the allyl cation hydrogen atom and the hydroxyl group as illustrated in **283**.

Chan and coworkers have examined a sequential Brønsted acid, halocyclization of cyclopropyl methanol substrates to generate tetrahydrofurans (Scheme 75) [81]. The one-pot, two-step cyclization protocol involved initially subjecting **286** to TfOH and H_2O followed by treating the resulting alkene with NIS to give substituted tetrahydrofurans **287**. This reaction requires that tertiary alcohols be used and was selective for the formation of the isomer having the larger substituent and the halogen *cis* to one another. The authors showed that the reaction proceeds via homoallylic alcohol **288**.

Vares and Rein have developed a powerful approach to tetrahydrofurans where they coupled an asymmetric Horner–Wadsworth–Emmons (HWE) reaction with a Pd-catalyzed ring closure to generate *cis*- and *trans*-tetrahydrofuran derivatives (Schemes 76 and 77) [82]. From *meso*-dialdehyde **289**, asymmetric HWE gave *E*-alkene isomer **291** with high levels of diastereoselectivity. Aldehyde reduction was followed by pivaloyl migration to afford cyclization precursor **292** in 63 % yield. Pd-catalyzed π-allyl formation and cyclization proceeded with overall retention of stereochemistry to give 2,5-*cis*-tetrahydrofuran **293** in 76 % yield.

2,5-*trans*-Tetrahydrofurans were generated by Vares and Rein through the use of the corresponding *Z*-olefin in the Pd-catalyzed cyclization. Asymmetric HWE using trifluoroethylphosphonate ester **294** gave *Z*-olefin **295**. Reduction of the remaining aldehyde gave a 1.3:1 mixture of secondary alcohol **296** and the corresponding primary alcohol. After separation, Pd-catalyzed cyclization resulted in the generation of 2,5-*trans*-tetrahydrofuran **297** as the *E*-olefin isomer.

Enders and coworkers have observed high levels of enantioselectivity from organocatalytic tandem Michael additions (Scheme 78). Their approach to tetrahydrofurans involved an initial nitrodiene-aldehyde condensation followed by an intramolecular oxy-Michael addition [83]. As illustrated, the sequence gave **301** in high diastereo- and enantioselectivity.

9 Miscellaneous Approaches to Tetrahydrofurans

Yamazaki and coworkers have described formal [3+2]-cycloadditions of ethene tricarboxylates with substituted propargyl alcohols to give methylene tetrahydrofurans (Scheme 79) [84]. Interesting was that the *E*/*Z* selectivity of the reaction depended upon the Lewis acid that was used. *Z*-Alkene isomers **304** were obtained in moderate to high yields when triply activated alkene **302** was treated with propargyl alcohol **303** in the presence of $ZnBr_2$, $InCl_3$, $FeCl_3$, or $AlCl_3$ salts. The corresponding *E*-isomer **308** was obtained when $SnCl_4$ was used. The authors speculate that protonation in the $SnCl_4$ reaction occurs from allenyl species **309** where tin is coordinated to oxygen. Protonation in the Zn-, In-, Fe-, and Al-catalyzed reactions is thought to occur from the *Z*-vinyl species **305**.

Xi and coworkers have discovered that zirconacyclopentadienes can be used to generate tetrahydrofurans when they are reacted with aldehydes [85]. The transformation requires the presence of CuCl and two equivalents of aldehyde (Scheme 80).

To explain the process, Xi et al. have proposed a mechanism where zirconacycle **312** reacts with the first equivalent of aldehyde to give **313**. Transmetallation with CuCl and reaction with the second equivalent of aldehyde provides cyclization precursor **315**. Acidic workup and cyclization provide the corresponding tetrahydrofuran (Scheme 81).

10 Conclusions

Significant progress has been made in the development of new or the optimization of old methods that target the synthesis of substituted tetrahydrofuran substrates. It is expected that these efforts will continue with a focus on the application of the methods described here along with the discovery of new methods that enable even simpler starting materials to be used.

References

1. Pan J-Y, Chen S-L, Yang M-H, Wu J, Sinkkonen J, Zou K (2009) Nat Prod Rep 26:1251–1292
2. Ferrier RJ (1998) Carbohydr Chem 30:268–315
3. Liaw C-C, Wu T-Y, Chang F-R, Wu Y-C (2010) Planta Med 76:1390–1404
4. Lorente A, Lamariano-Merketegi J, Albericio F, Alvarez M (2013) Chem Rev. ASAP
5. For a review that covers the period up to 2005, see Wolfe JP, Hay MB (2006) Tetrahedron 63:261–290
6. Larsen CH, Ridgway BH, Shaw JT, Woerpel KA (1999) J Am Chem Soc 121:12208–12209
7. Kim H, Wooten CM, Park Y, Hong J (2007) Org Lett 9:3965–3968
8. Mitchell TA, Romo D (2007) J Org Chem 72:9053–9059
9. Mead KT, Pillai SK (1993) Tetrahedron Lett 34:6997
10. Yang HW, Romo D (1997) J Org Chem 62:4–5
11. Mitchell TA, Zhao C, Romo D (2008) J Org Chem 73:9544–9551
12. Gharpure SJ, Shukla MK, Vijayasree U (2009) Org Lett 11:5466–5469
13. Opalka S, Steinbacher J, Lambiris B, McQuade D (2011) J Org Chem 76:6503–6517
14. Donohoe TJ, Harris RM, Williams O, Hargaden GC, Burrows J, Parker J (2009) J Am Chem Soc 131:12854–12861
15. Donohoe TJ, Williams O, Churchill GH (2008) Angew Chem Int Ed 47:2869–2871
16. Pohlhaus P, Johnson J (2005) J Org Chem 70:1057–1059
17. Smith AG, Slade MC, Johnson JS (2011) Org Lett 13:1996–1999
18. Pohlaus PD, Sanders SD, Parsons AT, Li W, Johnson JS (2008) J Am Chem Soc 130: 8642–8650
19. Yang G, Shen Y, Li K, Sun Y, Hua Y (2011) J Org Chem 76:229–233
20. Yang G, Sun Y, Shen Y, Chai Z, Zhou S, Chu J, Chai J (2013) J Org Chem 78. ASAP
21. Pohlhaus P, Johnson J (2005) J Am Chem Soc 127:16014–16015
22. Benfatti F, de Nanteuil F, Waser J (2012) Org Lett 14:386–389
23. Miyake T, Endo S, Moriyama T, Sakata K, Nishibayashi Y (2013) Angew Chem Int Ed 52: 1758–1762
24. Parsons AT, Johnson JS (2009) J Am Chem Soc 131:3122–3123
25. Campbell MJ, Johnson JS (2009) J Am Chem Soc 131:10370–10371
26. Sanders SD, Ruiz-Olalla A, Johnson JS (2009) Chem Commun: 5135–5137
27. Karadeolian A, Kerr MA (2010) J Org Chem 75:6830–6841
28. Shin C, Cavre SN, Pae AN, Cha JH, Koh HY, Chang MH, Choi JH, Cho YS (2005) Org Lett 7:3283–3285
29. Chavre SN, Choo H, Cha JH, Pae AN, Choi KI, Cho YS (2006) Org Lett 8:3617–3619
30. Masse CE, Panek JS (1995) Chem Rev 95:1293–1316
31. Micalizio GC, Roush WR (2000) Org Lett 2:461–464
32. Hodgson DM, Salik S (2012) Org Lett 14:4402–4405
33. Akiyama T, Funaki S, Fuchibe K (2006) Heterocycles 67:369–374

34. Malkov AV, Kysilka O, Edgar M, Kadlcikova A, Kotora M, Kocovsky P (2011) Chem Eur J 17:7162–7166
35. Tinsley JM, Mertz E, Chong PY, Rarig R-AF, Roush WR (2005) Org Lett 7:4245–4248
36. Tinsley JM, Roush WR (2005) J Am Chem Soc 127:10818–10819
37. Xu X, Han X, Yang L, Hu W (2009) Chem Eur J 15:12604–12607
38. Inoki S, Mukaiyama T (1990) Chem Lett: 67–70
39. Mishra J, Wipf P, Sinha S (2010) J Comb Chem 12:609–612
40. Morra NA, Pagenkopf BL (2011) Org Lett 13:572–575
41. Wu D, Forsyth CJ (2013) Org Lett 15:1178–1181
42. Nicolai S, Waser J (2011) Org Lett 13:6324–6327
43. Nicolai S, Erard S, González Fernández D, Waser J (2010) Org Lett 12:384–387
44. Carlsen PHJ, Katsuki T, Martin VS, Sharpless KB (1981) J Org Chem 46:3936–3938
45. Cheng H, Stark CBW (2010) Angew Chem Int Ed 49:1587–1590
46. Göksel H, Stark CBW (2006) Org Lett 8:3433–3436
47. Klein E, Rohahn W (1965) Tetrahedron 21:2353–2358
48. Morris CL, Hu Y, Head GD, Brown LJ, Whittinhgam WG, Brown RCD (2009) J Org Chem 74:981–988
49. Avedissian H, Sinha SC, Yazbak A, Sinha A, Neogi P, Sinha SC, Keinan E (2000) J Org Chem 65:6035–6051
50. Hu Y, Brown RCD (2005) Chem Commun: 5636–5637
51. Cecil ARL, Hu Y, Vicent MJ, Duncan R, Brown RCD (2004) J Org Chem 69:3368–3374
52. Champdoré M, Lasalvia M, Piccialli V (1998) Tetrahedron Lett 39:9781–9784
53. Donohoe TJ, Butterworth S (2003) Angew Chem Int Ed 115:978–981
54. Donohoe TJ, Winship PCM, Pilgrim BS, Walter DS, Callens CKA (2010) Chem Commun 46:7310–7312
55. Donohoe TJ, Butterworth S (2005) Angew Chem Int Ed 44:4766–4768
56. Donohoe TJ, Harris RM, Burrows J, Parker J (2006) J Am Chem Soc 128:13704–13705
57. Donohoe TJ, Winship PCM, Tatton MR, Szeto P (2011) Angew Chem Int Ed 50:7604–7606
58. Donohoe TJ, Lindsay-Scott PJ, Parker JS (2009) Tetrahedron Lett 50:3523–3526
59. Rychnovsky SD, Bartlett PA (1981) J Am Chem Soc 103:3963–3964
60. Lü B, Jiang X, Fu C, Ma S (2009) J Org Chem 74:438–441
61. Wang D-S, Chen Q-A, Lu S-M, Zhou Y-G (2012) Chem Rev 112:2557–2590
62. Feiertag P, Albert M, Nettekoven U, Spindler F (2006) Org Lett 8:4133–4135
63. Overman LE (2009) Tetrahedron 65:6432–6446
64. Cohen F, MacMillan DWC, Overman LE, Romero A (2001) Org Lett 3:1225–1228
65. Brown MJ, Harrison T, Herrinton PM, Hopkins MH, Hutchinson KD, Mishra P, Overman LE (1991) J Am Chem Soc 113:5365–5537
66. Chavre SN, Ullapu PR, Min S-J, Lee JK, Pae AN, Kim Y, Cho YS (2009) Org Lett 11: 3834–3837
67. Mihovilovic M, Bianchi D, Rudroff F (2006) Chem Commun: 3214–3216
68. Csákÿ AG, Medel R, Murcia MC, Plumet J (2005) Helv Chim Acta 88:1387–1396
69. Winkler JD, Asselin SM, Shepard S, Yuan J (2004) Org Lett 6:3821–3824
70. Aljarilla A, Murcia MC, Csákÿ AG, Plumet J (2009) Eur J Org Chem: 822–832
71. Liu Z, Rainier JD (2005) Org Lett 7:131–133
72. Oikawa M, Ikoma M, Sasaki M, Gill MB, Swanson GT, Shimamoto K, Sakai R (2009) Eur J Org Chem: 5531–5548
73. Benjamin NM, Martin SF (2011) Org Lett 13:450–453
74. Cooper HD, Wright DL (2013) Molecules 18:2438–2448
75. Nicolaou KC, Prasad CVC, Somers PK, Hwang C-K (1989) J Am Chem Soc 111:5330–5334
76. Bellur E, Frieifeld I, Böttcher D, Bornscheuer UT, Langer P (2006) Tetrahedron 62: 7132–7139
77. Bellur E, Görls H, Langer P (2005) Eur J Org Chem: 2074–2090
78. Freifeld I, Holtz E, Dahmann G, Langer P (2006) Eur J Org Chem: 3251–3258

79. Passiniemi M, Koskinen AMP (2011) Org Biomol Chem 9:1774–1783
80. Arthuis M, Beaud R, Gandon V, Roulland E (2012) Angew Chem Int Ed 51:10510–10514
81. Mothe SR, Kothandaraman P, Rao W, Chan PWH (2011) J Org Chem 76:2521–2531
82. Vares L, Rein T (2000) Org Lett 2:2611–2614
83. Enders D, Wang C, Greb A (2010) Adv Synth Catal 352:987–992
84. Morikawa S, Yamazaki S, Tsukada M, Izuhara S, Morimoto T, Kakiuchi K (2007) J Org Chem 72:6459–6463
85. Zhao C, Lu J, Li Z, Xi Z (2004) Tetrahedron 60:1417–1424
86. Pohlhaus PD, Sanders SD, Parsons AT, Li W, Johnson JS (2008) J Am Chem Soc 130: 8642–8650
87. Bates RH, Shotwell JB, Roush WR (2008) Org Lett 10:4343–4346
88. Huh CW, Roush WR (2008) Org Lett 10:3371–3374

Synthesis of Saturated Tetrahydropyrans

Matthew A. Perry, Scott D. Rychnovsky, and Nicholas Sizemore

Contents

Abstract Tetrahydropyran (THP) rings are important motifs in biologically active molecules. This review presents common strategies for THP synthesis based on

M.A. Perry • S.D. Rychnovsky (✉) • N. Sizemore
Department of Chemistry, University of California, 1102 Natural Sciences II, Irvine, CA 92697-2025, USA
e-mail: srychnov@uci.edu

J. Cossy (ed.), *Synthesis of Saturated Oxygenated Heterocycles I*, Topics in Heterocyclic Chemistry 35, DOI 10.1007/978-3-642-41473-2_2,

Table 1 Comparison of the frequency and median yields for the disconnections discussed in this survey

Disconnection	Strategy	*n*	Median % yield
O1–C2	Conjugate addition	14	75
	Nucleophilic substitution cyclization	10	84
	Alkene-mediated cyclization	8	81
	Epoxide opening	2	82
C2–C3	Prins cyclization	13	68
	Petasis–Ferrier union/rearrangement	3	77
	Panek annulation	4	78
C3–C4	Class 1 ring-closing metathesis	5	94
	Class 2 ring-closing metathesis	3	73
Others	Hetero-Diels–Alder	8	69
	Lactol reduction	4	95
	Lactone reduction/reductive acetylation	13	84

typical retrosynthetic disconnections. General mechanistic and stereochemical considerations for each disconnection are included. The various strategies related to THP ring formation are discussed in the context of natural product synthesis.

Keywords Etherification • Hetero-Diels–Alder • Oxocarbenium ion • Reductive acetylation • Ring-closing metathesis • Tetrahydropyran

1 Introduction

Tetrahydropyran (THP) rings are important motifs in natural products and medicinal chemistry programs. As such, several reviews on the synthesis of THP rings have been reported [1–3]. The focus of this review is to present the strategies for THP synthesis in a systematic way. This chapter is divided into sections based on the retrosynthetic disconnections typically used to construct THP rings. Each section begins with a brief introduction of the different strategies for each disconnection and includes discussion of general mechanistic and stereochemical considerations. Finally, examples of each disconnection are given with respect to the synthesis of a select group of natural products.

The single-bond disconnections presented in this review are O1–C2, C2–C3, and C3–C4. A section on the synthesis of dihydropyran (DHP) rings using hetero-Diels–Alder reactions is included since these products are easily reduced to THP products. The functionalization of lactones and lactols is also included because these methods have found widespread use in THP synthesis. Table 1 shows a statistical snapshot of the disconnections and strategies in terms of yields presented in this review.

Given the wealth of THP-containing natural products and reports of THP syntheses, we decided that a small subset of natural products would be chosen to reduce the data set to a manageable size. The natural products shown in Fig. 1 were

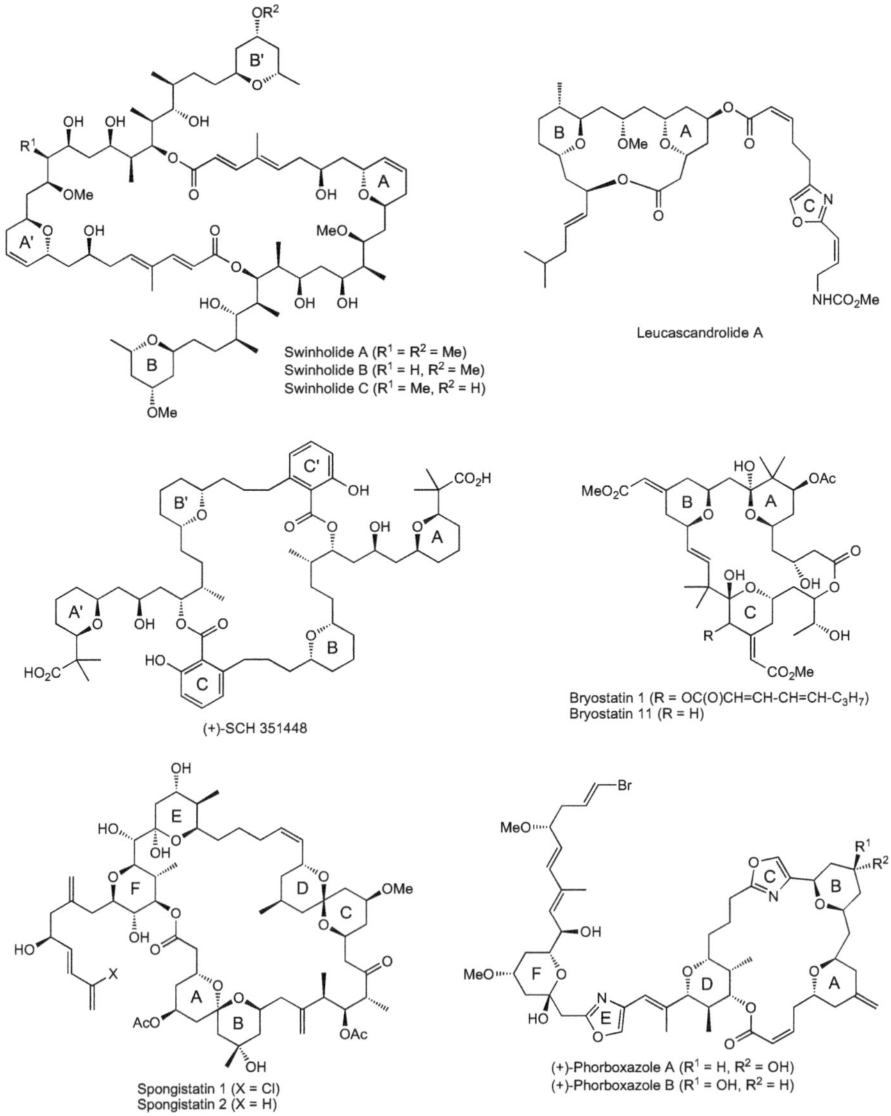

Fig. 1 The representative natural products selected for this survey

selected in order to give a broad range of substrate complexity and a variety of substitution patterns (including, but not limited to, 2,6-*cis* and 2,6-*trans* THP rings) [4–12]. These natural products have the added benefit of having been synthesized many times using various strategies, which allows for direct comparison across the different synthetic methods. This approach allowed for maximum utility within the chapter using a relatively small subset of THP-containing natural products.

2 O1–C2 THP-Forming Processes

Of the numerous connections for the construction of THP rings, O1–C2 bond formation has proven to be an efficient, stereochemically predictable, and reliable approach. Such methods encompass S_N2 and S_N1 nucleophilic addition, conjugate addition, metal-promoted, and dehydrative cyclizations as represented in Scheme 1. This section will only cover those processes that produce tetrahydropyrans (Scheme 1, Eqs. 1–4). As such, the common O1–C2 closure by dehydration of δ-hydroxy ketones to give dihydropyrans will not be discussed (Scheme 1, Eq. 5). The goal of this section is to highlight methods used for the stereoselective construction of tetrahydropyrans in the context of complex natural product synthesis.

2.1 Conjugate Addition

Conjugate addition has found widespread use in organic synthesis and, not surprisingly, in the construction of oxygen-containing heterocycles [13]. As shown in Scheme 2, intramolecular nucleophilic hydroxyl attack onto the electron-deficient β-carbon of an α,β-unsaturated carbonyl system proceeds through an *exo* (or *endo*) ring closure to provide the corresponding tetrahydropyran. The most common mode of ring closure involves 6-*exo-trig* cyclization of α,β-unsaturated hydroxy ketones or esters. Cyclization can be carried out under Brønsted basic or acidic conditions. Under basic conditions, 2,6-*trans* THPs are kinetically favored at low temperatures (−78 °C) and short reaction times, whereas 2,6-*cis* THPs are thermodynamically favored at higher temperatures and longer reaction times. The stereoselectivity observed for base-mediated reactions has been described by the difference in energy and HOMO/LUMO orbital overlap between the s-*cis* (**TS-A**, lower energy, better orbital overlap) and s-*trans* (**TS-B**, higher energy, decreased orbital overlap) TS conformations (Scheme 2, Eq. 1) [14–18]. Under acidic conditions, the transition state (**TS-D**) leading to the thermodynamic 2,6-*cis* disubstituted pyran is now kinetically favored based on a frontier molecular orbital (FMO) theory argument (Scheme 2, Eq. 2). Inspection of the FMO coefficients of the allylic cationic species and the orbital overlap with the oxygen lone pair indicates greater stereoelectronic stabilization in **TS-D** than **TS-C** [13, 19]. These arguments validate the observed selectivity for simple 2,6-substituted tetrahydropyrans, but the stereochemical outcome of more complex THPs requires conformational analysis of the resultant heterocycle.

Paterson et al. reported perhaps the simplest intramolecular oxyanion conjugate addition in the synthesis of the C1–C15 fragment of swinholide A [20, 21]. This particular cyclization constitutes one of the first examples of the less common *endo* conjugate addition to a dihydropyran. Cyclization of **13** under Lewis acid/Brønsted basic conditions provided the racemic dihydropyrone **14** in good yield (61 %) (Scheme 3). Attempts at reaction optimization by changing solvent had little effect,

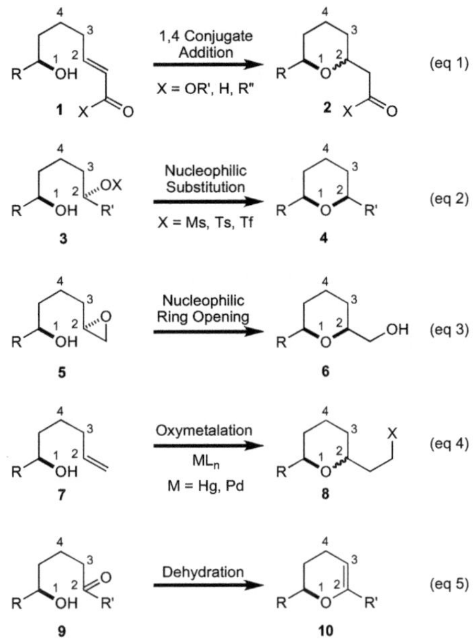

Scheme 1 General approaches to O1–C2 bond formation

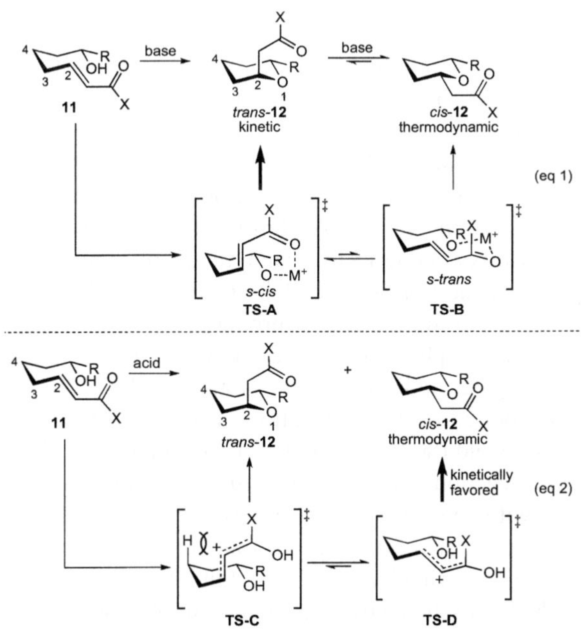

Scheme 2 Stereochemical consequences of reaction conditions in conjugate additions

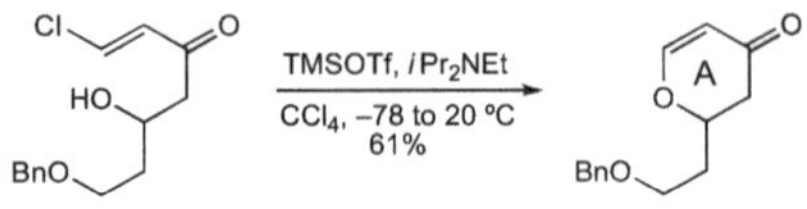

13 14

Scheme 3 *Endo* conjugate addition approach to the A ring of swinholide A [21]

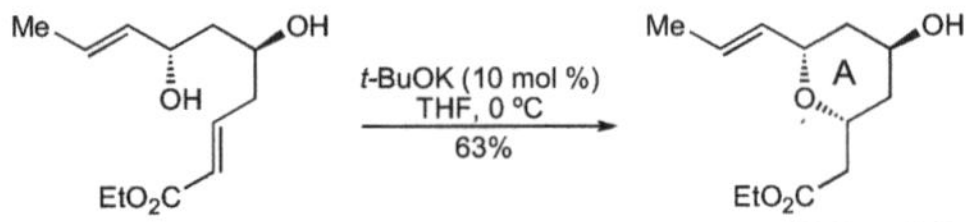

15 16 (dr = 10:1)

Scheme 4 Early installation of A ring fragment of leucascandrolide A [22]

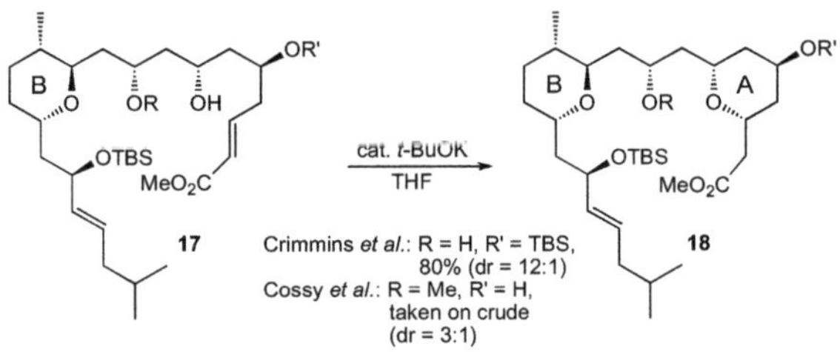

17 18

Scheme 5 Late-stage A ring assembly of leucascandrolide A [23, 24]

and alternative cyclization conditions gave only traces of the DHP (~20 %). This fragment was later used for a Ferrier-type rearrangement to give the 2,6-disubstituted DHP.

One of the many benefits of the conjugate addition approach stems from the ability of the product to self-catalyze the reaction. Carreira and Fettes reported on an oxa-conjugate addition in the synthesis of leucascandrolide A in 2002 [22]. Exposure of hydroxy enone **15** to a catalytic quantity potassium *tert*-butoxide under thermodynamic conditions provided the A ring pyran **16** in good yield (63 %) and with a dr of 10:1 favoring the 2,6-*cis* product (Scheme 4).

Both Crimmins and Siliphaivanh, as well as Cossy et al., utilized the catalytic base-promoted conjugate addition in the total synthesis of leucascandrolide A. Crimmins first reported on the cyclization of alcohol **17** to provide the A ring THP **18** in very good yield (80 %) and a dr of 12:1 (Scheme 5) [23]. In 2007, Cossy described the same reaction using a very similar substrate to provide the 2,6-*cis* disubstituted THP with low diastereoselectivity (dr = 3:1) [24]. These examples clearly illustrate the subtle effects that remote functionality bears on stereoselectivity.

De Brabander et al. described a conjugate addition approach to both the A and B ring of (+)-SCH 351448 (Scheme 6) [25]. Treatment of neopentyl alcohol **19** with a catalytic amount of base under equilibrating conditions provided 2,6-*cis* THP **20** in high yield and high diastereoselectivity (90 %, dr > 95:5). Similarly, homoallylic

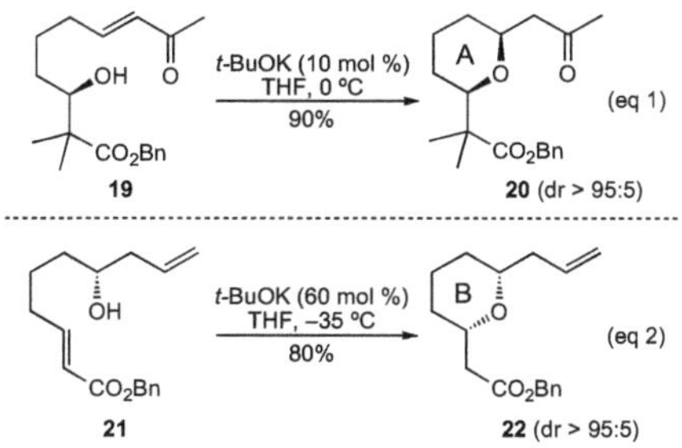

Scheme 6 Synthesis of key fragments for convergent synthesis of (+)-SCH 351448 [25]

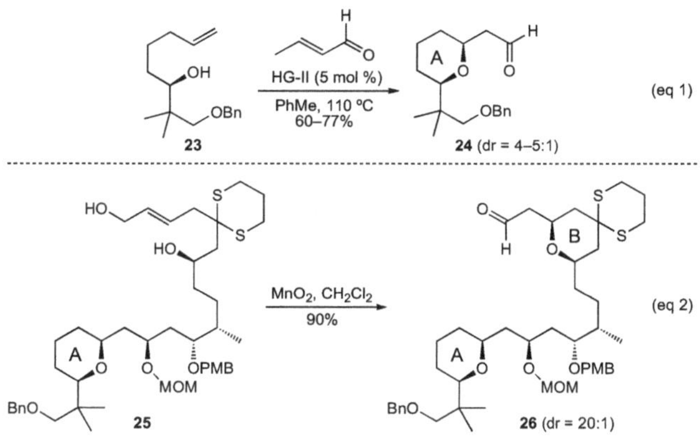

Scheme 7 Cross-metathesis and oxidation/conjugation addition cascades leading to (+)-SCH 351448 [26]

alcohol **21** afforded *cis*-substituted B ring **22** in 80 % yield and a dr > 95:5. Interestingly, in both cyclizations, the diastereoselectivity was shown to be exquisitely controlled through reaction conditions (kinetic conditions led to a dr of 7:93 in both cases).

An appealing feature of oxa-Michael cyclizations is their ability to participate in cascade sequences. Hong et al. utilized a tandem cross-metathesis/conjugate addition sequence in a formal synthesis of SCH 351448 [26]. Use of Hoveyda–Grubbs second generation catalyst (HG-II) with methylacrolein and alkenyl alcohol **23** afforded 2,6-*cis* THP aldehyde **24** in good yield (60–77 %) with moderate diastereoselectivity (dr = 4–5:1) (Scheme 7, Eq. 1). In the same report, Hong et al. described an allylic oxidation/conjugate addition sequence to access the tetrahydropyranyl B ring. Oxidation of allylic alcohol **25** using manganese dioxide provided an intermediate enal that was subsequently trapped by the pendant alcohol to provide dithiane-protected 2,6-*cis* disubstituted pyrone **26** in 90 % yield and a dr of 20:1 (Scheme 7, Eq. 2).

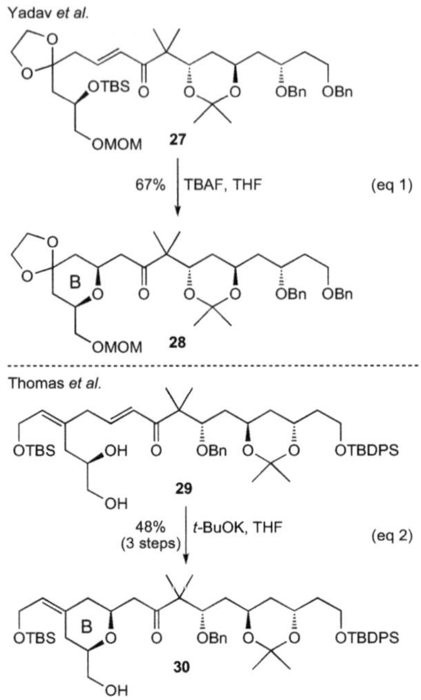

Scheme 8 Related approaches to the B ring of the bryostatins [27, 28]

The THP rings contained within the bryostatin family of natural products have also succumbed to synthesis by Michael-type cyclization. Yadav et al. reported the tandem desilylation and conjugate addition to give the B ring of bryostatin 1 [27]. TBAF-mediated deprotection of silyl ether **27** resulted in 67 % isolated yield of 2,6-*cis* THP **28** (Scheme 8, Eq. 1). Thomas and coworkers described a similar approach on a related substrate to that of Yadav et al. (Scheme 8, Eq. 2) [28]. Cleavage of TES and TMS ethers with HF/pyridine resulted in diol **29**, and a subsequent cyclization with catalytic base proceeded stereoselectively to 2,6-*cis*-4-methylene THP **30** (48 % yield over three steps).

Brønsted acid-mediated THP synthesis provides an effective alternative to the more commonly used Brønsted base protocol. Paterson and Keown were able to effect cyclization under mildly acidic conditions en route to spongistatin 1 (Scheme 9) [29]. Acetonide deprotection with acetic acid followed by conjugate addition of the resultant diol provided 2,6-*trans* and 2,6-*cis* THPs (*trans*-**2.32** and *cis*-**32**, respectively) in a dr of 2.5:1 and 95 % yield. Fortunately, treatment of the mixture with Triton methoxide provided the required *cis*-isomer in 70 % overall yield.

A Horner–Wadsworth–Emmons (HWE) olefination/conjugate addition sequence was reported by Roush et al. for the synthesis of the bis-THP subunit of spongistatin 1 (Scheme 10) [30]. This example demonstrates how substitution affects product distribution of conjugate additions under thermodynamic control.

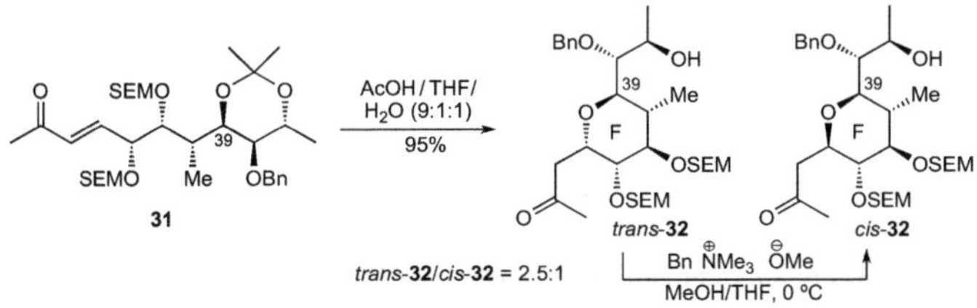

Scheme 9 Conjugate addition approach to the highly substituted F ring of spongistatin 1 [29]

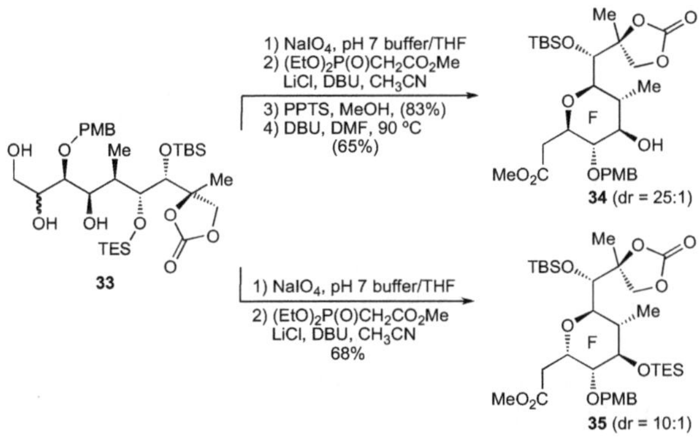

Scheme 10 Selective synthesis of the F ring subunit of spongistatin 1 [30]

The terminal diol of **33** was oxidatively cleaved, and the resulting crude aldehyde was subjected to an HWE reaction using excess LiCl at room temperature. Under these conditions, the TES group migrated from the C39 hydroxyl to the C41 hydroxyl, and the newly revealed alcohol underwent 1,4-addition to the enoate. This one-pot olefination/silyl migration/1,4-addition sequence furnished the undesired 2,6-*trans* THP **35** as the major diastereomer (dr = 10:1) in 68 % yield. Diastereoselectivity was reversed by employing a stepwise procedure involving HWE olefination, subsequent cleavage of the TES ether, followed by 1,4-addition under thermodynamic conditions to afford the 2,6-*cis* THP **34** as the major diastereomer (dr = 25:1) in 65 % yield.

Forsyth and coworkers provided one of the earliest examples of the desilylation/ conjugate addition sequence en route to the total synthesis of phorboxazole A (Scheme 11) [31, 32]. Subjecting silyl ether **36** to fluoride-mediated deprotection furnished the alkoxide, which upon conjugate addition to the enoate gave the THP-containing ester **37** in moderate yield and diastereoselectivity (46 %, dr = 4:1) after 3 days. The use of shorter reaction times provided a mixture of mono- and dihydroxy acrylate and THP products. Attempts at improving conversion to the desired THP including increased temperature, longer reaction times, and stronger bases were unsuccessful.

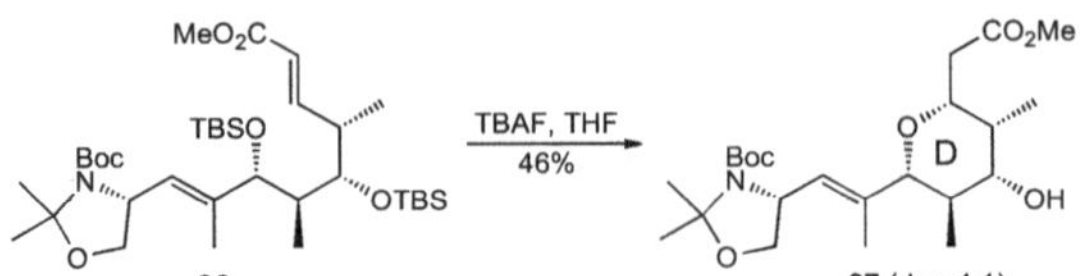

Scheme 11 Desilylation/conjugate addition approach to the D ring of phorboxazole A [31]

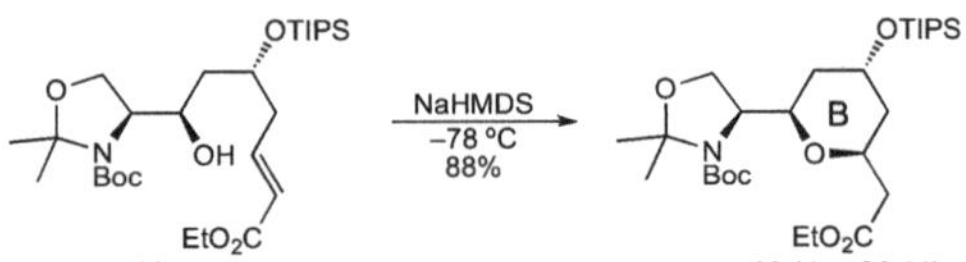

Scheme 12 Synthesis of B ring fragment of phorboxazole A [33]

Pattenden and Plowright described the synthesis of the B ring of phorboxazole by a simple conjugate addition (Scheme 12) [33]. Treatment of alcohol **38** with NaHMDS under typical kinetic control conditions (−78 °C) resulted in a selective intramolecular oxyanion conjugate addition to produce the thermodynamically favored 2,6-*cis* THP **39** in 88 % yield and a dr of 86:14. No attempt at equilibration under thermodynamic conditions to give a higher *cis*/*trans* ratio was reported. This example demonstrates the dramatic effects that subtle changes in substitution can have on stereochemical outcomes.

2.2 *Nucleophilic Substitution Cyclizations*

Cyclizations based on nucleophilic substitution represent the simplest strategy leading to tetrahydropyrans. The plethora of methods for the stereoselective installation of secondary alcohols allows efficient synthesis of the requisite hydroxy nucleophile as well as the leaving group (usually derived from a chiral alcohol). The 6-*exo-tet* cyclization proceeds with inversion at the electrophilic carbon center in accord with a Williamson reaction [34]. In this way, both the 2,6-*cis* and 2,6-*trans* THP are readily accessible based on the configuration of the reactive center (Scheme 13). Reactivity toward nucleophilic substitution follows the rate trend of primary > secondary > tertiary; most examples covered herein utilize secondary electrophiles with secondary alcohol nucleophiles. In contrast to S_N2-like processes, S_N1 processes involve the generation of stabilized allylic or benzylic carbocations for nucleophilic trapping. In this method, the stereochemical outcome of the cyclization is dictated by the stereogenic composition of the substrate and the experimental conditions.

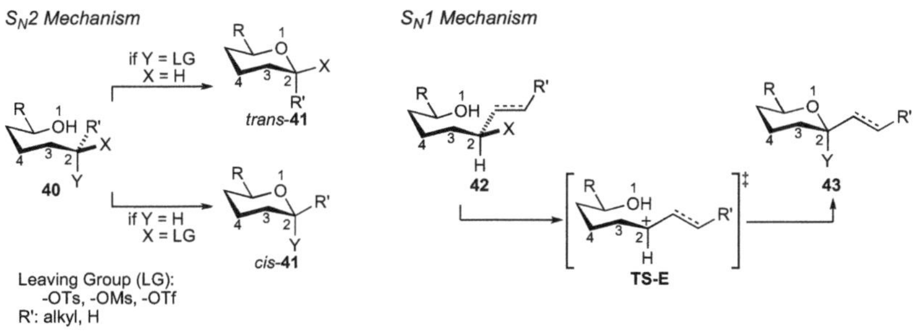

Scheme 13 Mechanistic considerations for nucleophilic substitution cyclizations

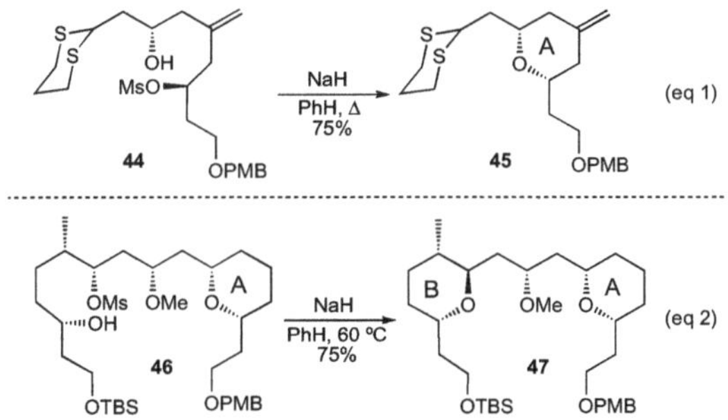

Scheme 14 Nucleophilic substitution approach to the bis-THP fragment of leucascandrolide A [35]

The ability of nucleophilic substitution to provide either 2,6-*cis* or 2,6-*trans* THP has been exploited in the synthesis of several natural products. Williams and coworkers reported the formation of both the A and B ring of leucascandrolide A by nucleophilic substitution (Scheme 14) [35]. Both examples relied on a methanesulfonate leaving group and secondary alcohol nucleophile. Subjecting mesylate **44** (or **46**) to sodium hydride deprotonation followed by heating resulted in THP product **45** (or **47**) in 75 % yield as a single diastereomer.

The relative ease of cyclization with primary electrophilic centers prompted Smith and coworkers to use this method in their scalable route to (+)-spongistatin 1 (Scheme 15) [36]. Sharpless asymmetric dihydroxylation of alkene **48** provided diol **49**, which underwent cyclization in the presence of sodium methoxide to afford THP **50** in 85 % yield over two steps as a single diastereomer.

The utility of sequential displacement cyclization strategies has been demonstrated in the construction of the phorboxazole bis-THP subunit. Milder conditions can be used for effective cyclization in cases where existing functionality can result in undesired side product formation. Cink and Forsyth reported an elimination pathway that resulted in formation of a conjugated diene when alcohol **51** was treated with sodium hydride [37]. Use of a less basic hindered amine in refluxing

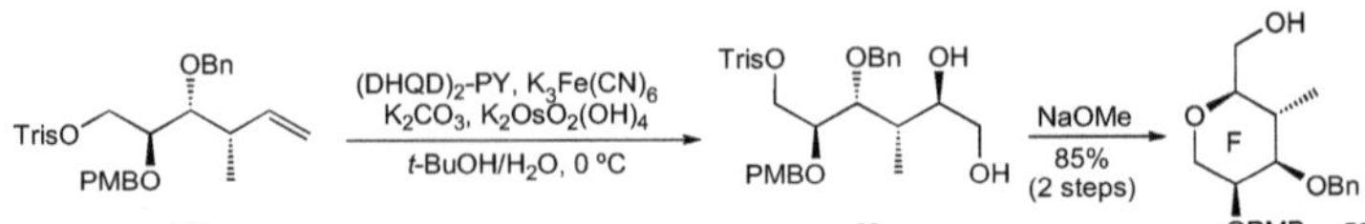

48 49 50

Tris = 2,4,6-triisopropylbenzenesulfonyl

Scheme 15 Dihydroxylation/displacement strategy to the F ring of spongistatin 1 [36]

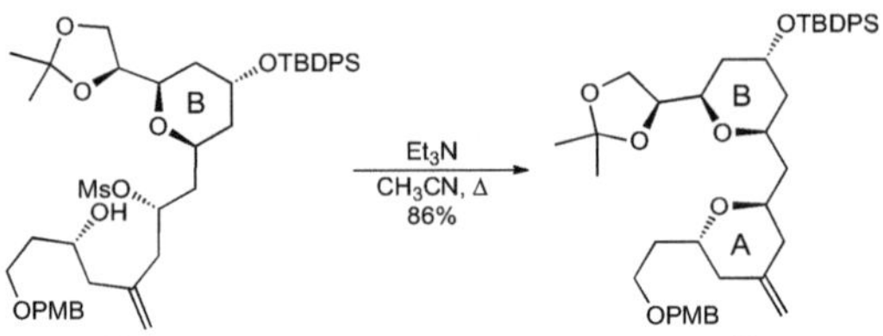

51 52

Scheme 16 Mild basic displacement to the A ring of phorboxazole A [37]

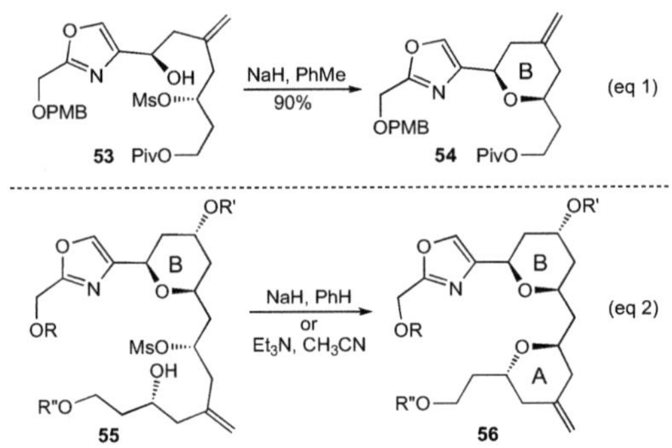

53 54 55 56

Williams *et al.*: R = PMB, R' = TBDPS, R" = Piv (89%)
Pattenden *et al.*: R = PMB, R' = TIPS, R" = H (78%)
Lin *et al.*: R = TBS, R' = TBS, R" = TBDPS (83%)

Scheme 17 Displacement strategies for the synthesis of the B ring in phorboxazoles [33, 38, 39]

acetonitrile proceeded smoothly to furnish bis-THP **52** in 86 % yield as a single diastereomer (Scheme 16).

Williams et al.'s synthesis of phorboxazole A utilized a direct displacement strategy to produce both THP rings in the bis-THP fragment of the natural product [38]. Sodium hydride deprotonation of alcohol **53** and displacement gave B ring intermediate **54** in high yield (90 %) as a single diastereomer (Scheme 17, Eq. 1). Following the same procedure, Williams et al. also effected cyclization of **55** to afford the A ring in 89 % yield (Scheme 17, Eq. 2). This method and substrate were subsequently used by both Pattenden and Plowright and Lin and coworkers to construct the bis-THP domain, with the only modification being the use of triethylamine as the base [33, 39]. Of note, substrate control can provide both 2,6-*cis* and 2,6-*trans* THPs by the same method.

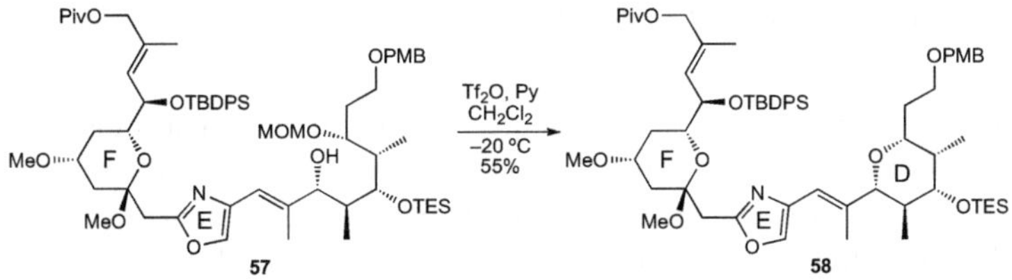

Scheme 18 Carbocation-induced D ring closure en route to phorboxazole A [40]

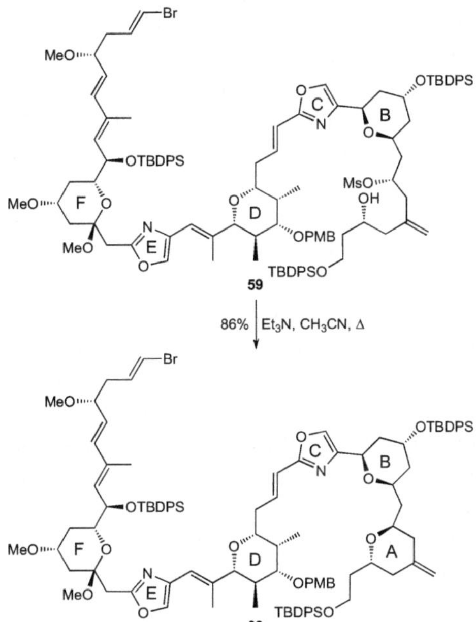

Scheme 19 Late-stage displacement strategy to the A ring of phorboxazole A [42]

In addition to the A and B rings, Williams et al. also used an interesting cationic process to build the D ring contained in phorboxazole A (Scheme 18) [40]. The transformation involved formation of the triflate from **57** followed by expulsion of triflic acid to generate an allylic carbocation. Subsequent internal trapping of the transoid allylic cation intermediate by the C22 methoxymethyl ether and concomitant dealkylation of the resultant oxonium species provided THP **58** in moderate yield (55 %) as the sole diastereomer.

White and coworkers reported perhaps the most noteworthy example of nucleophilic substitution in a complex setting. The late-stage closure of the A ring of phorboxazole A proceeded smoothly under mild conditions [41, 42]. Exposure of mesylate **59** to Et_3N afforded 2,6-*trans* THP **60** in high yield (86 %) as a single diastereomer (Scheme 19).

2.3 Alkene-Mediated Cyclizations

Electrophile-activated alkene additions are a common chemical transformation of carbon–carbon double bonds. This section will review some of the more common methods encountered in the context of macrolide natural product synthesis. Typically, THP formation occurs by a 6-*exo-trig* cyclization of δ-hydroxy alkenes in the presence of an appropriate metal salt. Activation of the π-system occurs through reversible formation of a π-complex or an onium intermediate that leads to the heterocycle by attack of the pendant oxygen nucleophile [43]. The stereochemical outcome of the cyclization is dictated by the nucleophilic attack occurring on the face opposite electrophilic π-complexation (Scheme 20). Facial discrimination of the alkene can therefore be achieved through either substrate control (chiral directing-group coordination) or chiral metal reagent/catalyst.

Metal-activated alkene additions can be classified as stoichiometric or catalytic processes. Stoichiometric processes for THP synthesis typically involve the use of mercury(II) salts and to a lesser extent iodo and seleno reagents. The progress of intramolecular oxymercuration is determined by the stability of the cationic intermediates. Product stereochemistry is under substrate control and usually leads to the thermodynamically more stable THP product. Catalytic variations generally involve palladium complexes [44], but other transition metals are becoming more common (e.g., Pt [45], Ag [46], Sn [47], Ce [48]). The oxidation state of Pd determines the catalyst reactivity. Palladium(0) complexes are nucleophilic and participate in tetrahydropyran synthesis through π-allyl cation intermediates, whereas Pd(II) complexes possess electrophilic character and progress through a reversible π-complex.

Masamune et al. reported on an oxymercuration approach to the synthesis of the B ring of bryostatin 1 (Scheme 21) [49]. Mercury acetate-mediated cyclization of **63** followed by acetylation of the C7 alcohol allowed for oxidative cleavage of the organomercurial intermediate to give a 1:1 diastereomeric mixture of 4-methylene THP **64** in good yield (64 % over three steps). The lack of diastereoselection can be attributed to the absence of a chelating directing group near the reacting alkene center.

Leighton et al. described an effective and mild palladium-catalyzed tandem alkene addition/carbonylation procedure in the synthesis of leucascandrolide A [50]. Intramolecular alkoxycarbonylation of diol **65** under Semmelhack conditions proceeded efficiently to provide the desired 2,6-*cis*-tetrahydropyran **66** in 75 % yield with a dr of >10:1 (Scheme 22). Reaction optimization showed that use of benzonitrile as a cosolvent leads to cleaner and more efficient reactions. The functional group tolerance and chemoselectivity of the reaction simplified the protecting group strategy.

Fettes and Carreira were the first to describe the use of a selenium-based reagent to effect alkoxymetallation of an alkene to give a 2,6-*trans* tetrahydropyran in the total synthesis of leucascandrolide A [22]. Various electrophiles (I_2, IBr, and $Hg(OAc)_2$) gave disappointing levels of diastereoselectivity (1:1), whereas

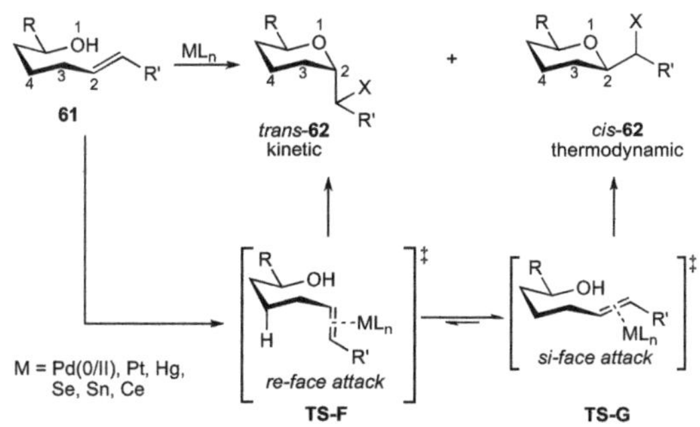

Scheme 20 Stereochemical outcomes of metal-mediated alkene cyclizations

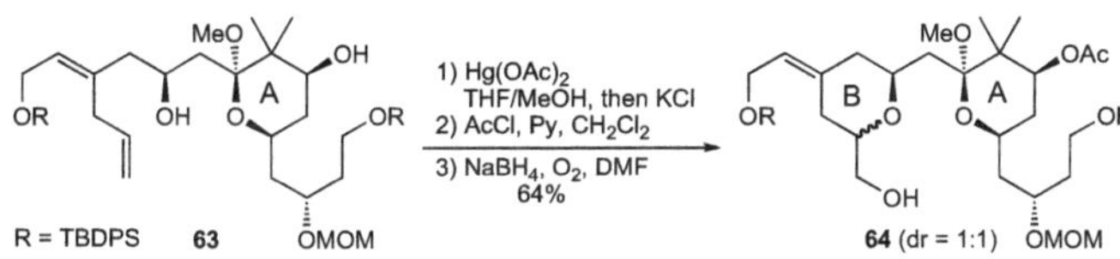

Scheme 21 Oxymercuration strategy to the B ring of the bryostatins [49]

Scheme 22 Pd-catalyzed alkoxycarbonylation to the bis-THP fragment of leucascandrolide A [50]

phenylselenyl chloride gave a diastereomeric ratio of 3:1 (*trans*/*cis*). That finding prompted the examination of bulkier, substituted phenylselenyl halides in order to increase the diastereoselectivity. Treatment of hydroxy alkene **67** with 2,4,6-triisopropylphenylselenyl bromide afforded the desired 2,6-*trans* tetrahydropyran **68** in 65 % yield with a dr of 88:12 (Scheme 23). The *trans* stereoselectivity can be rationalized by conformational analysis whereby the selenide complexes to the allylic strain minimized alkene from the less hindered *re* face (that of the methoxy group) forcing attack of the alcohol to occur from the *si* face.

Lucas and Burke reported on a Pd(0)-catalyzed asymmetric allylic etherification/desymmetrization of *meso* substrates by double cyclizations to the bis-THP subunit of phorboxazoles A and B [51]. Based on Trost and Toste's transition state model for the diphenylphosphino benzoic acid (DPPBA) ligand system [52], Burke rationalized that cyclization of tetraol **69** with (*R*,*R*)-DPPBA ligand would effect the double cyclization to give the *trans*-A ring and *cis*-B ring in bis-THP **70** (Scheme 24). Under optimized conditions, tetraol **69** was converted to **70** in 75 %

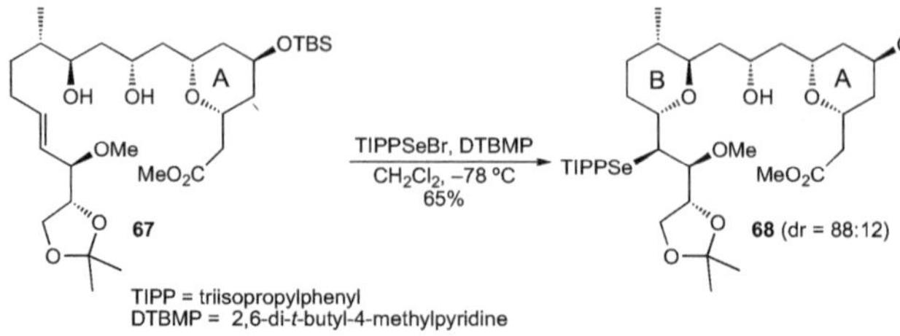

Scheme 23 Selenium-mediated alkene cyclization to construct the B ring of leucascandrolide A [22]

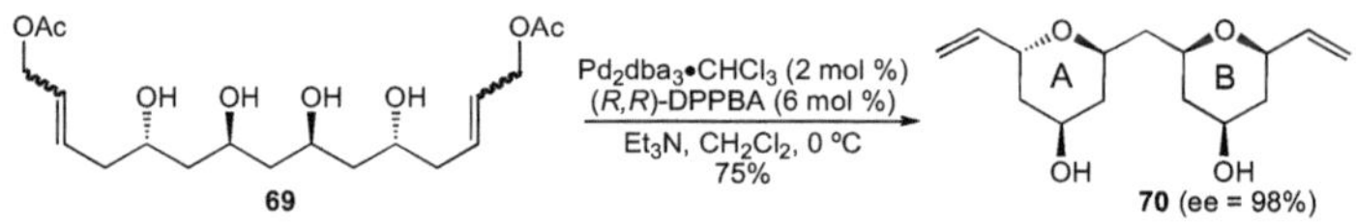

Scheme 24 Pd-catalyzed desymmetrization to the bis-THP fragment of the phorboxazoles [51]

yield with an er of 99:1 and dr of 6.4:1 (*trans–cis*/*cis–cis*). Fast ligand-controlled cyclization installs the *cis*-substituted B ring, inhibits formation of the other enantiomer, and leads to the high levels of enantioselectivity observed. The moderate diastereoselectivity arises from a competitive ligand-controlled cyclization to the *trans*-substituted A ring with the substrate-controlled cyclization to the *cis*-substituted A ring. Specifically, as a result of a matched interaction between ligand control and steric bias, the B ring allylic stereogenic center has been set in the *R* configuration. However, the second cyclization event represents a mismatched case of ligand control and inherent steric bias. Cyclization of the A ring under ligand control provided the desired *R,R* (*cis–trans*)-isomer with one equatorial and one axial vinyl group. Cyclization under substrate control provided the more stable *S,R* (*cis–cis*)-isomer with two equatorial vinyl groups. This example highlights the utility of ligand-controlled Pd-catalyzed desymmetrization to expeditiously produce useful chiral bis-THP building blocks.

White et al. also reported on the palladium(II)-catalyzed 1,2-alkoxycarbonylation in the synthesis of both the B and D rings of phorboxazole A (Scheme 25) [41]. The D ring THP **73** was prepared in high yield (86 %) from cyclization of hydroxy alkene **71**. The stereochemistry and efficiency of the reaction is best described by a minimization of steric interactions between the α-methyl substituent and the *exo* Pd substituent (e.g., Cl) of the complexed alkene intermediate **72** preceding cyclization. Unfortunately, an excess of palladium acetate was required to compensate for the reduction of Pd(II) to inactive Pd(0). For the synthesis of the B ring, use of a stoichiometric oxidant, *p*-benzoquinone, with catalytic palladium circumvented this issue. Under these improved conditions, trisubstituted B ring THP **75** was isolated as a single diastereomer in a moderate 58 % yield along with 15–20 % recovered starting material **74**.

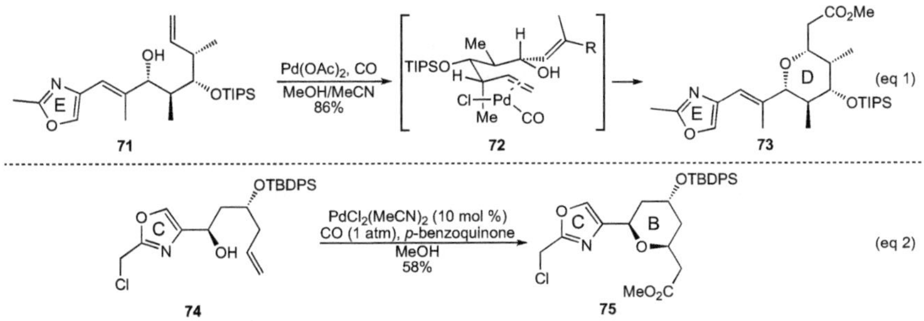

Scheme 25 Pd-catalyzed alkoxycarbonylation to the D and B ring fragments of phorboxazole A [41]

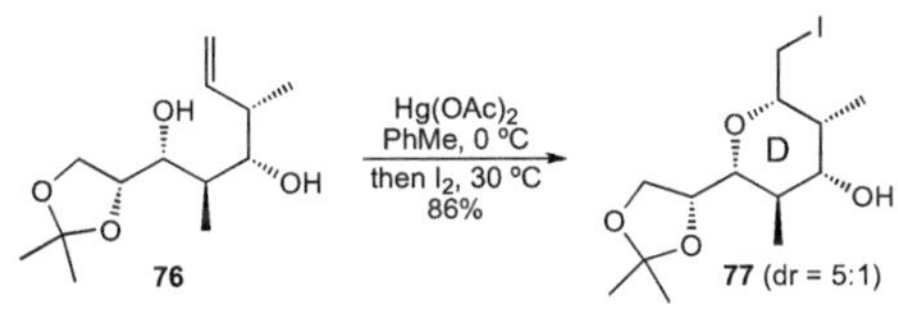

Scheme 26 Iodomercuration of the D ring of the phorboxazoles [39]

The ability of mercury salts to effect nucleophilic alkene addition is apparent in Lin and coworkers' synthesis of the phorboxazole D ring (Scheme 26) [39]. Initial attempts at cyclization proceeded by addition to a halo-activated alkene. The use of iodine under basic conditions afforded the desired THP in 46 % yield as a 2.6:1 ratio of *cis*/*trans* isomers. The use of bulkier NIS increased the diastereoselectivity to 7.7:1 with no increase in yield. A more efficient cyclization was realized by using mercury(II) acetate. Installation of the iodide was accomplished by the addition of iodine after cyclization was complete. These conditions gave 2,6-*cis* THP **77** in 86 % yield and moderate diastereoselectivity (5:1). This method also benefits from the displacement of the organomercurial intermediate in a single-pot procedure, thereby mitigating the isolation of potentially toxic mercury-containing substrates.

2.4 Nucleophilic Substitution Cyclizations by Epoxide Opening

Analogous to the nucleophilic substitution methods, nucleophilic epoxide opening provides access to substituted THP adducts in a stereospecific manner. Seminal studies by Nicolaou et al. established the structural requirements necessary for regio- and stereocontrolled synthesis of THP adducts [53, 54]. Regioselectivity is highly dependent on the configuration of the epoxide (Scheme 27); δ-hydroxy *trans*-epoxides show high selectivity for THP formation over oxepane formation

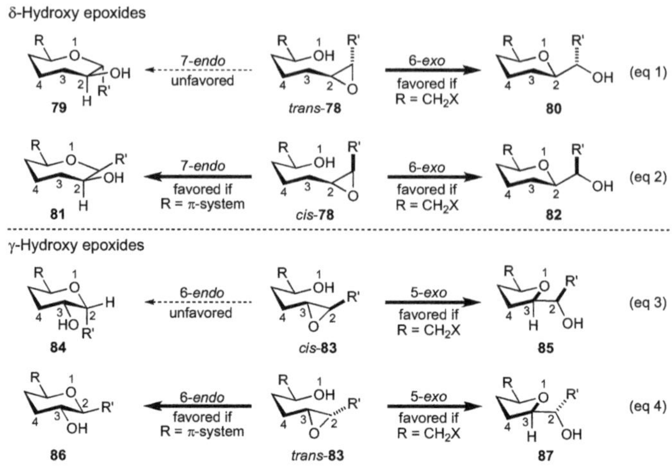

Scheme 27 Stereochemical consequences of intramolecular epoxide opening

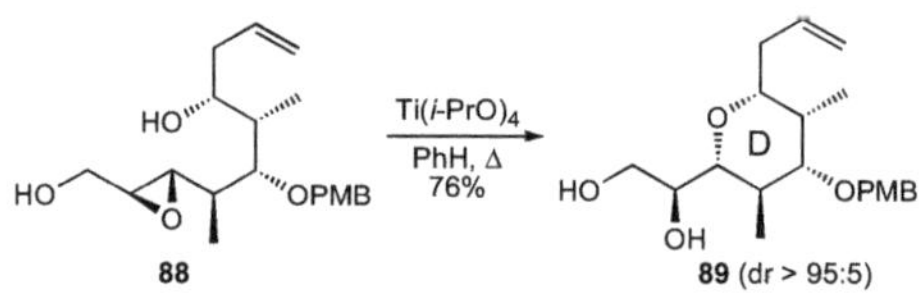

Scheme 28 Lewis acid-mediated epoxide-opening cyclization to D ring of phorboxazole A [57]

irrespective of the R substituent (Eq. 1). For *cis*-epoxides, saturated R substituents prefer 6-*exo* cyclization, whereas unsaturated R substituents prefer 7-*endo* cyclizations (Eq. 2). In the case of γ-hydroxy *cis/trans*-epoxides, either can undergo 6-*endo-tet* or 5-*exo-tet* cyclization (Eqs. 3–4). *Cis*-epoxides give 5-*exo* products (THF), whereas *trans*-epoxides are dependent on the identity of the R substituent. If R = alkane, *trans*-epoxides give 5-*exo* products (THF), whereas if R = π-system, 6-*endo* products are preferred (THP). This method benefits from the availability of several stereoselective epoxidation methods, the introduction of useful functional groups for further elaboration, and the ability to form polycyclic frameworks by an iterative process. However, the constraints imposed by the pendent epoxide substituents require a high level of substrate design with regard to stereochemistry and substitution of the cyclization precursor. Blanc and Toste, as well as Morimoto and coworkers, have described similar methods that do not follow the general trend described above [55, 56].

Ye and Pattenden described an epoxide-opening/cyclization approach to the synthesis of the pentasubstituted C22–C26 THP of phorboxazole A (Scheme 28) [57]. The use of a *trans*-epoxide with a hydroxymethylene side chain allowed selective formation of a 2,6-*cis* THP by 6-*exo* cyclization. Subjecting hydroxy epoxide **88** to Lewis acid in refluxing benzene provided desired THP **89** in 76 % yield with high diastereoselectivity (dr > 95:5).

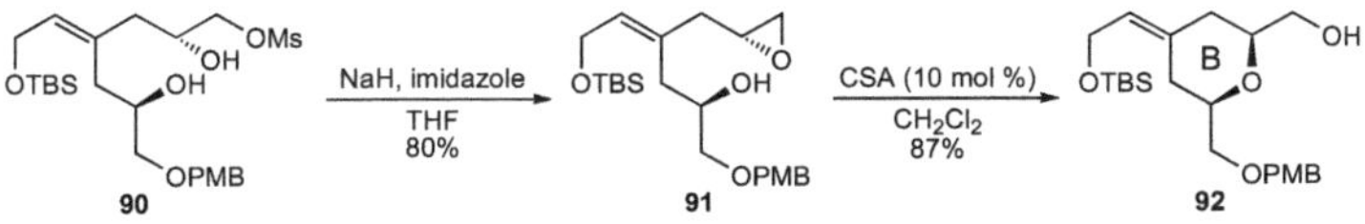

Scheme 29 Brønsted acid-catalyzed epoxide opening to the B ring of the bryostatins [58]

Hale et al. reported on the synthesis of the bryostatin B ring by acid-catalyzed nucleophilic epoxide opening (Scheme 29) [58]. Their goal was to convert *O*-mesylate epoxide precursor **90** directly into THP **92** by treatment with two equivalents of sodium hydride and imidazole. However, the only product isolated was epoxy alcohol **91** in 80 % yield. To effect the desired 6-*exo-tet* ring closure, epoxide **91** was treated with a catalytic amount of camphorsulfonic acid. Tetrahydropyran **92** was acquired as the sole product in 87 % yield as a single diastereomer.

2.5 *Summary*

The above examples represent some of the most efficient, reliable, and powerful methods for the construction of tetrahydropyrans by oxygen–carbon bond-forming processes. The conjugate addition approach can provide both 2,6-*cis* and 2,6-*trans* THPs by simple modification of the reaction conditions. Additionally, the precursors are amenable to the use of tandem processes that maximize step economy. Nucleophilic substitution methods benefit from the plethora of stereoselective alcohol-forming transformations but are limited chiefly to primary and secondary electrophiles. Epoxide-opening/cyclization processes are effective within a very specific context and thus lack generality. Electrophile-induced alkene addition reactions are efficient and can be stereoselective provided the substrate or catalyst contains the structural elements necessary for π-facial discrimination of the alkene. This method also benefits from the continued growth of palladium and other transition metal-mediated chemistry. While several approaches are available for the synthesis of tetrahydropyrans by O–C bond formation, the advantages and disadvantages of each approach must be evaluated when considering a 1,2 disconnection.

3 C2–C3 THP-Forming Processes

A common strategy for the synthesis of tetrahydropyran rings is C2–C3 disconnection (Scheme 30). The natural order of reactivity typically generates an oxocarbenium ion at C2 while C3 acts as the nucleophile. This strategy is manifested in three general classes of reactions: Prins cyclizations (Eq. 1), Petasis–Ferrier union/rearrangements (Eq. 2), and Panek annulation (Eq. 3).

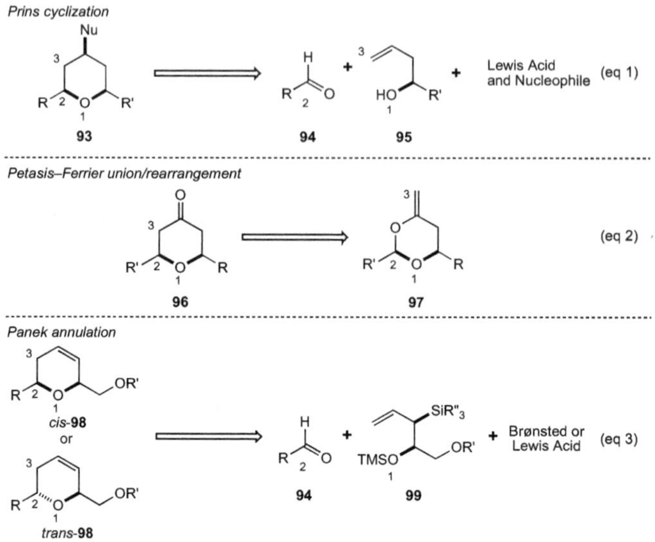

Scheme 30 Common retrosynthetic approaches that forge C2–C3 bonds

3.1 Prins Cyclization Strategies

Prins cyclization is a two-component coupling reaction between homoallylic alcohol **94** and aldehyde **95** that proceeds by Lewis acid-induced formation of oxocarbenium ion **101** (Scheme 31) [59–63]. Prins cyclizations proceed through chair-like transition state **TS-A**, where resulting secondary carbocation **102** is stabilized by orbital overlap [64]. The stereochemical outcome at C4 is dictated by the nature of the nucleophilic trapping agent, which is usually a Lewis acid counterion (Scheme 32) [65]. Dissociated ion pairs (e.g., $SnBr_5^-$) undergo equatorial nucleophilic attack to give the all *cis* arrangement (**93**), whereas tight ion pairs (e.g., Br^- from TMSBr) show a preference for axial attack, leading to the nucleophile at C4 being *trans* to the C2 and C6 groups (**103**).

Prins cyclization can suffer from degradation of optical purity through a variety of pathways [66]. The primary mode of racemization/epimerization in Prins cyclizations is by oxonia-Cope rearrangement (Scheme 33) [67, 68]. There are three primary examples of problematic substrates: *syn*-crotyl alcohols (Eq. 1), aryl-substituted homoallylic alcohols (Eq. 2), and allyl–alkyl or allyl–aryl carbinols (Eq. 3). Prins cyclization transition states of *syn*-crotyl alcohols exhibit significant diaxial interactions between the C3 hydrogen and the C5 substituent (Scheme 33, Eq. 1). As a result, a [3,3]-sigmatropic rearrangement (oxonia-Cope) can occur through a boat conformation. Resulting oxocarbenium ion **107** can then undergo a Prins cyclization where the C5 substituent can adopt an equatorial conformation. The overall result is a net epimerization of the C5 substituent while maintaining a 2,4,6-*cis* relationship in the THP ring (**106** vs. **108**). Substrate racemization can also

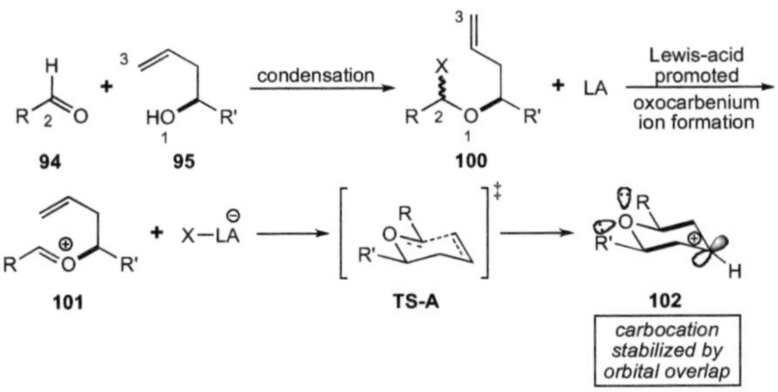

Scheme 31 General mechanism to the carbocationic intermediate of the Prins cyclization

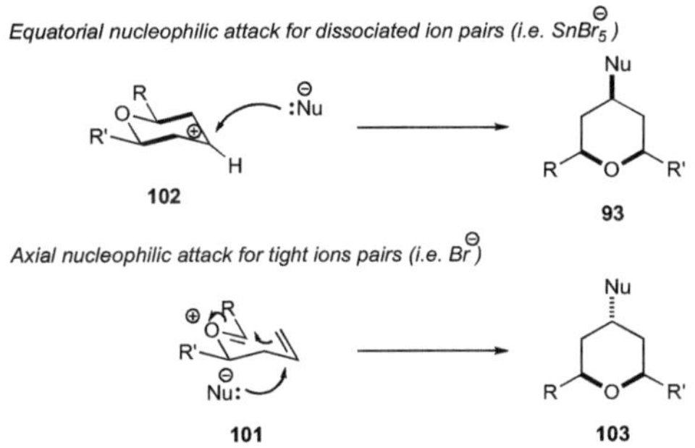

Scheme 32 Diastereoselective nucleophilic trapping in Prins cyclizations

occur with the use of benzylic homoallylic alcohols (Scheme 33, Eq. 2). In these cases, solvolysis of benzylic alcohol (−)-**109** leads to the formation of stable benzylic carbocation **112** and loss of chiral information [67]. Electron-donating aromatic groups are particularly problematic due to the increased stabilization of the resultant carbocation. THP rings that are generated by Prins cyclization can also undergo racemization through a sequence of equilibrating condensation, hydrolysis, and oxonia-Cope rearrangement pathways (Scheme 33, Eq. 3) [69]. The mechanism for racemization relies on an allyl transfer process. Alcohol (−)-**115** can condense with aldehyde **94** to generate oxocarbenium ion **116** with liberation of a molecule of water. Oxocarbenium ion **116** can then undergo a 2-oxonia-Cope rearrangement to afford oxocarbenium ion **117**. Addition of water to oxocarbenium ion **117** leads to a fragmentation process generating aldehyde **94′** and alcohol (+)-**118**. The formation of **94′** now permits a symmetric 2-oxonia-Cope rearrangement to occur. Alcohol (−)-**115** can condense with aldehyde **94′**, undergo a 2-oxonia-Cope rearrangement, and fragment by addition of water to provide epimeric (+)-**115**. Similarly, alcohol (+)-**118** can react with aldehyde **94** to generate epimeric (−)-**118**. This pathway is largely dictated by the reaction conditions such as solvent and temperature. Structural features (e.g., electron-withdrawing groups, alkene substitution) can reduce the reversibility of the 2-oxonia-Cope rearrangement, thereby eliminating racemization in the Prins cyclization.

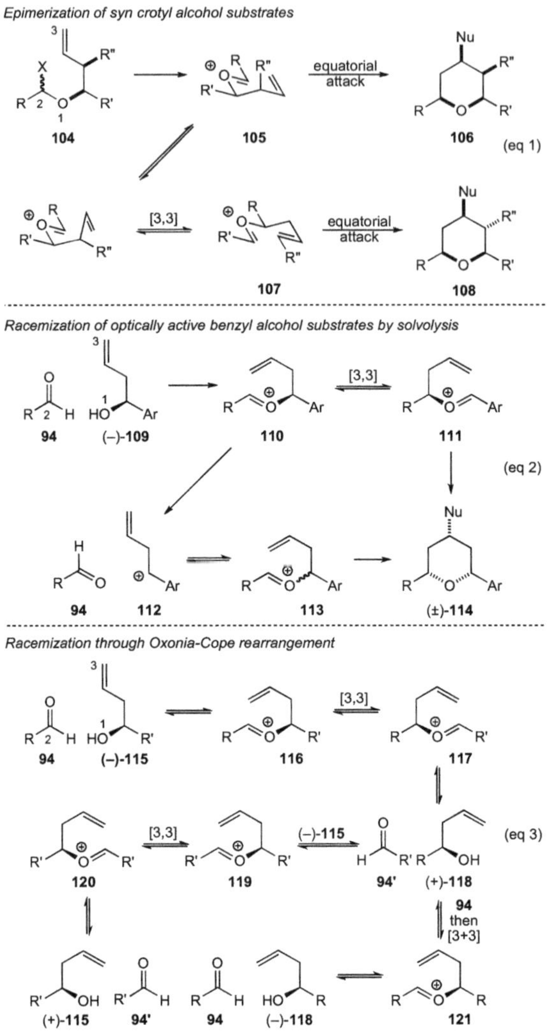

Scheme 33 Potential pathways for the degradation of stereochemistry in Prins cyclizations

The alkene geometry of the homoallylic alcohol dictates the C3 stereochemistry relative to the nucleophilic trap in Prins cyclizations (Scheme 34) [70]. For example, *E*-alkene **122** undergoes facile Prins cyclization through a chair-like transition state where the C3 substituent is in an equatorial position. Subsequent equatorial nucleophilic attack gives rise to 3,4-*trans* THP **125**. In contrast, *Z*-alkene **126** can undergo Prins cyclization leading to 3,4-*cis* product **129**; however, diaxial interactions from the axially disposed substituent often suppress this pathway in favor of an envelope transition state leading to THF **130** [71].

The reactive nature of carbocationic THP intermediates in Prins cyclizations is not limited to intermolecular nucleophilic trapping. The cyclic carbocation **133**

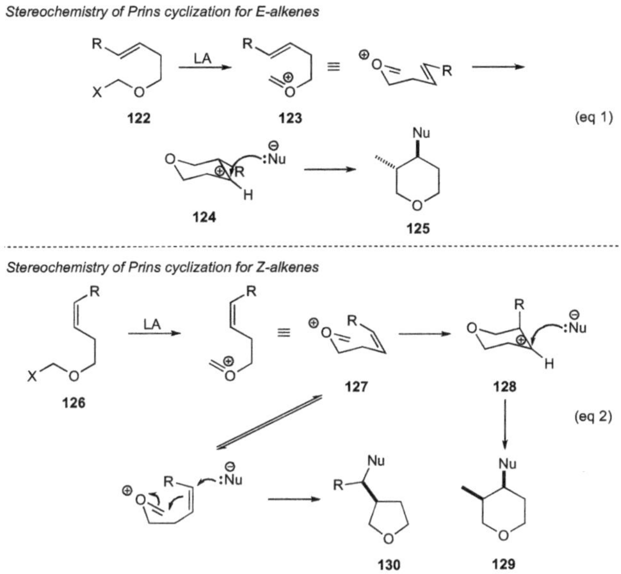

Scheme 34 Relative stereochemistry in the Prins cyclization is dictated by alkene geometry

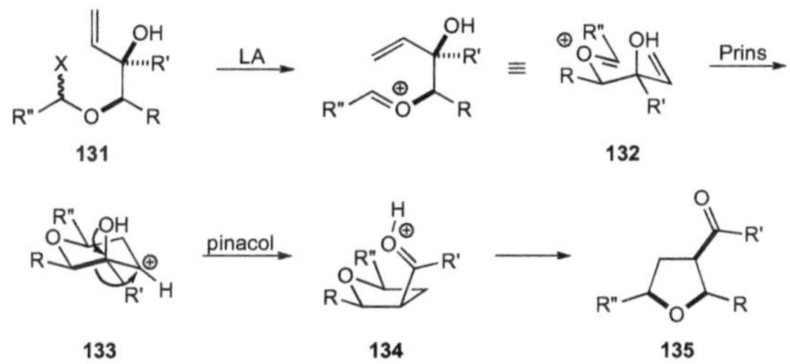

Scheme 35 Allylic 1,2-diols undergo Prins–pinacol reactions

formed when employing allylic 1,2-diols in Prins cyclization readily undergoes ring contraction through a pinacol rearrangement to give 3-acyl THF **135** (Scheme 35) [72–74]. The difficulties associated with attenuating the Prins–pinacol pathway must be considered when planning a synthetic strategy to access THP rings bearing an alcohol.

The 2,6-*cis* THP ring of leucascandrolide A has been approached by classical Prins methods as well as been a testing ground for novel extensions of this reaction. Kozmin and coworkers performed a vinylogous transesterification of alcohol **136** to afford Prins cyclization precursor **138** (Scheme 36) [75, 76]. Subsequent treatment of enol ether **138** with trifluoroacetic acid initiated a Prins cyclization, and basic ester hydrolysis provided the 2,4,6-*cis* THP alcohol **139** in 77 % as a single diastereomer.

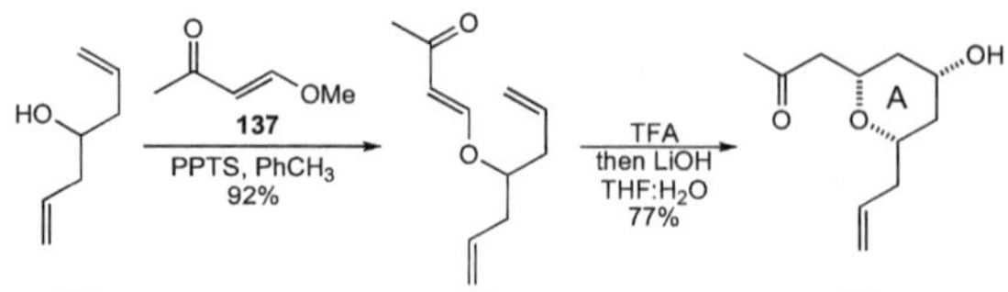

Scheme 36 Brønsted acid-mediated Prins cyclization to A ring of leucascandrolide A [75]

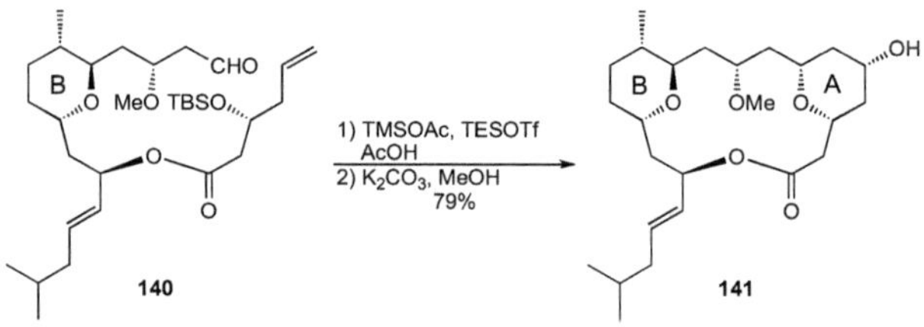

Scheme 37 Late-stage Prins macrocyclization toward leucascandrolide A [77]

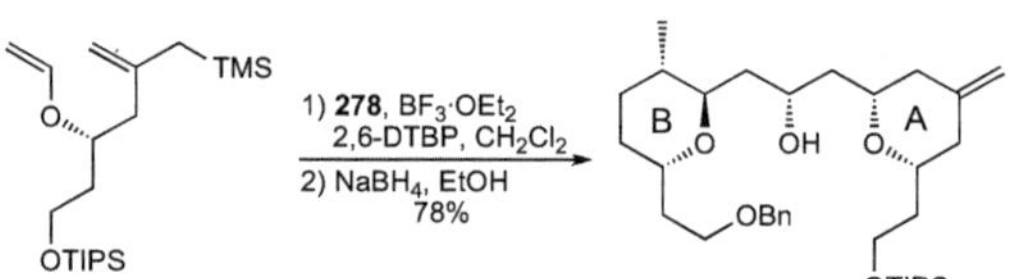

Scheme 38 Convergent assembly of the bis-THP portion of leucascandrolide A by MAP method [78]

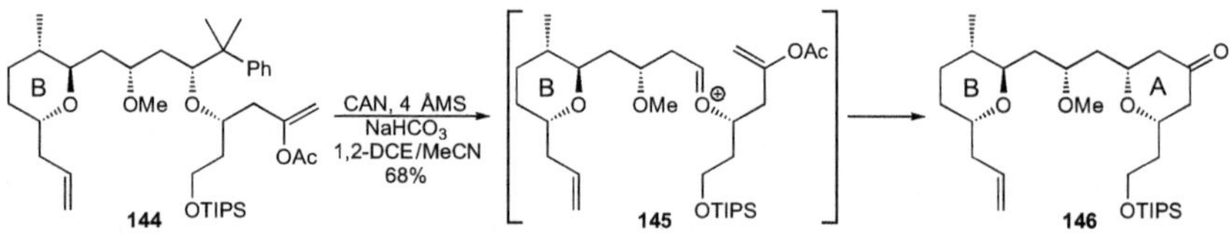

Scheme 39 A ring closure of leucascandrolide A by ETIC method [80]

Yadav et al. utilized a Prins reaction in the context of a macrocyclization strategy toward leucascandrolide A (Scheme 37) [77]. Lewis acid activation of aldehyde **140** leads to oxocarbenium ion formation and Prins cyclization with nucleophilic acetate trapping. Direct base hydrolysis afforded macrocyclic fragment **141** as a single diastereomer in good yield (79 % over two steps).

Kopecky and Rychnovsky also targeted leucascandrolide A to demonstrate their Mukaiyama Aldol-Prins (MAP) method (Scheme 38) [78]. The MAP reaction generates an oxocarbenium intermediate from an enol ether and an aldehyde, which is capable of undergoing a Prins cyclization with the pendant allylsilane. Aldehyde **278** (from Scheme 73, Eq. 1) in the presence of enol ether **142** and boron trifluoride etherate underwent the MAP cascade without incident. The reaction mixture was treated with sodium borohydride to reduce any of the unreacted aldehyde and simplify purification. Desired alcohol **143** was then isolated in 78 % yield as a 5.5:1 mixture of diastereomers.

While use of Lewis or Brønsted acids is the most common way to generate the requisite oxocarbenium ion for a Prins cyclization, single electron pathways offer an attractive alternative. The mild, functional group tolerant conditions of electron transfer initiated cyclization (ETIC) method pioneered by Floreancig and coworkers were highlighted in their synthesis of leucascandrolide A [80]. Oxidative benzylic carbon–carbon bond cleavage of ether **144** was achieved by treatment with ceric ammonium nitrate to give oxocarbenium ion **145** (Scheme 39). Subsequent attack of the enol acetate through a chair-like transition state led to 2,6-tetrahydropyran-4-one **146** (68 % yield as a single diastereomer).

Indium Lewis acids have garnered attention due to their mild reactivity and air and water stability. Both Li et al. and Chan and Loh have shown that In(III) complexes are suitable Lewis acids for Prins cyclizations [81, 82]. These reports prompted Loh and coworkers to embark on a synthesis of (+)-SCH 351488 that utilized this strategy (Scheme 40) [83]. Condensation of homoallylic alcohol **147** and aldehyde **148** in the presence of indium tribromide and TMSBr gave 4-bromo THP **149** in 65 % overall yield as an inconsequential mixture of diastereomers (2,4-*cis*/2,4-*trans* = 75:25). Complete retention of the homoallylic alcohol stereochemistry is responsible for the key 2,6-*cis* relationship in the product. Initial attempts to apply these same conditions to the B ring resulted in acetonide deprotection and no THP formation. Subsequent optimization revealed that indium triflate and TMSCl were competent additives to effect cyclization. Careful temperature control was required to suppress an undesired Prins side reaction. The combination of homoallylic alcohol **150** and aldehyde **151** in the presence of the appropriate Lewis acids at −78 °C, followed by warming to –40 °C for 4 h, led to the desired monomer precursor **152** in 42 % yield.

Rychnovsky et al. used their MAP strategy to construct the A ring of (+)-SCH 351488, while the B ring was constructed using a reductive acetylation/Prins cyclization protocol (Scheme 41) [84]. Use of enol ether **153** and aldehyde **154** in the presence of titanium tetrabromide afforded the ring-closed product with the appended side chain. The resulting bromide was reduced under radical conditions, and SEM protection of the alcohol gave advanced alkene **155** in 55 % yield over three steps. The B ring was constructed using α-acetoxy ether **156**, derived from reductive acetylation of the corresponding ester, as a Prins precursor. Treatment of **156** with tin tetrabromide led to THP formation. Reduction of the bromide and acidic removal of the silyl moiety afforded 2,6-*cis* THP alcohol **157** in 55 % yield over two steps. These examples demonstrate the utility of masked oxocarbenium ions (aldehyde equivalents) in expanding the scope of this method.

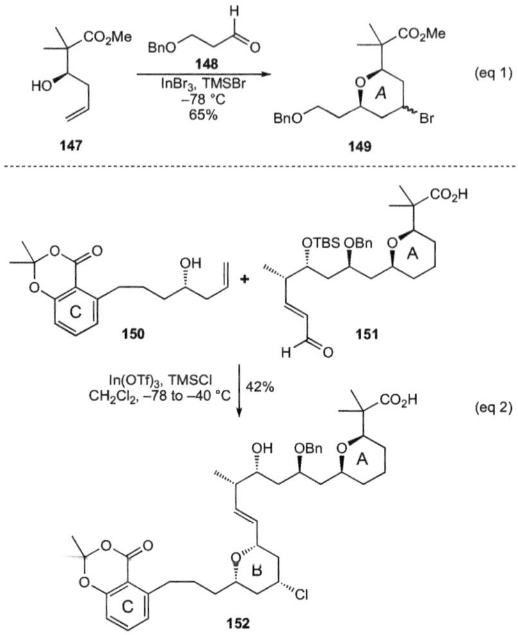

Scheme 40 Indium(III)-promoted Prins cyclization in the progress toward (+)-SCH 351488 [83]

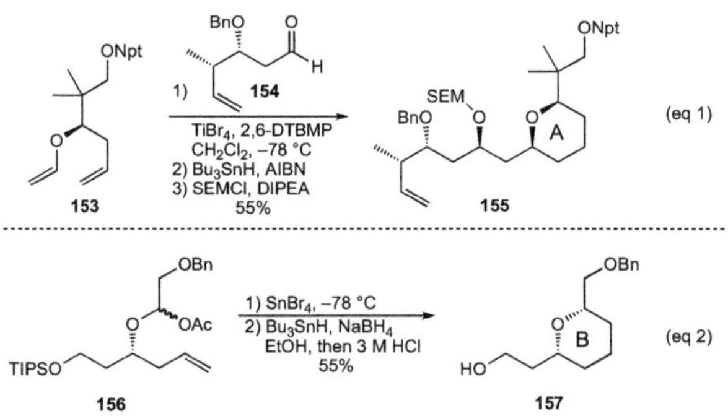

Scheme 41 MAP and Prins approaches to key intermediates for (+)-SCH 351488 [84]

The synthesis of bryostatins and related congeners contains some of the most complex examples of Prins cyclizations in natural products synthesis. Simultaneous reports by both Wender and coworkers and Keck et al. illustrate the power and tolerance of Prins cyclization on complex bryostatin analogs (Scheme 42)

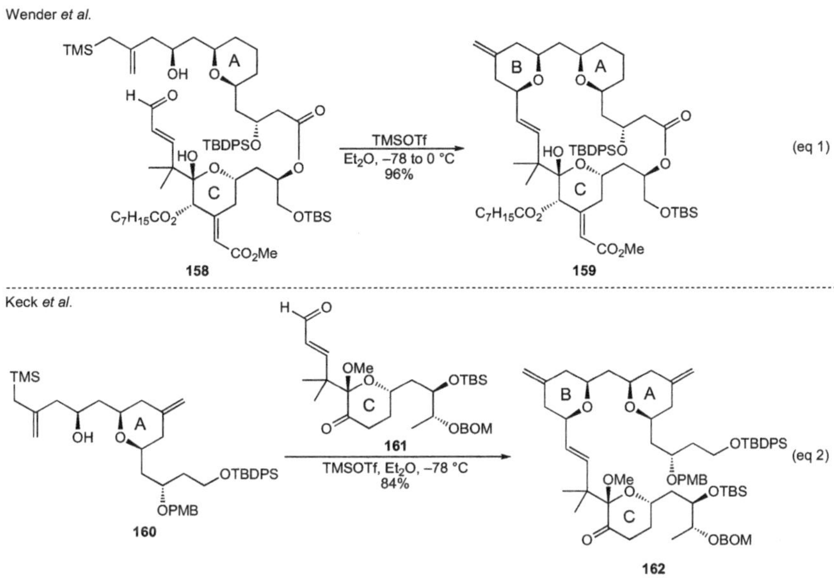

Scheme 42 B ring assembly of bryostatin analogs in intra- and intermolecular contexts [85, 86]

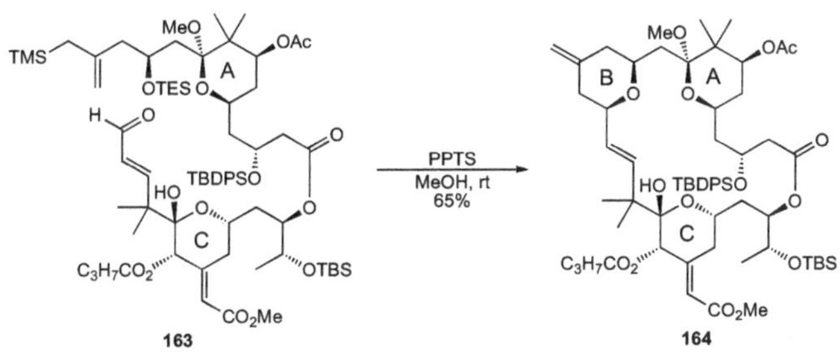

Scheme 43 Highly functionalized Prins macrocyclization en route to bryostatin 9 [87]

[85, 86]. Wender et al. implemented a Prins reaction in the context of a late-stage macrocyclization event. Treatment of allylsilane aldehyde **158** with TMSOTf gave excellent conversion to 20-membered macrocycle **159** (96 % yield). Keck et al. also disclosed a related intermolecular strategy incorporating an allylsilane and an enal. In the presence of TMSOTf, allylsilane **160** and enal **161** led to macrocyclic precursor **162** in an impressive 84 % yield. The highly functionalized substrates used in these examples underscore the mildness of the reaction conditions.

A subsequent report by Wender and Schrier implemented a similar late-stage macrocyclization in the synthesis of bryostatin 9 (Scheme 43) [87]. The Lewis acid, TMSOTf, was abandoned in favor of pyridinium *p*-toluenesulfonate (PPTS) in order to mitigate side reactions due to the ketal functionality in the A ring. As a result, treatment of allylsilane aldehyde **163** with PPTS in MeOH unveiled the

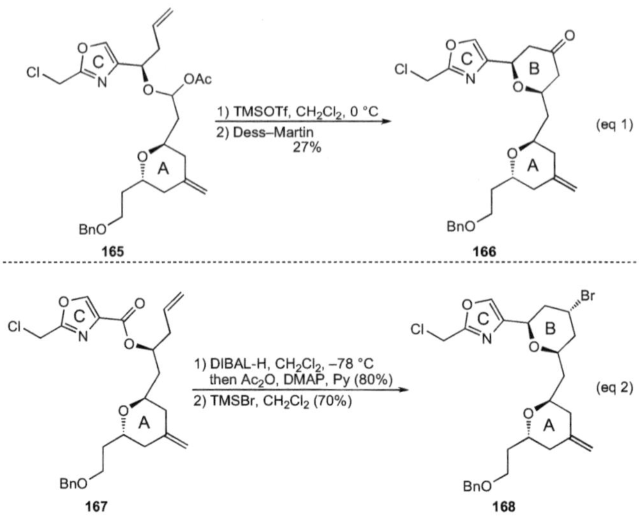

Scheme 44 Reductive acetylation approaches to the bis-THP fragment of phorboxazole B [88]

TES-protected alcohol, which underwent Prins cyclization to form the B ring in 20-membered macrocycle **164** in 65 % yield. This advanced intermediate was carried forward to bryostatin 9 in four synthetic steps. These convergent strategies represent concise approaches to the bryostatin family of natural products and highlight the utility of Prins cyclizations in complex settings.

The flexible nature of a reductive acetylation/Prins cyclization protocol was demonstrated by Rychnovsky et al. in the synthesis of the bis-THP fragment of phorboxazole B (Scheme 44) [88]. The initial strategy of using α-acetoxy ether **165** as a Prins precursor proved problematic due to unwanted reactivity from the oxazole moiety translating to a 27 % yield of 2,6-*cis*-4-one THP **166**. Switching the relative location of the α-acetoxy and homoallylic moieties attenuated the unwanted pathway, a strategy that required minimal functionalization of advanced intermediates. The second-generation Prins precursor **167** underwent facile cyclization to give 4-bromo THP **168** in 70 % yield.

3.2 Petasis–Ferrier Union/Rearrangement

Synthesis of Petasis–Ferrier union/rearrangement substrates usually involves the condensation of bis-silylated β-hydroxy acid **169** and aldehyde **94′** to afford dioxanone **170** (Scheme 45, Eq. 1) [89, 90]. Carbonyl olefination (typically using Cp_2TiMe_2) followed by treatment with a Lewis acid (often alkyl aluminum

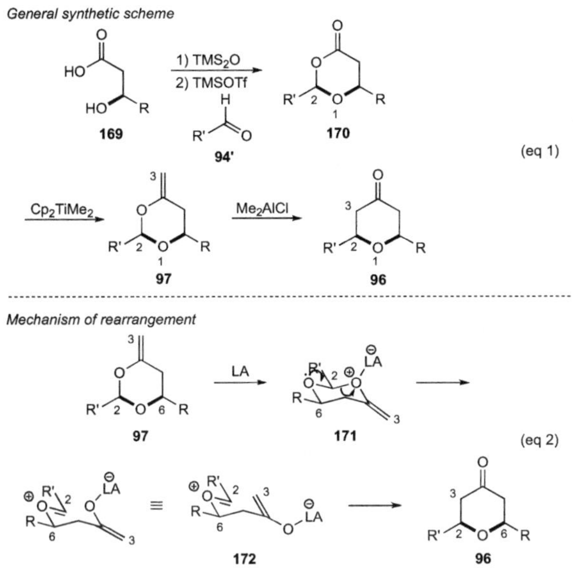

Scheme 45 Synthesis and mechanism of the Petasis–Ferrier union/rearrangement from β-hydroxy acids

reagents) gives either 2,6-*cis*-tetrahydropyran-4-ones **96** or 2,6-*cis*-tetrahydropyran-4-ols depending on the Lewis acid used. As in both Prins cyclizations and Panek annulations, this 2,3 disconnection relies on Lewis acid-mediated oxocarbenium ion formation. In the Petasis–Ferrier reaction, the Lewis basic oxygen (attached to C4) coordinates to the Lewis acid to open the acetal, simultaneously revealing an oxocarbenium ion and an enolate (e.g., **172**). Upon bond rotation, enolate attack onto the oxocarbenium ion furnishes THP **96** (Scheme 45, Eq. 2). The use of *i*-Bu_3Al leads to reduction to the alcohol, whereas Me_2AlCl affords the ketone. One tactical advantage is that the Petasis–Ferrier strategy allows for the construction of highly substituted THPs (i.e., 2,3,5,6-tetrasubstituted) in a predictive manner.

When trisubstituted alkenes are present in the substrate, both *E*- and *Z*-alkenes converge to the same major diastereomer (Scheme 46). This observation is rationalized as follows; upon bond rotation, *Z*-enolate **175** proceeds through a chair-like transition state that places the alkene substituent in an equatorial position. (*E*)-Enolate **176** must adopt a boat-like transition state in order to reduce diaxial interactions. Both transition states lead to a *trans* relationship between the C2 and C3 substituents in product **177**.

The Petasis–Ferrier method to construct 2,6-*cis*-4-one THP rings has been extensively developed and implemented by Smith and coworkers en route to a number of natural products [90]. This strategy has proven especially useful in the preparation of highly substituted THP rings present in (+)-phorboxazole B (the B

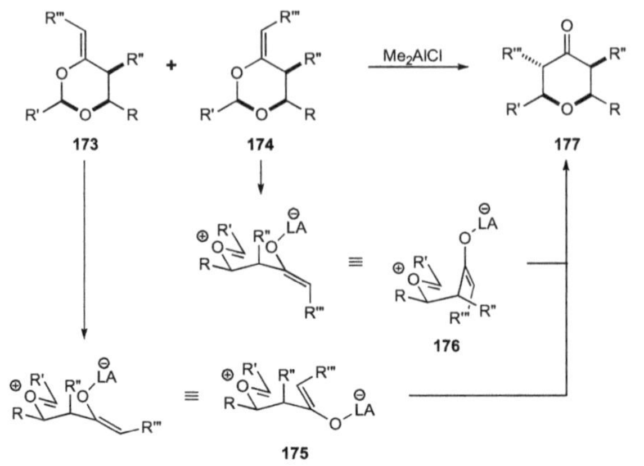

Scheme 46 Convergent Petasis–Ferrier reaction with trisubstituted alkenes

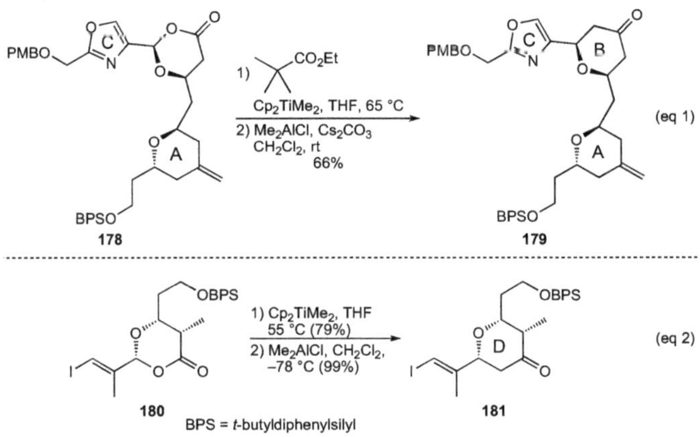

Scheme 47 Petasis–Ferrier approaches to the B and D rings of phorboxazole A [91]

and D rings) and (+)-spongistatin 2 (the F ring). The resulting ketone moiety on the THP ring allows for subsequent functionalization by standard enolate chemistries.

Smith and coworkers envisaged both the B and D rings of (+)-phorboxazole A arising from Petasis–Ferrier union/rearrangements (Scheme 47) [91]. The olefination of dioxanone **178** proceeds in the presence of dimethyltitanocene (Petasis reagent). Ethyl pivalate acts as a slow in situ scavenger to mitigate any side reactivity from Petasis reagent. The resulting enol acetal was then treated with dimethylaluminum chloride to afford rearranged 2,6-*cis* THP product **179**. The use of cesium carbonate was crucial for suppressing cleavage of the PMB group and allowed for construction of the B ring product in 66 % yield over two steps as a single diastereomer. A similar olefination union/rearrangement strategy was brought to bear on dioxanone **180**. Olefination proceeded in 79 % and the

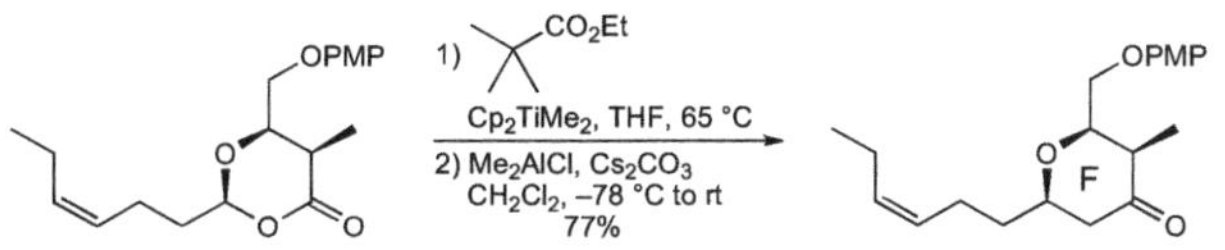

Scheme 48 Petasis–Ferrier union/rearrangement to F ring fragment of spongistatin 1 [36, 92]

rearrangement occurred in excellent yield (99 %) to afford ketone **181**. This product was further elaborated to the 2,3,5,6-tetrasubstituted THP by substrate-controlled stereoselective lithium enolate alkylation with methyl iodide to set the 3,5-*trans* relationship.

The highly substituted F ring of (+)-spongistatin 1 was also constructed using a Petasis–Ferrier union/rearrangement strategy by Smith and coworkers (Scheme 48) [36, 92, 93]. Using the conditions developed for the B ring synthesis in (+)-phorboxazole B (ethyl pivalate and cesium carbonate as additives), dioxanone **182** underwent smooth conversion to desired ketone **183** in 77 % yield as a single diastereomer. In this case, the ketone was stereoselectively hydroxylated at the C5 position using Davis oxaziridine reagent.

3.3 Panek [4+2]-Annulation Strategies

Panek annulation describes the Brønsted or Lewis acid-mediated formal [4+2]-cyclization of a *syn*-allylsilane and an aldehyde (e.g., *syn*-**99** and **94**) leading to a dihydropyran product (**185** or **186**) [94, 95]. The reaction is highly diastereoselective and the relative configuration (2,6-*cis* or 2,6-*trans*) depends on the R′ substituent employed (Scheme 49). This method complements the Prins cyclization, RCM, and lactone/lactol functionalization strategies due the stereodivergent nature of the reaction. Panek and coworkers subsequently developed a stereoselective annulation method employing *anti*-allylsilanes to afford *cis*-DHP rings [96]. These protocols take advantage of the established methods for asymmetric synthesis of the allylsilane substrates.

The stereochemical outcome of Panek annulation when using *syn*-allylsilanes is rationalized as follows (Scheme 49). Trimethylsilyl ether *syn*-**99**, in the presence of a Brønsted or Lewis acid, condenses on aldehyde **94** to give oxocarbenium ion *syn*-**184**. When the substituent on the pendant oxygen (R′) is small and electron-donating (i.e., Me), anchimeric-assisted stabilization of the methyl ether onto the oxocarbenium ion leads to a twist-boat conformation that places the silyl group pseudo-axial to provide 2,6-*trans* DHP **185**. In contrast, when R′ is large and electron withdrawing (i.e., SO_2Mes), destabilizing 1,2-diaxial interactions lead to a chair conformation and ultimately 2,6-*cis* DHP **186**.

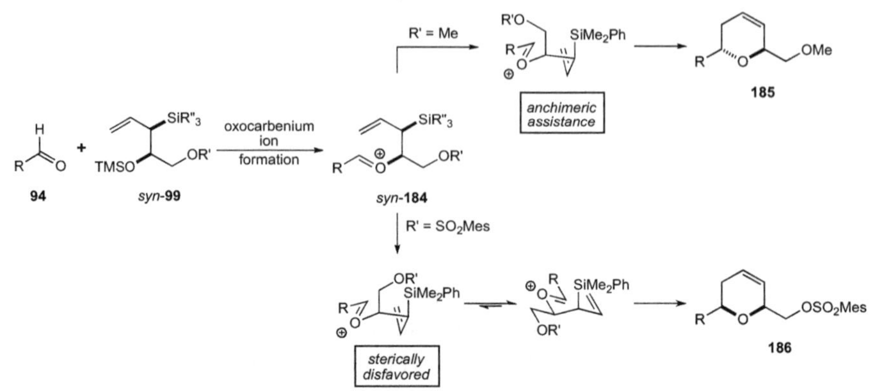

Scheme 49 Rationale for the observed stereoselectivities in Panek annulation of *syn*-allylsilanes

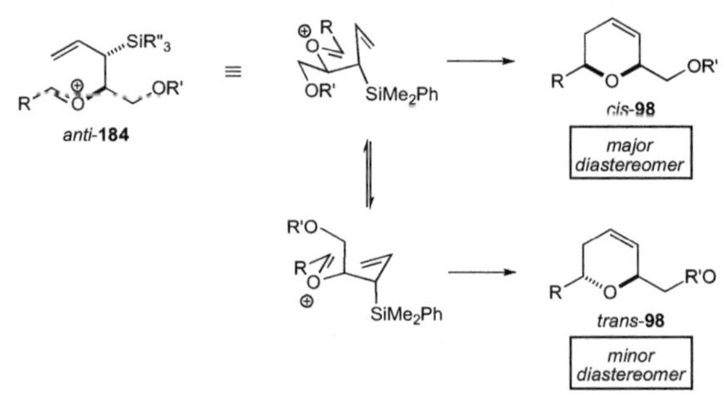

Scheme 50 Panek annulation with *anti*-allylsilanes leads to 2,6-*cis* DHPs

Sterically demanding aldehydes can give inconsistent results with *syn*-allylsilane annulation. Building upon the work of Roush and Dilley [97] concerning the annulation of *anti*-β-hydroxysilanes and aldehydes to form 2,6-*cis* DHP products, Panek developed an *anti*-allylsilane annulation protocol [96]. The rationale for selectivity, analogous to Roush's, is shown in Scheme 50. Upon formation of the oxocarbenium ion derived from an *anti*-allylsilane (e.g., *anti*-**184**), equilibrium is established between the boat and chair conformations. The favored boat conformation places only the silyl group pseudo-axial, whereas the chair conformation is destabilized by the axial disposition of both the silyl group and OR′ side chain. As a result, Panek annulation with *anti*-allylsilanes affords 2,6-*cis* DHP *cis*-**98** as the major diastereomer, regardless of the electronic or steric effects of R′.

Su and Panek's synthesis of leucascandrolide A highlights the versatility of this annulation strategy (Scheme 51) [98]. Treatment of *syn*-allylsilane **187** with triflic acid in the presence of aldehyde **188** led to desired DHP **189** in high yield and diastereoselectivity (82 %, dr = 12:1). DHP **189** was subsequently oxymercurated to install the required C4 oxygen without incident (76 % yield) and further elaborated to aldehyde **191**. The 2,6-*trans* B ring of the natural product was installed using the previously described *anti*-crotylsilane protocol [94]. Thus,

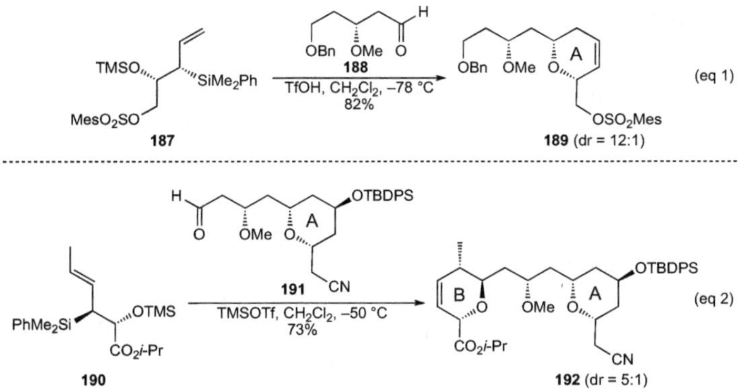

Scheme 51 [4+2]-annulation using both allyl- and crotylsilanes en route to leucascandrolide A [98]

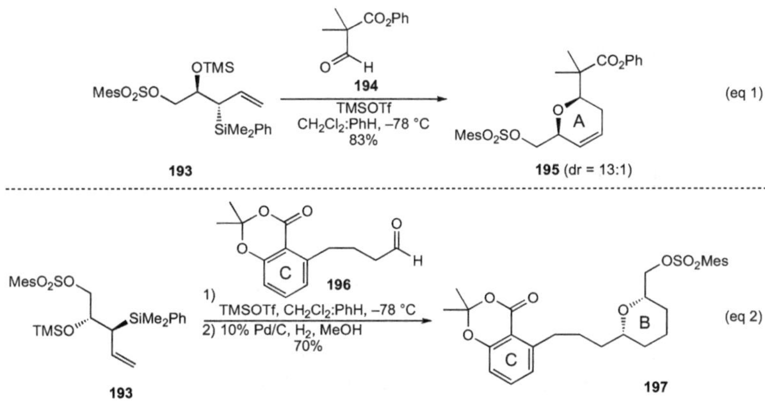

Scheme 52 A common precursor provides the A and B rings of (+)-SCH 351448 [96]

treatment of *anti*-crotylsilane **190** with TMSOTf in the presence of aldehyde **191** afforded the desired 2,6-*trans* DHP **192** in good yield and moderate selectivity (73 %, dr = 5:1). Simple hydrogenation with Pd/C proceeded in high yield (89 %) to provide the B ring of leucascandrolide A.

Zhu and Panek utilized the *anti*-allylsilane annulation protocol to construct both the A and B rings of (+)-SCH 351448 (Scheme 52) [96]. The 2,6-*cis* configuration of both rings in the natural product dictated that sulfonate allylsilane **193** be employed as the nucleophile. Treatment of hindered aldehyde **194** and silane **193** with TMSOTf afforded 2,6-*cis* DHP **195** in 83 % yield with a dr of 13:1. This example is particularly remarkable given the steric demand of the neopentyl electrophile. Subsequent use of silane **193** with aliphatic aldehyde **196** under identical conditions, followed by reduction of the resultant 2,6-*cis* DHP, led to the desired 2,6-*cis* THP **197** in good yield (70 % over two steps) as a single diastereomer. The use of the same silane to construct both rings allowed for a highly convergent route to this macrodiolide.

3.4 Summary

C2–C3 Disconnections remain attractive retrosynthetic strategies for the construction of THP rings. Prins cyclization protocols are the predominate method for such disconnections in the context of natural product synthesis. The wide utility and scope of this transformation must be balanced by careful selection of substrates, since a variety of side reactions or loss of stereochemical information is possible (*vide supra*). This method has proved extremely reliable for the installation of 2,6-*cis* THP rings. The related Petasis–Ferrier method also provides access to 2,6-*cis* THP rings, with the added advantages of increased THP substitution patterns and the C4 ketone as a functional handle. Panek annulation allows rapid access to either 2,6-*cis* or 2,6-*trans* DHP rings by judicious substrate selection (R′ for allyl substrates and *syn*/*anti* for crotylsilanes). These DHP products can easily be converted to THP rings through simple reduction or be elaborated further by manipulation of the alkene moiety. Each of these methods has proven useful in complex natural product synthesis and will continue to remain useful for constructing THP motifs.

4 C3–C4 THP-Forming Processes

Ring-closing metathesis (RCM) is a powerful method for carbon–carbon bond formation that is often employed in the synthesis of complex natural products [99, 100]. Oxygen-containing heterocycles, such as tetrahydropyran (THP) rings, can be envisioned to arise from such transformations (Scheme 53) [101]. When considering such disconnections, the C3–C4 ring closure is preferable to the slower and more problematic C2–C3 metathesis [102]. RCM strategies for THP synthesis are useful due to the mild conditions, excellent functional group compatibility, retention of stereochemical information, and high yields. The main drawback is that stereochemistry must be installed prior to the reaction and further reduction is necessary to reach the THP oxidation state.

There are two main classes of RCM reactions that are important in the synthesis of THP rings (Scheme 54). Class 1 involves the ring closure of ether **199** bearing allylic and homoallylic functionalities to afford 3,4-dihydropyrans **198**, which upon simple reduction leads to THP **202**. Class 2 involves an RCM of homoallylic acrylate substrate **203** to provide unsaturated lactone **204**, which can be further functionalized to THP **202**. All of the RCM metathesis examples in this survey utilize either the first-generation Grubbs catalyst (**G-I**) or the second-generation Grubbs catalyst (**G-II**), shown in Fig. 2.

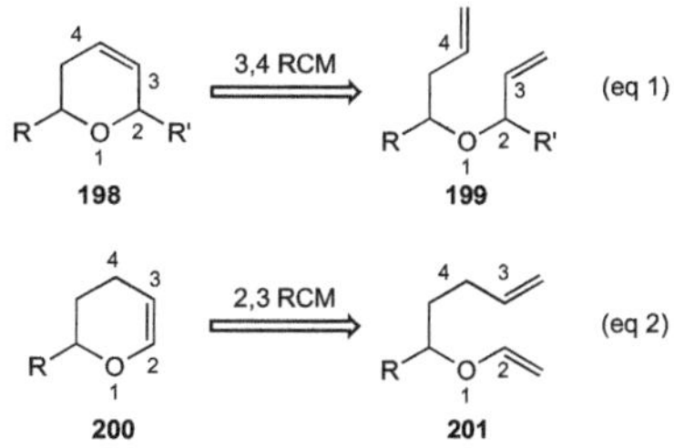

Scheme 53 Possible retrosynthetic disconnections for THP synthesis using ring-closing metathesis (RCM)

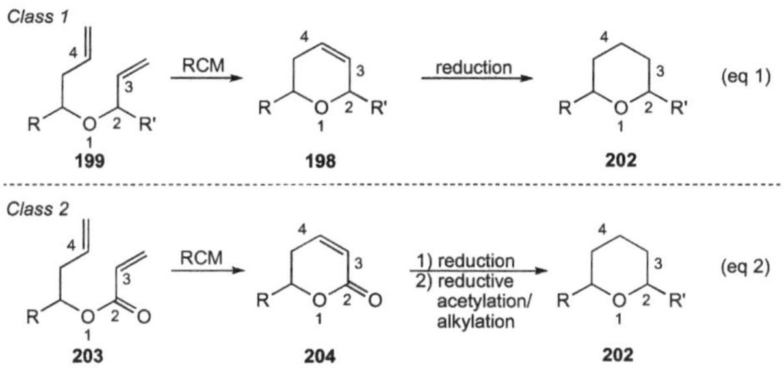

Scheme 54 RCM leading to DHPs (class 1) and RCM leading to α,β-unsaturated lactones (Class 2)

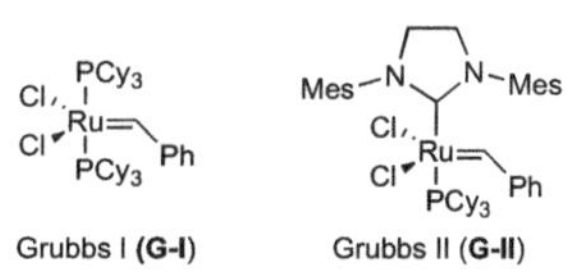

Fig. 2 Common ruthenium catalysts used for RCM

4.1 Class 1 Ring-Closing Metathesis

A general synthetic strategy for the synthesis of Class 1 RCM substrates is shown in Scheme 55. Easily accessible chiral allylic alcohols, such as **205**, can be alkylated with haloacetic acids and derivatized to the oxazolidinone **206**. The chiral auxiliary can undergo standard enolate alkylation to provide the homoallylic ether **207** in high diastereoselectivity. The major advantage to this approach is that either 2,6-*cis* or 2,6-*trans* stereochemistry can be accessed from the allylic alcohol [103]. Unfortunately, this approach requires extra synthetic operations to append and remove the stoichiometric chiral auxiliary, as well as further functionalization in order to access the RCM substrate **199**.

Perhaps the most striking example of the power of the Class 1 RCM reaction comes from Vanier and Crimmins in their enantioselective total synthesis of the

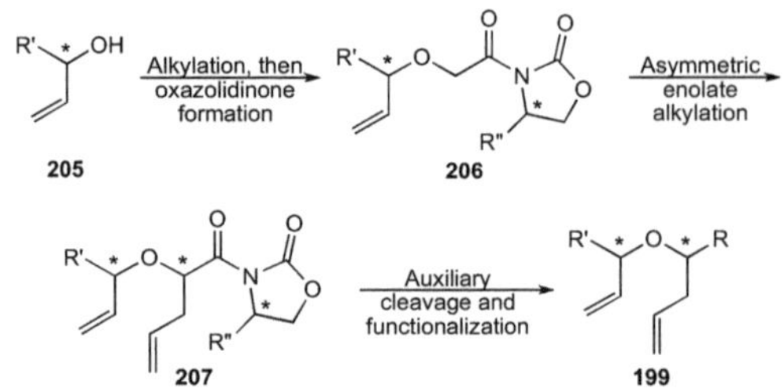

Scheme 55 General strategy for stereoselective synthesis of Class 1 RCM substrates

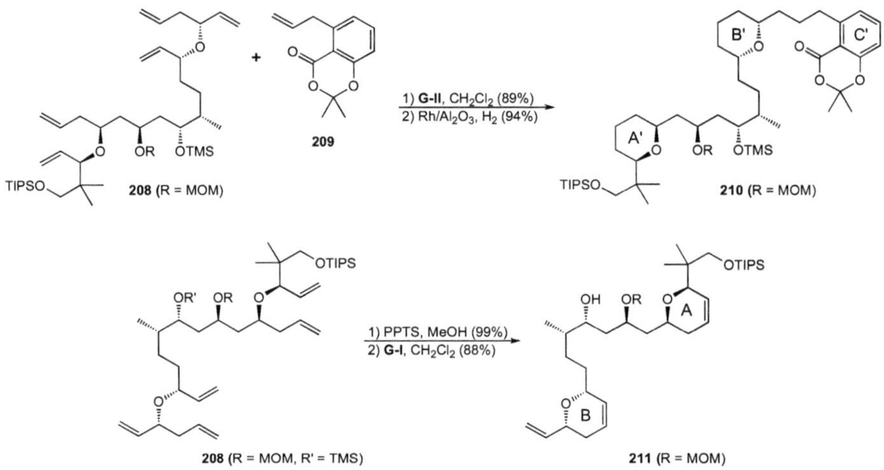

Scheme 56 Differentiated monomer units of (+)-SCH 351448 from tandem Class 1 RCM [104]

dimeric diolide (+)-SCH 351448 (Scheme 56) [104]. The acyclic polyene monomer precursor **208**, in the presence of catalyst **G-II** and dioxenone **209**, undergoes a tandem double ring-closing metathesis/cross-metathesis in 89 % yield. This reaction forms three carbon–carbon bonds by closing both rings and appending the required salicylic moiety. Rhodium-catalyzed hydrogenation proceeds smoothly to afford monomer **210** in 94 % yield. The same precursor **208** was also subjected to double ring-closing metathesis after acid-induced cleavage of the silyl ether. In this example, Vanier and Crimmins used catalyst **G-I** to close both dihydropyran rings, which provided alcohol **211** in 88 % yield. These differentiated monomer units **210** and **211** were then coupled in a stepwise fashion to give a remarkably convergent approach to this selective activator of LDL-R.

Another variation on the Class 1 RCM reaction employs allylic ethers bearing a pendent enol ether. Upon hydrolysis or deprotection, this strategy affords 4-tetrahydropyranones. This method is complementary to the Petasis–Ferrier rearrangement (Sect. 3.2), with the advantage that it also allows access to 2,6-*trans*-substituted tetrahydropyrans. Nakagawa and Crimmins employed this

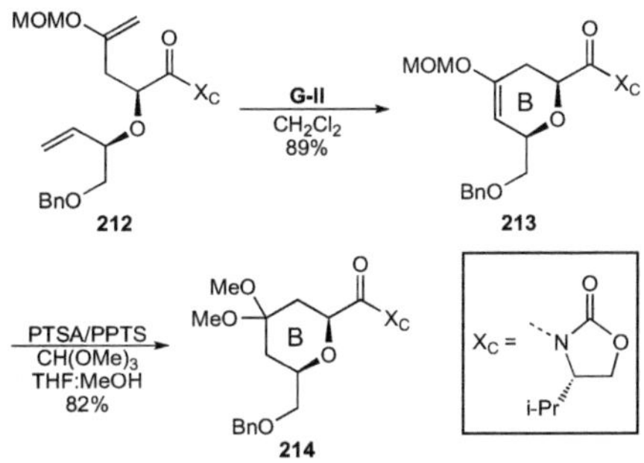

Scheme 57 Enol ether Class 1 RCM to access the B ring of the bryostatins [105]

variation to form the B ring in their northern fragment synthesis of the bryostatins (Scheme 57) [105]. Enol ether **212** in the presence catalyst **G-II** affords the ring-closed product **213** in 89 % yield. Subsequent acid-catalyzed MOM cleavage and dimethyl acetal protection gave the 2,6-*cis* THP **214** in 82 % yield.

4.2 Class 2 Ring-Closing Metathesis

Class 2 RCM reactions, involving the use of homoallylic acrylate esters to form unsaturated lactones, have also found synthetic utility in the context of THP-containing natural products. These acrylate substrates are rapidly accessed from the straightforward esterification of a homoallylic alcohol. While the second class does not formally yield a THP, the lactone affords the appropriate handles for a reductive acetylation/alkylation protocol, which is a powerful method for THP functionalization (Sect. 6.2).

The bis-THP segment of the cytostatic phorboxazole natural products has been shown to be a portion suitable to a Class 2 RCM strategy. Greer and Donaldson demonstrated in 2000 that the B ring could be constructed from a Class 2 RCM reaction of the acrylate **215** by the action of catalyst **G-I** in the presence of titanium tetraisopropoxide to give lactone **216** in 73 % yield (Scheme 58) [106].

Subsequent studies by Yadav and coworkers established that assembling the lower A ring of the bis-THP fragment by a Class 2 RCM reaction was also a viable strategy [107]. When the corresponding acrylate **217** was treated to the same conditions described by Greer and Donaldson, lactone **218** was formed in 94 % yield. Subsequent reduction and alkylation proceeded in good yield to provide bis-THP **219** with the required 2,6-*trans* relationship on the newly formed A ring (Scheme 59).

Cossy and coworkers also employed a Class 2 RCM to form the B ring in their formal total synthesis of the anticancer macrolide leucascandrolide A (Scheme 60) [108]. Using catalyst **G-II** followed by in situ reduction afforded the lactone **221**

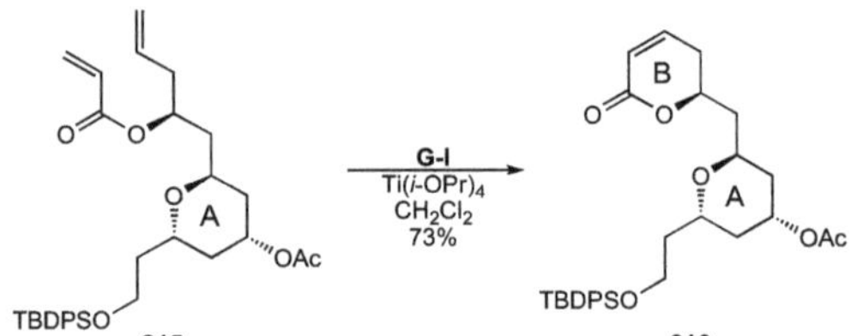

Scheme 58 Class 2 RCM strategy to the A ring fragment of the phorboxazoles [106]

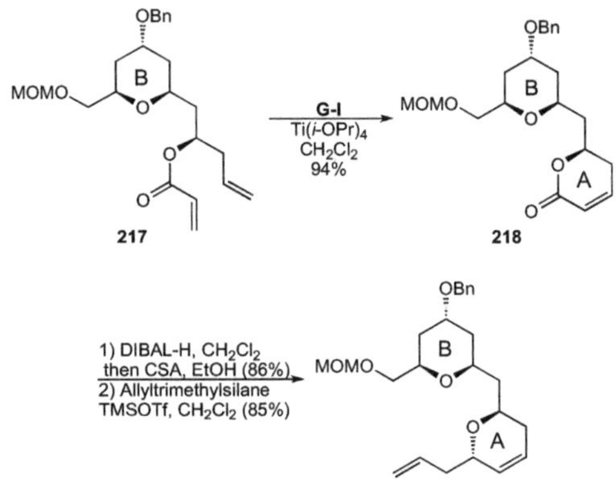

Scheme 59 Class 2 RCM strategy to the B ring fragment of the phorboxazoles [107]

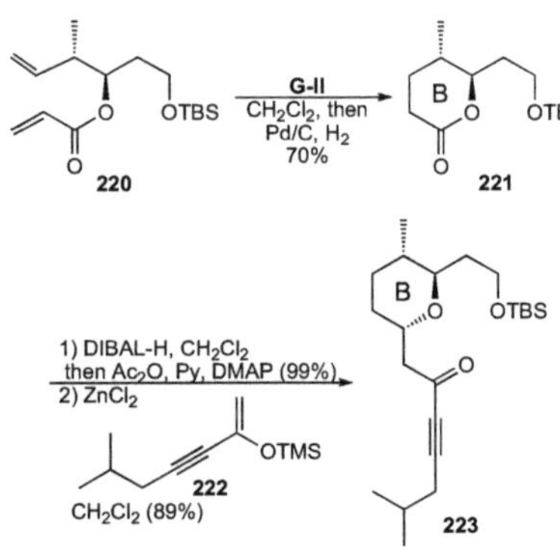

Scheme 60 Class 2 RCM/reductive acetylation/alkylation sequence to B ring fragment of leucascandrolide A [108]

from acrylate **220** in 70 % yield. The resulting lactone **221** was then subjected to reductive acetylation conditions and alkylated with silyl enol ether **222** in the presence of zinc chloride to afford the 2,6-*trans* THP intermediate **223**.

4.3 Summary

These representative examples demonstrate the versatility of RCM reactions for THP synthesis in the context of complex natural products. Both Class 1 and Class 2 substrates afford a reliable way to construct 6-membered oxygen-containing rings in high yields. The Class 1 RCM is versatile from the standpoint of having a number of ways to stereoselectively install the allylic and homoallylic moieties, whereas the strength of Class 2 RCM lies in the versatile synthetic handles afforded in the unsaturated lactone products. While neither of these reactions directly yield THP rings, the established functionalization methods certainly make RCM an attractive strategy. These methods will no doubt be a common strategy in the future.

5 O1–C6 and C2–C3 DHP-Forming Processes

The Diels–Alder (DA) reaction is a powerful method for the efficient and stereoselective formation of highly functionalized six-membered rings. The pivotal role of DA reactions in the construction of carbocyclic frameworks can be translated to the synthesis of moderately complex heterocycles [2]. The hetero-Diels–Alder (HDA) reaction generally proceeds with high regio- and diastereo-selectivity (setting up to 4 contiguous stereogenic centers in a single step) and moderate to excellent yield [109]. From a retrosynthetic perspective, a THP is assembled by bond formation at O1–C6 and C2–C3 followed by reduction of the resultant alkene at C4–C5. For DHP synthesis, there are two modes of reactivity: the normal-electron demand (Scheme 61, Eq. 1) and inverse-electron demand systems (Scheme 61, Eq. 2). The normal-electron demand system uses a carbonyl compound dienophile with a conjugated diene, whereas inverse demand systems have an alkene dienophile and an α,β-unsaturated carbonyl as the diene. Cycloadduct stereochemistry is dependent on transition state geometries [110]. There are four different transition structures that arise from dienophile orientation (*endo* vs. *exo*) and diene geometry (*E*- vs. *Z*-isomers). The examples discussed herein concern only normal demand systems using a carbonyl and (*E*)-olefin dienes and can therefore be described by the transition states (i.e., **TS-A** and **TS-B**) leading to **226**.

There are two main HDA approaches to stereocontrolled THP synthesis. The first uses a chiral auxiliary to direct π-facial selectivity and generally proceeds through an *endo* transition state to give 2,6-*cis* cycloadducts. The second approach uses coordination of a chiral Lewis acid to activate the carbonyl while directing the approach of the diene to one face of the carbonyl dienophile. Several catalysts have been developed for the asymmetric HDA reaction with Jacobsen's Cr(III) complexes [111, 112] exhibiting good to excellent enantioselectivities with a variety of substrates [113].

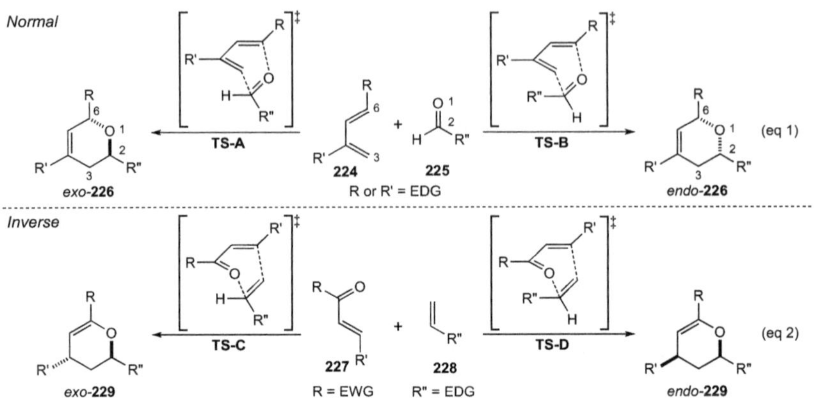

Scheme 61 Stereochemical consequences of hetero-Diels–Alder (HDA) approaches to THPs

5.1 Hetero-Diels–Alder (HDA) Reactions

Danishefsky's diene [114] has found use in DHP synthesis but is limited by the loss of stereochemistry from formation of the enone double bond. In the absence of chirality, the Lewis acid-activated HDA reaction of **230** and **231** leads to (±)-dihydropyrone **232** as reported in Greer and Donaldson's synthesis of phorboxazole A (Scheme 62, Eq. 1) [106]. This particular method is useful for subsequent DHP elaboration (e.g., conjugate addition regioselective α-substitution of the enone) but fails to capitalize on the ability of the HDA reaction to install multiple stereogenic centers in a convergent, stereoselective fashion. Similarly, Yamamura et al. reported the use of enantioenriched aldehyde **233** and diene **231** provided the B ring of bryostatin 3 in good yield (72 %) and high enantioselectivity (ee > 90 %) (Scheme 62, Eq. 2) [115].

Chiral acetonides and siloxydienes allow for the synthesis of 2,6-*cis* THP rings. Cink and Forsyth reported a convergent HDA in the synthesis of phorboxazole A (Scheme 63, Eq. 1) [37]. Use of *ent*-**233** with diene **235** followed by cleavage of the resultant silyl enol ether provided the desired tetrahydropyrone **236** as the major diastereomer (dr = 16:4:1) in 60 % yield over two steps. Burke and coworkers later capitalized on this strategy in the synthesis of the bryostatin 1 B ring (Scheme 63, Eq. 2) [116]. Subjecting chiral acetonide **237** and the complex diene **238** to Lewis acidic conditions provided the 2,6-*cis* DHP **239** in 67 % yield with a dr of 15:4:1. The desired major product was easily separated and taken forward to the natural product. This complex example shows the utility and functional group compatibility of the HDA reaction.

Chiral Lewis acid-catalyzed HDA reactions have found application in the asymmetric synthesis of THP-containing natural products. Chiral chromium complexes, especially the adamantyl-Cr(III) complexes discovered by Jacobsen et al., have been applied to the use of unactivated aldehyde dienophiles with various diene partners [111]. Paterson and coworkers employed this variation in the synthesis of

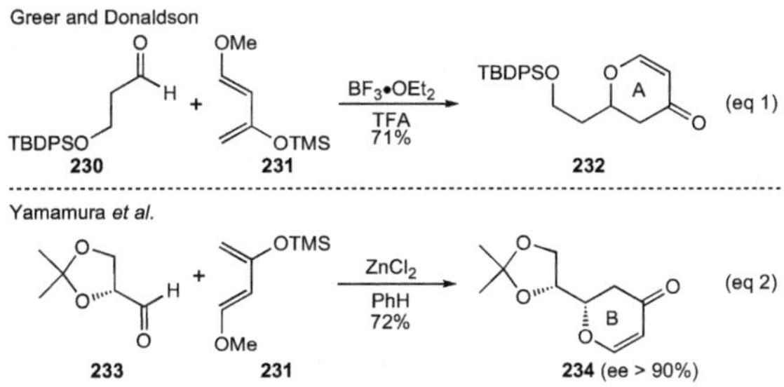

Scheme 62 Danishefsky's diene in HDA reactions toward phorboxazole and bryostatin natural products [106, 115]

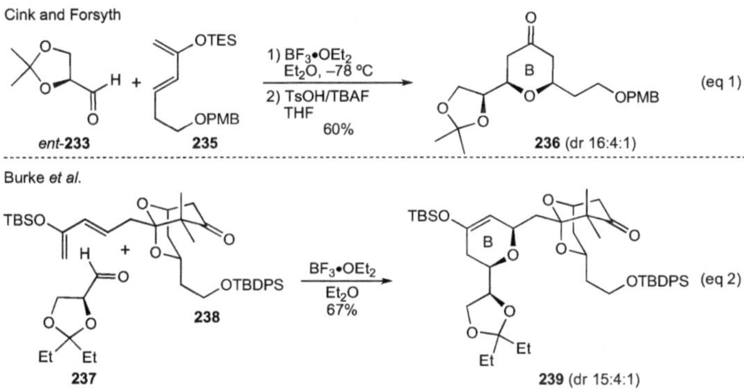

Scheme 63 Chiral acetonide dienophiles in HDA reactions toward phorboxazole and bryostatin natural products [37, 116]

leucascandrolide A and phorboxazole A as shown in Scheme 63. In a simple context, the use of aldehyde **240**, siloxydiene **235**, and catalytic Lewis acid *ent*-**cat. A** followed by acidic cleavage of the resultant silyl enol ether provided 2,6-*cis* THP **241** for leucascandrolide A (Scheme 64, Eq. 1) [117]. This example exhibited excellent enantio- and diastereoselectivity with concomitant high yield. One of the more challenging examples of the asymmetric catalytic HDA reactions was reported in Paterson's total synthesis of phorboxazole A (Scheme 64, Eq. 2) [118]. In the context of a convergent route, the union of aldehyde **242** and siloxydiene **243** proceeded in moderate yield (44 %, 90 % brsm) and low diastereoselectivity (dr = 1.5:1 of *endo*-cycloadducts) to provide the B ring THP **244**.

Wender et al. reported an interesting example that appears to exhibit a cooperative or matched-matched case for the chiral Lewis acid-catalyzed HDA of an aldehyde bearing a chiral auxiliary and Danishefsky's diene (Scheme 65) [119]. Subjecting chiral aldehyde **245** and siloxydiene **231** to Jacobsen's catalyst resulted in the formation of 2,3-DHP **246** in 88 % yield with excellent diastereoselectivity (dr = 33:1).

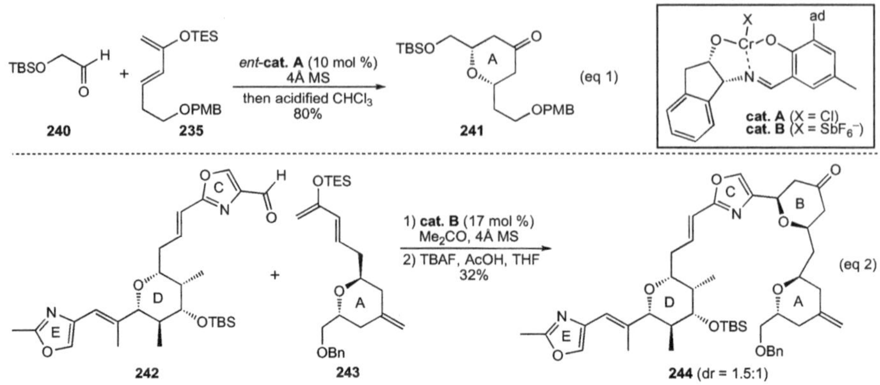

Scheme 64 Early- and late-stage applications of Cr(III)-catalyzed HDA reactions to leucascandrolide A and phorboxazole A [117, 118]

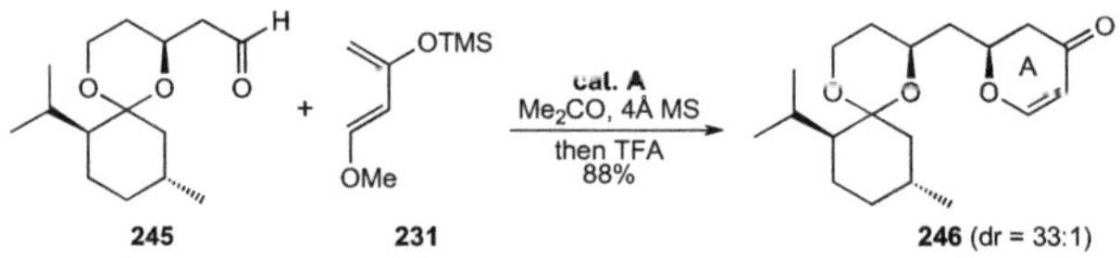

Scheme 65 Highly diastereoselective Cr(III)-catalyzed HDA reaction with chiral dienophile to access A ring fragment of bryostatins [119]

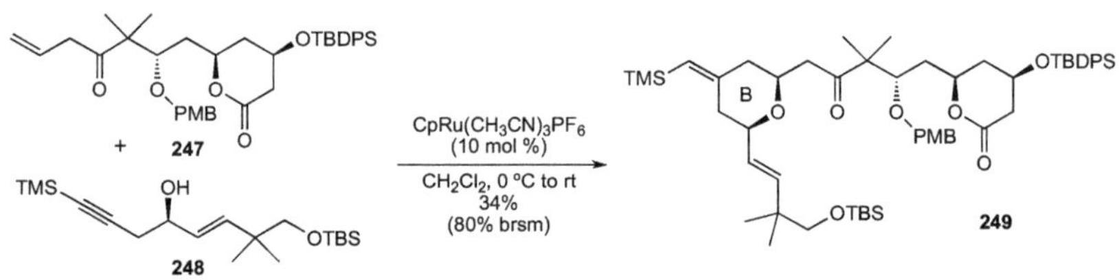

Scheme 66 Ru-catalyzed tandem alkene–alkyne coupling/Michael addition en route to bryostatins [121]

While the asymmetric HDA has proven to be an effective method for the construction of DHP/THP motifs, additional methods based on a metal-mediated/metal-catalyzed coupling and cyclization strategy are burgeoning. The study of cationic ruthenium in ene-type addition of alkenes to acetylenes has been studied previously [120], and Trost et al. recently reported an alternative approach to HDA THP cycloadducts by a similar process [121]. The first step of the transformation involves addition of the Ru-alkene complex of **247** to acetylene **248** followed by reductive elimination to give an enone. The enone subsequently undergoes a conjugate addition with the pendant alcohol to afford the desired 2,6-*cis*-tetrahydropyran **249**.

The tandem Ru-catalyzed alkene–alkyne coupling/conjugate addition sequence occurred in modest yield (34 %) yet showed remarkable chemoselectivity and functional group tolerance (80 % brsm) (Scheme 66).

5.2 *Summary*

These representative examples demonstrate the usefulness of HDA reactions for THP synthesis in the context of complex natural products. Both chiral auxiliary and asymmetric catalysis methods offer a reliable way to construct 6-membered oxygen heterocycles in high yield and stereoselectivity. The chiral auxiliary approach benefits from the wide array of available chiral auxiliaries as well as a number of ways to stereoselectively install chiral moieties contained within the natural product. Removal of the auxiliary imposes additional synthetic considerations. The strength of the asymmetric catalysis tactic comes from the ability to generate chiral products from achiral substrates with high levels of stereocontrol. While these approaches produce DHP rings, there are several methods available for their transformation to highly substituted THP products making HDA an attractive method.

6 C2 Functionalization of Lactols and Lactones

Functionalization of lactols or lactones is a useful method for introducing substitution at the C2 position of tetrahydropyran rings [122]. Retrosynthetically, this transformation can be envisioned as arising from a Lewis acid-mediated alkylation of an oxocarbenium ion by an organometallic nucleophile (Scheme 67, Eq. 1). The stereochemical outcome of the reaction is predicted by axial attack of the nucleophile on the half-chair conformation of oxocarbenium ion **252**, to provide kinetically favored 2,6-*trans* THP **253** (Scheme 67, Eq. 2) [123]. The electronics of the newly formed side chain must be taken into account when considering such disconnections. For example, side chains bearing an electron-withdrawing group (EWG) can sufficiently lower the pK_a of the hydrogen on C2. As a result, C2 isomerization can occur through acid- or base-catalyzed mechanisms leading to the thermodynamically favored 2,6-*cis*-**254** (Scheme 67, Eq. 3) [124]. Another minor drawback is that substrate functionality must be tolerant of Lewis acids. Despite these small concerns, such lactone/lactol functionalization strategies for THP synthesis remain useful due to the number of methods to access lactols and lactones, the mild reaction conditions, predictive stereochemical outcomes, and high yields.

Lactol **257** (or **259**) is commonly accessed by the spontaneous closure of alcohol **256** (or **258**) on an aldehyde (or ketone) six atoms away (Scheme 68, Eq. 1–2) [125]. Note that lactols derived from ketones can lead to a 2,6-*cis* arrangement for R and R′ using a hydride nucleophile in the presence of a Lewis acid. α-Alkoxy ether **260** is typically synthesized by ozonolysis of the 5-alken-1-ol **7** followed by treatment with dimethyl sulfide and aqueous acid (Scheme 68, Eq. 3). Lactone

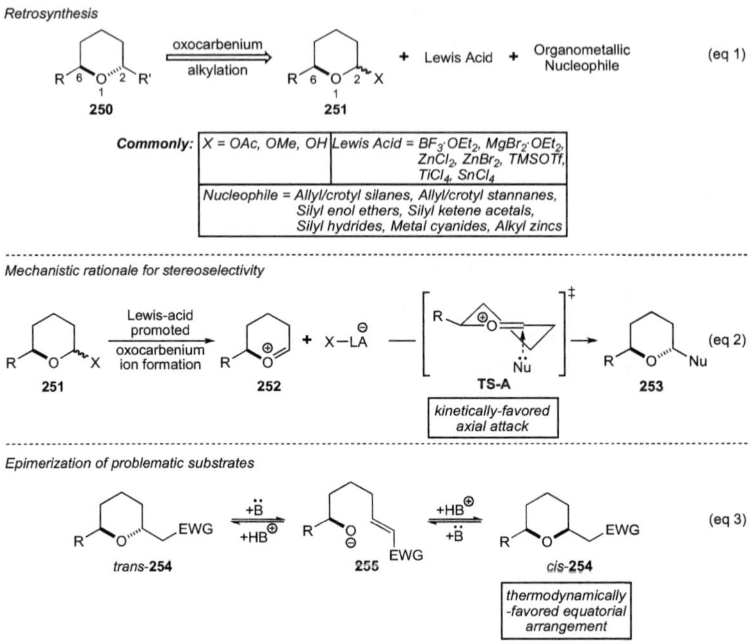

Scheme 67 Reductive C2 functionalization: mechanistic and structural implications

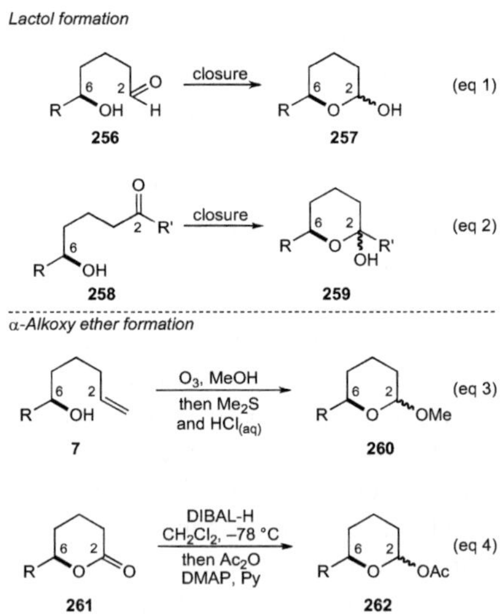

Scheme 68 Common methods for the synthesis of reductive C2 functionalization precursors

reduction with DIBAL-H at low temperatures and subsequent trapping with acetic anhydride leads to the synthetically useful α-acetoxy ether **262** (Scheme 68, Eq. 4) [126, 127]. Of the available methods, the relative stability of the α-acetoxy ether intermediate and the variety of available methods for lactone formation make the reductive acetylation strategy preferable.

6.1 Lactol Reduction

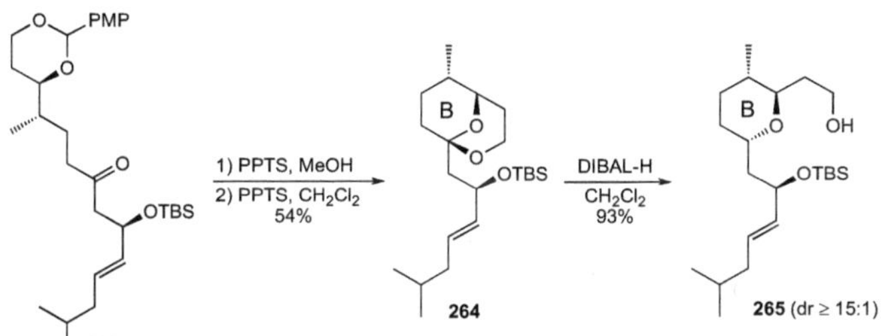

263 264 265 (dr ≥ 15:1)

Scheme 69 Chelate-controlled reduction to the B ring fragment of leucascandrolide A [23]

An interesting example of mixed ketal reduction can be found in Crimmins and Siliphaivanh's synthesis of leucascandrolide A (Scheme 69) [23]. The 1,3-protected diol **263** is deprotected in protic solvent in the presence of acid, followed by the acid-catalyzed formation of mixed ketal **264**. Bridged bicyclic ketals lead to either 2,6-*cis* or 2,6-*trans* arrangements depending on the reagents employed (Lewis acid and triethylsilane, or DIBAL-H, respectively). Chelate-controlled hydride delivery with DIBAL-H affords the desired 2,6-*trans* THP **265** in high yield and good diastereoselectivity (93 %, dr $\geq$ 15:1).

The two 2,6-*cis* THP rings present in the monomeric unit of the dimeric diolide (+)-SCH 351448 offered Backes and Koert an opportunity to exploit stereoselective lactol reduction (Scheme 70) [128]. Lewis acid activation of ketones **266** and **268** with $BF_3 \cdot OEt_2$, followed by closure with the pendant silyl ether, provides lactol intermediates. Subsequent oxocarbenium ion formation and hydride delivery by triethylsilane afforded the 2,6-*cis* THP rings **267** and **269** in high yields (98 % and 93 %, respectively) as single diastereomers.

Evans' synthesis of (+)-phorboxazole B demonstrates the functional group tolerance of lactol reduction in a complex setting (Scheme 71) [129]. Selective deprotection of the triethylsilyl ether **270** provides the lactol **271**, as a 92:8 mixture of closed/open forms that was carried on directly. Subsequent reduction of lactol **271** using the standard conditions afforded the bis-THP fragment **272** in good yield and high selectivity (96 %, dr = 95:5).

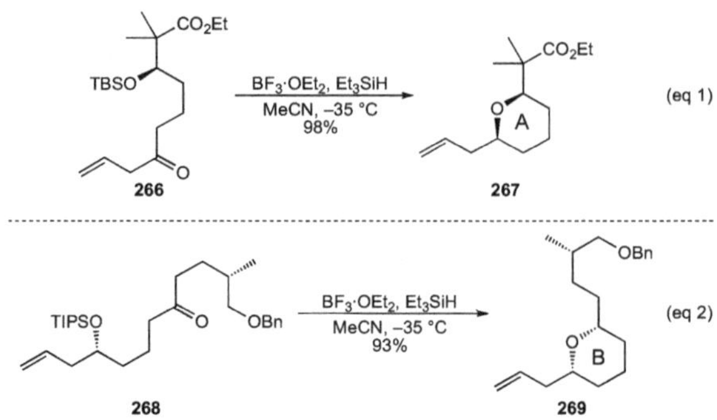

Scheme 70 Condensation/hydride reduction sequence to key fragments of (+)-SCH 351448 [128]

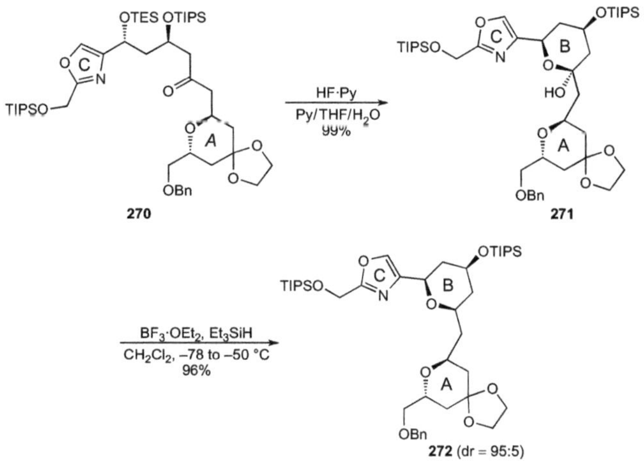

Scheme 71 Lactol formation/reduction protocol to access bis-THP fragment of phorboxazole B [129]

6.2 *Lactone Reduction and Reductive Acetylation*

Both Paterson et al. and Keck and Lundquist have used the α-methoxy ether **273** derived from ozonolysis of the requisite 5-alken-1-ol to functionalize the B ring of the swinholide family of natural products (Scheme 72) [20, 130]. Initial attempts by Paterson et al. to introduce a complex side chain using siloxydiene **274** were plagued by low yields. However, treatment of α-methoxy ether **273** with allyltrimethylsilane in the presence of a catalytic amount of trimethylsilyl triflate (10 mol %) afforded allyl THP **276** in good yields (96 % and 84 %, respectively) as a single diastereomer.

Perhaps no other natural product has highlighted the synthetic utility of a reductive acetylation/alkylation strategy for constructing 2,6-*trans* THP rings as

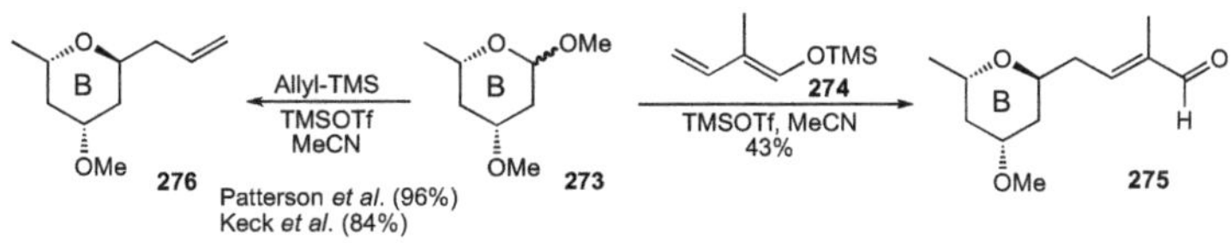

Scheme 72 Allylation of α-methoxy ether to access the B ring of the swinholides [20, 130]

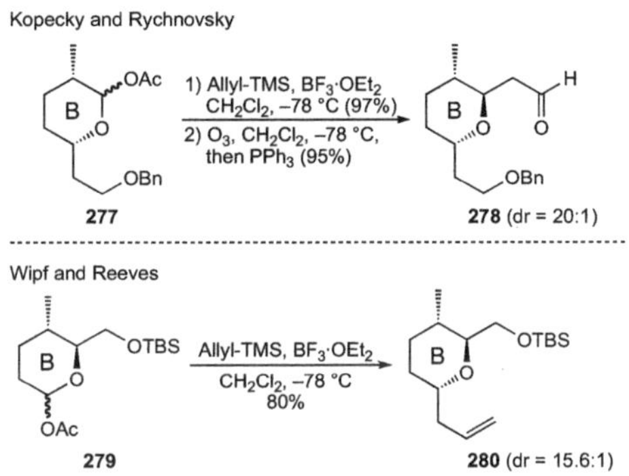

Scheme 73 Allylation of α-acetoxy ethers to functionalize either side of the B ring fragment of leucascandrolide A [78, 79]

much as leucascandrolide A. The B ring of this cytotoxic macrolide has been assembled using variations of this method no less than seven times. Kopecky and Rychnovsky were the first to demonstrate the allylation of α-acetoxy ether **277** with allyltrimethylsilane in the presence of boron trifluoride etherate in high yield and diastereoselectivity (97 %, dr $\geq$ 20:1) [78]. The resulting alkene was cleaved by ozonolysis to furnish aldehyde **278**, which was necessary for coupling the two THP fragments together (Scheme 73, Eq. 1; for coupling, see Scheme 38). Wipf and Reeves later utilized this strategy for alkylating the opposite side of the B ring to provide allyl fragment **280** (80 %, dr = 15.6:1), as shown in Scheme 73, Eq. 2 [79].

Reductive acetylation/alkylation has also provided access to advanced bis-THP intermediates of the type **284** in several reported pursuits of leucascandrolide A (Scheme 74) [76, 117, 131]. These complex examples highlight the remarkable functional group tolerance of this reaction type. The elaborate α-acetoxy ethers **283** demonstrate that a number of protecting groups (benzyl, silyl derivatives, and esters) survive the reaction conditions. Unsaturated silyl enol ethers (i.e., **281** and **282**) are effective nucleophiles and provide advanced intermediates like **284** in excellent yields and diastereoselectivities (80–90 %, dr $\geq$ 25:1 for all cases). A similar strategy, coupled with ring-closing metathesis, was also used by Cossy and coworkers for this natural product (Scheme 60) [108].

The rapid and highly convergent assembly of the bis-THP fragment **287** in Evans and Andrews' synthesis of leucascandrolide A relied on a reductive acetylation/alkylation strategy (Scheme 75) [132]. Treatment of α-acetoxy ether **285** with

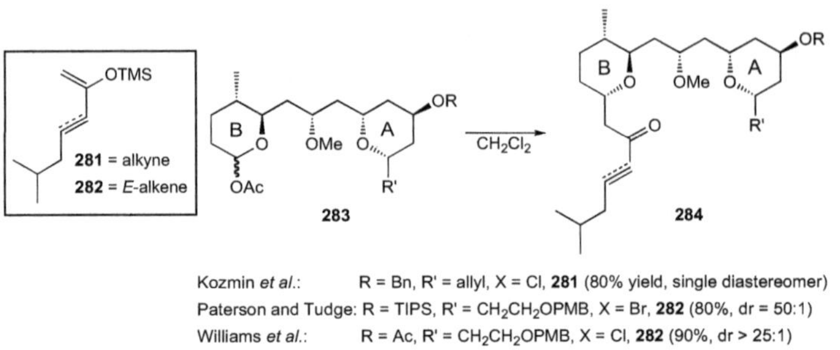

Scheme 74 Complex alkylations of α-acetoxy ethers employing silyl enol ether nucleophiles in the context of leucascandrolide A [76, 117, 131]

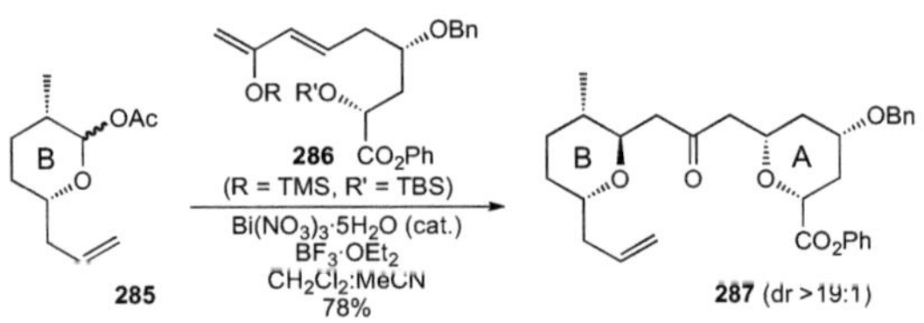

Scheme 75 Convergent strategy to bis-THP fragment of leucascandrolide A through α-acetoxy ether alkylation/conjugate addition sequence [132]

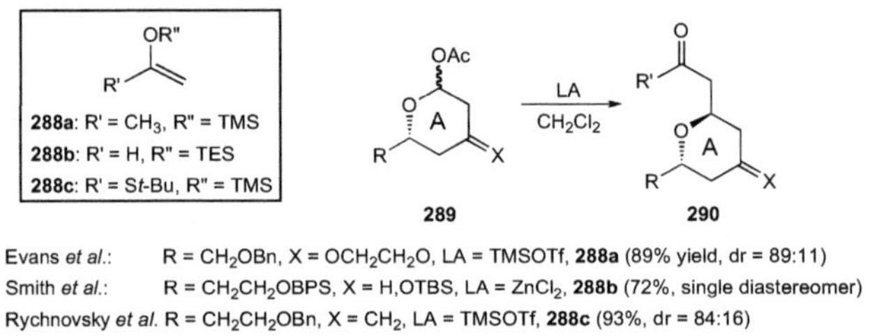

Scheme 76 A ring fragment functionalization of phorboxazoles [88, 91, 129]

boron trifluoride etherate in the presence of siloxydiene **286** at −40 °C leads to an enone intermediate, which upon warming to 0 °C with a catalytic amount of bismuth(III) nitrate pentahydrate closes in an oxa-Michael fashion to give bis-THP **287** in high yield and diastereoselectivity (78 %, dr $\geq$ 19:1). The use of bismuth as a Brønsted acid was important not only for the promotion of the conjugate addition but also to reduce epimerization of the B ring to the thermodynamically favored 2,6-*cis* configuration (*vide supra*).

The A ring of the cytotoxic macrolide phorboxazole family of natural products has also proven to be accessible through a number of reductive acetylation/alkylation protocols (Scheme 76) [88, 91, 129]. These examples employ silyl enol ethers and thioacetate nucleophiles **288a–288c** and α-acetoxy ether derivatives **289** to access

THP derivatives **290** that contain the pendant oxygen functionality required to construct the B ring contained in the natural products. In all cases, the products are obtained in high yields and good diastereoselectivities (72–93 %, dr = 84:16 to dr = 100:0).

6.3 Summary

Lactols and α-alkoxy ethers have proven to be useful intermediates for the synthesis of THP-containing natural products. These representative examples demonstrate that while most commonly used to access 2,6-*trans* motifs, careful substrate design and judicious choice of nucleophile also allows for the synthesis of 2,6-*cis* THP rings. The mild conditions for these functional group tolerant transformations allow for rapid construction of complex fragments in good yield and with high levels of predictable diastereoselectivity. In addition, the variety of methods for synthesizing lactols and lactones increases the overall utility of such approaches.

7 Conclusion

The examples described herein demonstrate the variety, efficiency, reliability, and stereofidelity of THP-forming processes in complex natural product synthesis over the last few decades. The seemingly straightforward construction of tetrahydropyrans through O1–C2 bond formation includes methods ranging from the long-standing S_N2/S_N1 and oxa-Michael-based cyclizations to the more recent transition metal-catalyzed reactions. The recent developments in epoxide-opening cyclizations allow for selective formation of THPs over the more favored THFs as well as the ability to construct fused tetrahydropyran units simultaneously by a cascade sequence. The less obvious C2–C3 bond-forming processes include the venerable Prins cyclization, Petasis–Ferrier rearrangement, and Panek annulation strategies. The reliability of the Prins in the formation of 2,6-*cis* THP rings and the ability to trap various nucleophiles at the C4 position have made this transformation a cornerstone in THP synthesis. Similarly, the Petasis–Ferrier method provides access to 2,6-*cis*-tetrahydropyran-4-ones, with the added advantages of increased THP substitution patterns and the C4 ketone for further functionalization. Panek annulation provides either 2,6-*cis* or 2,6-*trans* DHP rings that can be rapidly converted to highly functionalized THPs but require careful substrate selection. Only upon the advent of alkene metathesis was the C3–C4 disconnection made practical. The RCM strategy benefits from the variety of asymmetric methods to install allylic and homoallylic alcohols (Class 1) and/or the synthetic handles afforded in the unsaturated lactone products for further functionalization (Class 2). The improvement in asymmetric hetero-Diels–Alder reactions, due to the development of chiral organocatalytic and Lewis acid-catalyzed procedures, has provided functionalized THP scaffolds with high regio-, diastereo-, and enantioselectivity (up to four contiguous chiral centers in a single step). The fabrication of THPs by lactone/lactol functionalization remains an attractive method due to the ease of access to lactols and lactones, mild reaction conditions, and the ability to construct 2,6-*trans* THPs. The scope of such ring-forming strategies has been established by the total synthesis of a large number of structurally complex natural products. Furthermore, emerging concepts and methods will continue to facilitate developments in this field.

References

1. Clarke PA, Santos S (2006) Eur J Org Chem: 2045–2053
2. Tietze LE, Kettschau G (1997) Top Curr Chem 189:1–120
3. Boivin TLB (1987) Tetrahedron 43:3309–3362
4. Carmely S, Kashman Y (1985) Tetrahedron Lett 26:511–514
5. Kobayashi M, Tanaka J, Katori T, Kitagawa I (1990) Chem Pharm Bull 38:2960–2966
6. D'Ambrosio M, Guerriero A, Debitus C, Pietra F (1996) Helv Chim Acta 79:51–60
7. Hegde VR, Puar MS, Dai P, Patel M, Gullo VP, Das PR, Bond RW, McPhail AT (2000) Tetrahedron Lett 41:1351–1354
8. Pettit GR, Herald CL, Doubek DL, Herald DL, Arnold E, Clardy J (1982) J Am Chem Soc 104:6846–6848
9. Pettit GR, Kamano Y, Herald CL (1987) J Org Chem 52:2848–2854
10. Pettit GR, Chicacz ZA, Gao F, Herald CL, Boyd MR, Schmidt JM, Hooper JNA (1993) J Org Chem 58:1302–1304
11. Pettit GR, Cichacz ZA, Gao F, Herald CL, Boyd MR (1993) J Chem Soc Chem Commun: 1166–1168
12. Searle PA, Molinski TF (1995) J Am Chem Soc 117:8126–8131
13. Fuwa H (2012) Heterocycles 85:1255
14. Banwell MG, Bissett BD, Bui CT, Pham HTT, Simpson GW (1998) Aust J Chem 51:9–18
15. Betancort JM, Martín VS, Padrón JM, Palazón JM, Ramírez MA, Soler MA (1997) J Org Chem 62:4570–4583
16. Ramírez MA, Padrón JM, Palazón JM, Martín VS (1997) J Org Chem 62:4584–4590
17. Banwell, MG, Bui CT, Pham HTT, Simpson GW (1996) J Chem Soc Perkin 1: 967–969
18. Schneider C, Schuffenhauer A (2000) Eur J Org Chem 2000:73–82
19. Houk KN, Strozier RW (1973) J Am Chem Soc 95:4094–4096
20. Paterson I, Cumming JG, Ward RA, Lamboley S (1995) Tetrahedron 51:9393–9412
21. Paterson I, Smith JD, Ward RA (1995) Tetrahedron 51:9413–9436
22. Fettes A, Carreira EM (2002) Angew Chem Int Ed 41:4098–4101
23. Crimmins MT, Siliphaivanh P (2003) Org Lett 5:4641–4644
24. Laurent F, Reymond S, Capdevielle P, Cossy J (2007) Org Lett 9:2461–2464
25. Bhattacharjee A, Soltani O, De Brabander JK (2002) Org Lett 4:481–484
26. Park H, Kim H, Hong J (2011) Org Lett 13:3742–3745
27. Yadav JS, Bandyopadhyay A, Kunwar AC (2001) Tetrahedron Lett 42:4907–4911
28. Ball M, Bradshaw BJ, Dumeunier R, Gregson TJ, MacCormick S, Omori H, Thomas EJ (2006) Tetrahedron Lett 47:2223–2227
29. Paterson I, Keown LE (1997) Tetrahedron Lett 38:5727–5730
30. Micalizio GC, Pinchuk AN, Roush WR (2000) J Org Chem 65:8730–8736
31. Lee CS, Forsyth CJ (1996) Tetrahedron Lett 37:6449–6452
32. Forsyth CJ, Ahmed F, Cink RD, Lee CS (1998) J Am Chem Soc 120:5597–5598

33. Pattenden G, Plowright AT (2000) Tetrahedron Lett 41:983–986
34. Williamson AW (1852) J Chem Soc 4:229–239
35. Williams DR, Plummer SV, Patnaik S (2003) Angew Chem Int Ed 42:3934–3938
36. Smith AB, Sfouggatakis C, Risatti CA, Sperry JB, Zhu W, Doughty VA, Tomioka T, Gotchev DB, Bennett CS, Sakamoto S, Atasoylu O, Shirakami S, Bauer D, Takeuchi M, Koyanagi J, Sakamoto Y (2009) Tetrahedron 65:6489–6509
37. Cink RD, Forsyth CJ (1997) J Org Chem 62:5672–5673
38. Williams DR, Clark MP, Berliner MA (1999) Tetrahedron Lett 40:2287–2290
39. Li D-R, Zhang D-H, Sun C-Y, Zhang J-W, Yang L, Chen J, Liu B, Su C, Zhou W-S, Lin G-Q (2006) Chem Eur J 12:1185–1204
40. Williams DR, Kiryanov AA, Emde U, Clark MP, Berliner MA, Reeves JT (2003) Angew Chem Int Ed 42:1258–1262
41. White JD, Kuntiyong P, Lee TH (2006) Org Lett 8:6039–6042
42. White JD, Lee TH, Kuntiyong P (2006) Org Lett 8:6043–6046
43. Chamberlin AR, Mulholland RL, Kahn SD, Hehre WJ (1987) J Am Chem Soc 109:672–677
44. Muzart J (2005) Tetrahedron 61:5955–6008
45. Qian H, Han X, Widenhoefer RA (2004) J Am Chem Soc 126:9536–9537
46. Yang C-G, Reich NW, Shi Z, He C (2005) Org Lett 7:4553–4556
47. Coulombel L, Favier I, Dunach E (2005) Chem Commun: 2286–2288
48. Marotta E, Foresti E, Marcelli T, Peri F, Righi P, Scardovi N, Rosini G (2002) Org Lett 4:4451–4453
49. Blanchette MA, Malamas MS, Nantz MH, Roberts JC, Somfai P, Whritenour DC, Masamune S, Kageyama M, Tamura T (1989) J Org Chem 54:2817–2825
50. Hornberger KR, Hamblett CL, Leighton JL (2000) J Am Chem Soc 122:12894–12895
51. Lucas BS, Burke SD (2003) Org Lett 5:3915–3918
52. Trost BM, Toste FD (1999) J Am Chem Soc 121:4545–4554
53. Nicolaou KC, Prasad CVC, Somers PK, Hwang C-K (1989) J Am Chem Soc 111:5330–5334
54. Nicolaou KC, Prasad CVC, Somers PK, Hwang CK (1989) J Am Chem Soc 111:5335–5340
55. Blanc A, Toste FD (2006) Angew Chem Int Ed 45:2096–2099
56. Morimoto Y, Nishikawa Y, Ueba C, Tanaka T (2006) Angew Chem Int Ed 45:810–812
57. Ye T, Pattenden G (1998) Tetrahedron Lett 39:319–322
58. Hale KJ, Hummersone MG, Bhatia GS (2000) Org Lett 2:2189–2192
59. Prins HJ (1919) Chem Weekblad 16:1072–1073
60. Prins HJ (1919) Chem Weekblad 16:1510–1526
61. Snider BB, Trost BM, Fleming I (eds) (1991) In: Comprehensive organic synthesis. Additions to C-X π-bonds, part 2, vol 2. Pergamon, Oxford, pp. 527–561
62. Pastor IM, Yus M (2007) Curr Org Chem 11:925–957
63. Olier C, Kaafarani M, Gastaldi S, Bertrand MP (2010) Tetrahedron 66:413–445
64. Alder RW, Harvey JN, Oakley MT (2002) J Am Chem Soc 124:4960–4961
65. Jasti R, Vitale J, Rychnovsky SD (2004) J Am Chem Soc 126:9904–9905
66. Jasti R, Rychnovsky SD (2006) J Am Chem Soc 128:13640–13648
67. Crosby SR, Harding JR, King CD, Parker GD, Willis CL (2002) Org Lett 4:577–580
68. Marumoto S, Jaber JJ, Vitale JP, Rychnovsky SD (2002) Org Lett 4:3919–3922
69. Jasti R, Rychnovsky SD (2006) Org Lett 8:2175–2178
70. Winstead RC, Simpson TH, Lock GA, Schiavelli MD, Thompson DW (1986) J Org Chem 51:275–277
71. Yang X-F, Mague JT, Li C-J (2001) J Org Chem 66:739–747
72. Martinet P, Mousset G, Colineau M (1969) Acad Sci Paris Ser C 268:1303–1306
73. Martinet P, Mousset G (1970) Bull Soc Chim Fr: 1071–1076
74. Overman LE, Pennington LD (2003) J Org Chem 68:7143–7157
75. Kozmin SA (2001) Org Lett 3:755–758
76. Wang Y, Janjic J, Kozmin SA (2002) J Am Chem Soc 124:13670–13671
77. Yadav JS, Pattanayak MR, Das PP, Mohapatra DK (2011) Org Lett 13:1710–1713

78. Kopecky DJ, Rychnovsky SD (2001) J Am Chem Soc 123:8420–8421
79. Wipf P, Reeves JT (2002) Chem Commun: 2066–2067
80. Jung HH, Seiders JR, Floreancig PE (2007) Angew Chem Int Ed 46:8464–8467
81. Yang J, Viswanathan GS, Li C-J (1999) Tetrahedron Lett 40:1627–1630
82. Chan K-P, Loh T-P (2005) Org Lett 7:4491–4494
83. Chan K-P, Ling YH, Loh T-P (2007) Chem Commun: 939–941
84. Cheung LL, Marumoto S, Anderson CD, Rychnovsky SD (2008) Org Lett 10:3101–3104
85. Wender PA, Dechristopher BA, Schrier AJ (2008) J Am Chem Soc 130:6658–6659
86. Keck GE, Kraft MB, Truong AP, Li W, Sanchez CC, Kedei N, Lewin NE, Blumberg PM (2008) J Am Chem Soc 130:6660–6661
87. Wender PA, Schrier AJ (2011) J Am Chem Soc 133:9228–9231
88. Vitale JP, Wolckenhauer SA, Do NM, Rychnovsky SD (2005) Org Lett 7:3255–3258
89. Petasis NA, Lu S-P (1996) Tetrahedron Lett 37:141–144
90. Smith AB III, Fox RJ, Razler TM (2008) Acc Chem Res 41:675–687
91. Smith AB, Razler TM, Ciavarri JP, Hirose T, Ishikawa T, Meis RM (2008) J Org Chem 73:1192–1200
92. Smith AB, Sfouggatakis C, Gotchev DB, Shirakami S, Bauer D, Zhu W, Doughty VA (2004) Org Lett 6:3637–3640
93. Smith AB III, Zhuang L, Brook CS, Boldi AM, McBriar MD, Moser WH, Murase N, Nakayama K, Verhoest PR, Lin Q (1997) Tetrahedron Lett 38:8667–8670
94. Huang H, Panek JS (2000) J Am Chem Soc 122:9836–9837
95. Su Q, Panek JS (2004) J Am Chem Soc 126:2425–2430
96. Zhu K, Panek JS (2011) Org Lett 13:4652–4655
97. Roush WR, Dilley GJ (2001) Synlett 2001:955–959
98. Su Q, Panek JS (2005) Angew Chem Int Ed 44:1223–1225
99. Grubbs RH, Miller SJ, Fu GC (1995) Acc Chem Res 28:446–452
100. Grubbs RH (2004) Tetrahedron 60:7117–7140
101. Deiters A, Martin SF (2004) Chem Rev 104:2199–2238
102. Brown RCD, Satcharoen V (2006) Heterocycles 70:705–736
103. Crimmins MT, Diaz CJ, Emmitte KA (2004) Heterocycles 62:179–183
104. Crimmins MT, Vanier GS (2006) Org Lett 8:2887–2890
105. Nakagawa-Goto K, Crimmins TM (2011) Synlett 2011:1413–1418
106. Greer PB, Donaldson WA (2000) Tetrahedron Lett 41:3801–3803
107. Yadav JS, Prakash SJ, Gangadhar Y (2005) Tetrahedron: Asymmetry 16:2722–2728
108. Ferrié L, Reymond S, Capdevielle P, Cossy J (2007) Org Lett 9:2461–2464
109. Weinreb SM, Staib RR (1982) Tetrahedron 38:3087–3128
110. Fringuelli F, Taticchi A, Fringuelli F, Taticchi A (eds) (2002) The Diels–Alder reaction: selected practical methods. Wiley, Chichester
111. Dossetter AG, Jamison TF, Jacobsen EN (1999) Angew Chem Int Ed 38:2398–2400
112. Schaus SE, Brånalt J, Jacobsen EN (1998) J Org Chem 63:403–405
113. Pellissier H (2009) Tetrahedron 65:2839–2877
114. Danishefsky S, Kitahara T (1974) J Am Chem Soc 96:7807–7808
115. Ohmori K, Suzuki T, Miyazawa K, Nishiyama S, Yamamura S (1993) Tetrahedron Lett 34:4981–4984
116. Voight EA, Seradj H, Roethle PA, Burke SD (2004) Org Lett 6:4045–4048
117. Paterson I, Tudge M (2003) Tetrahedron 59:6833–6849
118. Paterson I, Luckhurst CA (2003) Tetrahedron Lett 44:3749–3754
119. Wender PA, Baryza JL, Bennett CE, Bi FC, Brenner SE, Clarke MO, Horan JC, Kan C, Lacôte E, Lippa B, Nell PG, Turner TM (2002) J Am Chem Soc 124:13648–13649
120. Trost BM, Indolese A (1993) J Am Chem Soc 115:4361–4362
121. Trost BM, Dong G (2010) J Am Chem Soc 132:16403–16416
122. Kraus GA, Molina MT, Walling JA (1986) J Chem Soc Chem Commun: 1568–1569
123. Lewis MD, Cha JK, Kishi Y (1982) J Am Chem Soc 104:4976–4978

124. Liu L, Donaldson WA (1996) Synlett 1996:103–104
125. Coote SC, Smith LHS, Procter DJ (2011) Science of synthesis, knowledge updates. Product Class 17: Acyclic hemiacetals, lactols, and carbonyl hydrates, vol 2010, iss. 3. Georg Thieme Verlag, Stuttgart, pp 417–473
126. Kopecky DJ, Rychnovsky SD (2003) Org Synth 80:177–183
127. Sizemore N, Rychnovsky SD (2012) Org Synth 89:143–158
128. Backes JR, Koert U (2006) Eur J Org Chem 2006:2777–2785
129. Evans DA, Fitch DM, Smith TE, Cee VJ (2000) J Am Chem Soc 122:10033–10046
130. Keck GE, Lundquist GD (1999) J Org Chem 64:4482–4491
131. Williams DR, Patnaik S, Plummer SV (2003) Org Lett 5:5035–5038
132. Evans PA, Andrews WJ (2008) Angew Chem Int Ed 47:5426–5429

Synthesis of Saturated Six-Membered Ring Lactones

Kalanidhi Palanichamy and Krishna P. Kaliappan

Contents

Abstract The δ-lactone moiety is a common structural framework found in many natural products that exhibit interesting biological properties. This review details a collection of various selected methods developed for the synthesis of saturated δ-lactones. Synthesis of unsaturated δ-lactones and manipulation thereof to saturated δ-lactones are not covered in this review.

Keywords Baeyer–Villiger oxidation • Hydroxy acid • Iodolactonization • Lactol • δ-Lactone

K. Palanichamy • K.P. Kaliappan (✉)
Department of Chemistry, Indian Institute of Technology Bombay, Powai, Mumbai 400076, Maharashtra, India
e-mail: kpk@chem.iitb.ac.in

J. Cossy (ed.), *Synthesis of Saturated Oxygenated Heterocycles I*, Topics in Heterocyclic Chemistry 35, DOI 10.1007/978-3-642-41473-2_3,

1 Introduction

Lactones constitute one of the most important classes of compounds in organic chemistry, and they are generally classified into four main categories such as β-lactones, γ-lactones, δ-lactones, and macrolactones [1, 2]. Among these lactones, the δ-lactone moiety is a common structural subunit present in many natural products, including those isolated from insects, fungi, marine organisms, and plants [3–5]. Natural products containing the δ-lactone framework have been shown to exhibit a wide range of biological activity [6–11]. For example, discodermolide **1** [4, 12, 13] (Fig. 1), a polypropionate-derived marine natural product isolated from the Caribbean sponge *Discodermia dissoluta*, exhibits potent immunosuppressive activity comparable with FK506 and rapamycin. Similarly, mevinolin **2** (lovastatin) [14, 15] and compactin **3** [16, 17] are highly potent inhibitors of 3-hydroxy-3-methylglutaryl coenzyme A (HMG CoA) reductase and are effective in lowering blood plasma cholesterol levels. The β-hydroxy-δ-lactone moiety is the key structural feature of these inhibitors and essential for inhibition as it closely resembles the HMG portion of the enzyme. Another interesting compound with the δ-lactone backbone is the (+)-Prelog–Djerassi lactonic acid **4**, which was isolated independently by Prelog [18] and Djerassi [19] as a key degradation product of the macrolide antibiotics pikromycin and narbomycin and methymycin and neomethymycin, respectively. It has not only provided essential information in the structural elucidation of these antibiotics but has also served as a synthon in their synthesis. Pseudolaric acids A, **5**, and B, **6**, were isolated in 1965 from the bark of *Pseudolarix kaempferi Gordon* [20–24], and the extract from root bark of *P. kaempferi* has been used as a Chinese herbal medicine called *tujinpi* for the treatment of fungal infections. Pseudolaric acid B is a potent antifungal, antifertility, and cytotoxic agent and has also shown significant activity against multidrug-resistant cancer cell lines. Similarly, (−)-iridolactone **7** [25, 26] (Fig. 1) isolated from *Pedicularis* and *Vitex* plants is used as folk medicine in Asia. Ricciocarpins A, **8**, and B, **9**, are two novel sesquiterpene lactones isolated from an axenic culture of the European liverwort, *Ricciocarpos natans* (Ricciaceae) [27–30]. These two natural products have been shown to exhibit high molluscicidal activity against the water snail *Biomphalaria glabrata*, a vector of schistosomiasis. Mevalonolactone **10** [31, 32], the lactone form of mevalonic acid, is an important intermediate in biosynthetic pathways leading to sterols, carotenoids, terpenes, and other isoprenoids. Invictolide **11** was isolated by Rocca et al. in 1983 as one of the three lactones that function as the queen recognition pheromone of the red imported fire ant *Solenopsis invicta* Buren [3]. There are also several insect pheromone δ-lactones such as lactone **12**, the proposed pheromone from *Vespa orientalis* [33], and δ-lactone **13** present in the white butterfly *Idea leuconoe* [34]. (3*R*,4*S*,5*S*,9*S*)-3,5,9-Trihydroxy-4-methylundecanoic acid δ-lactone **14** [35, 36], a novel pentaketide lactone, is an example of δ-lactones obtained through

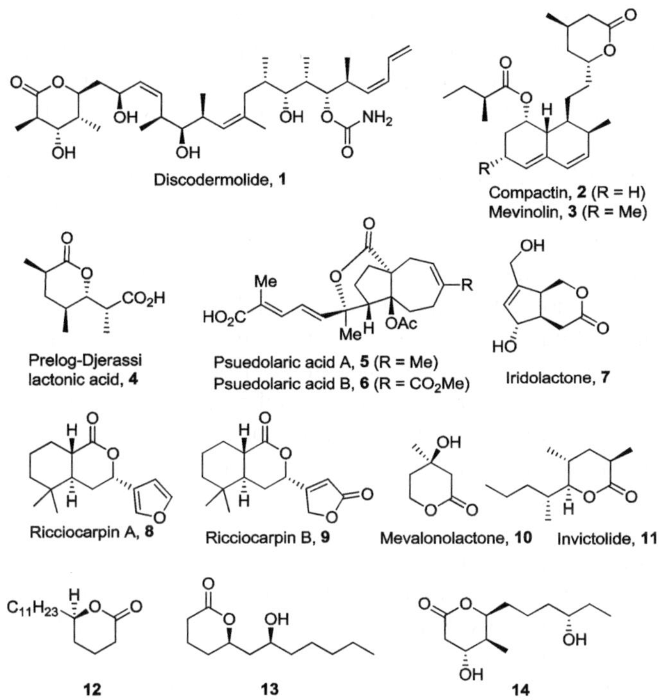

Fig. 1 Examples of natural products with δ-lactone framework

biogenetic engineering by expression of a truncated version of the gene of a polyketide synthon in a heterologous host *Saccharopolyspora erythraea*. Only some selected examples of biologically active natural products with the δ-lactone framework are represented in Fig. 1.

The biological importance of δ-lactones has sparked enormous interest among the synthetic chemists and led to the development of a wide variety of powerful methods for their synthesis. In this chapter, a brief review on selected approaches for the construction of the saturated δ-lactone framework is presented, focusing mainly on the ring-closing steps. We have to point out that methods for the synthesis of unsaturated δ-lactones and their derivatization to saturated δ-lactones are beyond the scope of this chapter.

2 Lactonization of δ-Hydroxy Acid Derivatives

Lactonization is the most widely used method for the formation of lactones, and there are numerous examples under this category for the construction of δ-lactones from the corresponding δ-hydroxy acid derivatives. For example, in 1962, Pappo and Jung reported the synthesis of δ-lactones from cyclohexenones via δ-hydroxy acid [37] (Scheme 1). The protocol involves dihydroxylation of enone **15** with osmium tetroxide, oxidative cleavage of diol with lead tetraacetate, and reduction of the resulting secoaldehyde-acid **16** with sodium borohydride followed by acid treatment to afford the desired lactone **18**.

The same transformation has also been accomplished in one pot by Chavdarian and Heathcock [38] (Scheme 2). Ozonolysis of cyclohexenone **19** in methanol at −60 °C, followed by the addition of excess sodium borohydride at 0 °C, afforded δ-lactone **21** in 45 % yield. The generality of this method was demonstrated using different cyclohexenones to obtain the corresponding δ-lactones.

Coke and Richon have constructed the δ-lactone framework of *n*-hexadecalactone, the proposed pheromone isolated from *Vespa orientalis*, through lactonization of a hydroxy acid intermediate [39] (Scheme 3). The optically pure amino alcohol **22** obtained by resolution was converted to the optically active epoxide **23** by quaternization, followed by Hofmann elimination. Addition of the dianion of propiolic acid to epoxide **23**, and subsequent reduction of the resulting acetylenic hydroxy acid with hydrogen and palladium, provided the saturated hydroxy acid **25**, which spontaneously cyclized to afford δ-lactone **12**. In a similar way, the enantiomer of amino alcohol **22** was also transformed into the antipode of lactone **12**. Furthermore, using this method, any terminal epoxide can easily be converted to the corresponding saturated δ-lactone in two steps.

Another approach to the synthesis of δ-lactones from δ-hydroxy masked acids has been demonstrated by Khan and Paterson [40] (Scheme 4). $ZnBr_2$-catalyzed phenylthioalkylation of ketene bis(trimethylsilyl)acetals **26** with appropriate α-chlorosulphides **27** afforded the corresponding alkylated product **28**, which on hydrolysis underwent lactonization to provide the δ-lactone **29**.

A general approach for the synthesis of δ-lactones from δ-hydroxy esters and amides has been described by Yamaguchi et al. [41]. The strategy involves addition of lithium enolates generated from the esters or amides to oxetanes **30** to afford the corresponding δ-hydroxyesters or δ-hydroxyamides **31** (Scheme 5). Hydrolysis of these intermediates and subsequent lactonization was achieved under either acidic or basic conditions to furnish δ-lactone **32**.

In another report on the synthesis of insect pheromone **12** by Gerth and Giese, lactonization of a δ-hydroxy ester has been used to build the δ-lactone skeleton [42] (Scheme 6). The radical generated from iodide **33** was treated with ethyl acrylate to afford ester **34**, which upon saponification and lactonization furnished lactone **36**. Ester **34** was also converted into the lactonic pheromone of the oriental hornet **12** through a series of transformations. Cleavage of the acetonide group of ester **34**, monotosylation, and protection of the secondary alcohol yielded tosylate **37**. Homologation of tosylate **37** followed by a one-pot deprotection and lactonization furnished pheromone **12**.

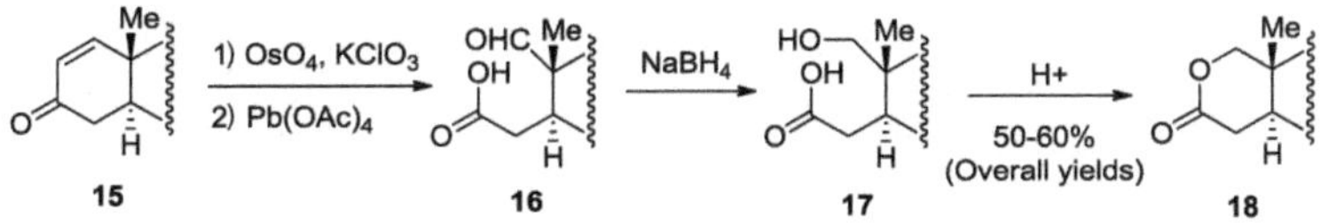

Scheme 1 Pappo and Jung synthesis of δ-lactones

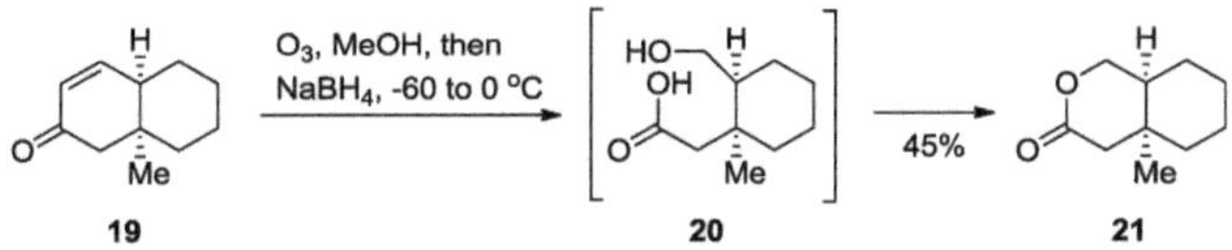

Scheme 2 Heathcock and Chavdarian synthesis of δ-lactones

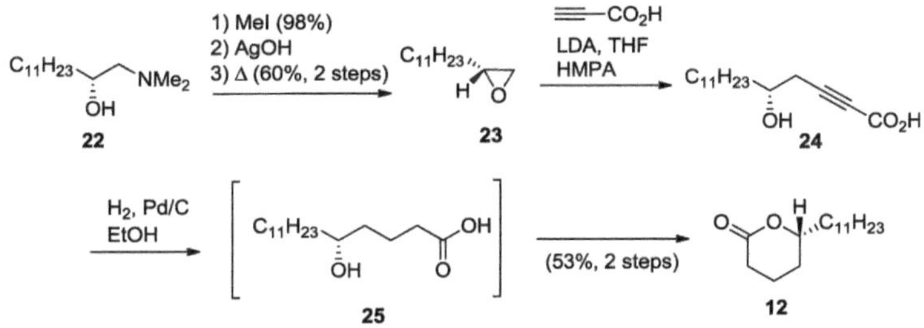

Scheme 3 Coke and Richon synthesis of the proposed pheromone from *Vespa orientalis*

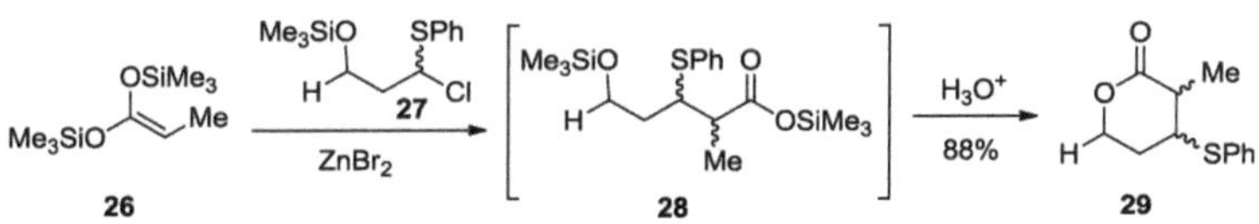

Scheme 4 Paterson and Khan synthesis of δ-lactones

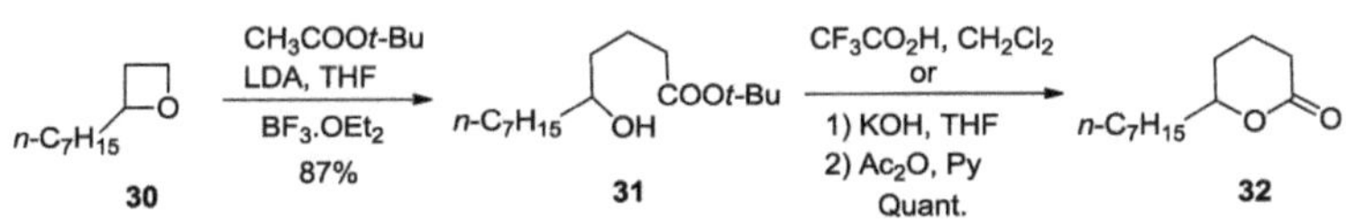

Scheme 5 Yamaguchi et al. synthesis of δ-lactones

In another example, Oda et al. have utilized an acid-catalyzed lactonization of a δ-hydroxy amide derivative in the synthesis of insect pheromone **12** [43] (Scheme 7). The condensation of the diacid dichloride derived from keto diacid **39**, with (*R*)-(+)-[1,1′-binaphthyl]-2,2′-diamine, afforded the coupled product **41**,

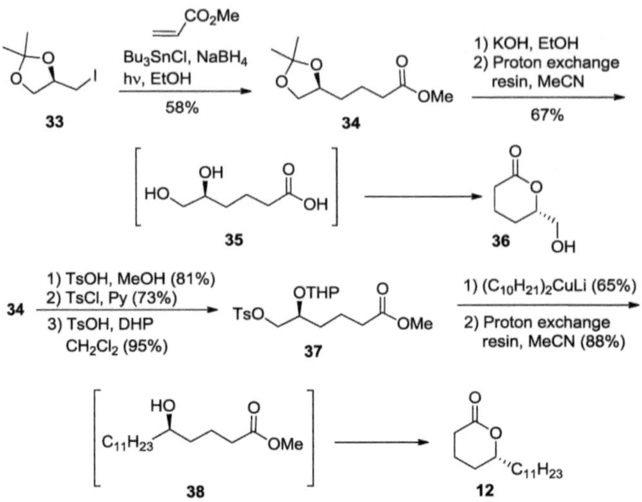

Scheme 6 Giese and Gerth synthesis of δ-lactones

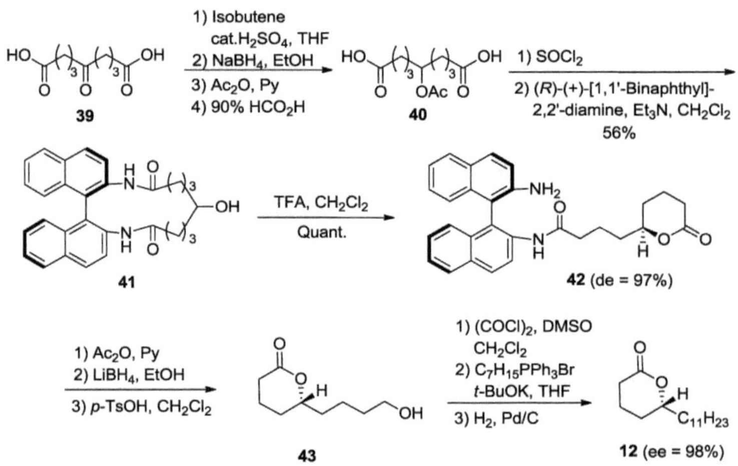

Scheme 7 Oda et al. synthesis of (*R*)-(+)-5-hexadecanolide

which upon treatment with trifluoroacetic acid gave the chiral lactone **42**. The *S* configuration of lactone **42** was established by converting it to the naturally occurring pheromone (*R*)-(+)-5-hexadecanolide **12**. Thus, acylation of the free amine, reduction of the lactone, and subsequent treatment with acid gave lactone **43**, which upon oxidation of the primary alcohol to aldehyde, Wittig reaction, and hydrogenation provided the pheromone **12**.

An example related to the formation of δ-lactone from a δ-hydroxy ortho ester derivative has been demonstrated by Carretero and Ghosez in the synthesis of (−)-argentilactone [44], an α,β-unsaturated δ-lactone [45] (Scheme 8). A normal

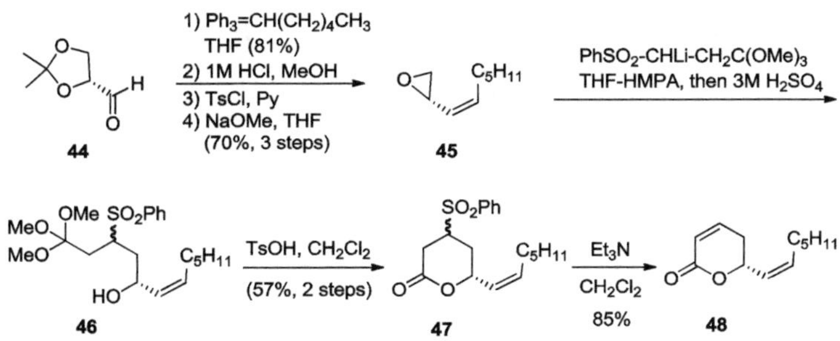

Scheme 8 Ghosez and Carretero synthesis of argentilactone

Wittig reaction on aldehyde **44** afforded the *cis*-olefin, which was then converted into the chiral vinyl epoxide **45** through a three-step sequence. Reaction of methyl 3-phenylsulphonylorthopropionate with epoxide **45** followed by acidic work-up and treatment with *p*-toluenesulfonic acid gave the saturated lactone **47**, which was converted into (−)-argentilactone **48** by treatment with triethylamine.

A novel synthesis of (+)-integerrinecic acid lactone, the necic acid component of the macrolactone pyrrolizidine alkaloid integerrimine **53**, has been reported by White and Jayasinghe, wherein lactonization of a δ-hydroxy acid was used for the ring closure [46] (Scheme 9). Reduction of epoxide **49** derived from *R*-(+)-β-citronellol and subsequent protection gave bis-3,5-DNB ester **50**. Oxidative cleavage of the double bond afforded the carboxylic acid which upon saponification of the esters and acidification resulted in spontaneous lactonization to provide lactone **52**.

Hiyama et al. have reported an enantioselective synthesis of β-hydroxy δ-lactones as simplified analogs of compactin and mevinolin [47] (Scheme 10). The β,δ-diketo ester **54** derived from Taber's chiral alcohol **57** was subjected to a stereoselective reduction with sodium borohydride in the presence of Et_2BOMe to afford a *syn* diol, which upon saponification and then heating in dry toluene led to the chiral lactone **56**.

An acid-catalyzed lactonization of δ-hydroxy esters has been used in the diastereoselective synthesis of *cis*-4,5-substituted δ-lactones by Saigo et al. [48] (Scheme 11). The ring-opening aldol-type reaction of 2-methoxy-2-(trimethylsiloxy)cyclobutanecarboxylic ester **58** with aldehydes **59** in the presence of Lewis acid afforded the corresponding adducts **60** and **61**. Subsequent treatment of these products with a catalytic amount of *p*-TsOH gave *cis*- and *trans*-4,5-substituted lactones in favor of the *cis*-isomer **62**.

Roe and Thomas have reported the transacylation of activated azetidinones to construct the δ-lactone portion of the macrocyclic lankacidin antitumor antibiotics [49] (Scheme 12). Treatment of azetidinone **64** with $BF_3{\cdot}OEt_2$ facilitated a smooth transacylation to give lactone **65** in good yield. In the case of R^1, possessing hydroxyl groups, careful choice of the protecting groups is necessary to avoid any side reactions, such as the formation of any medium ring lactones.

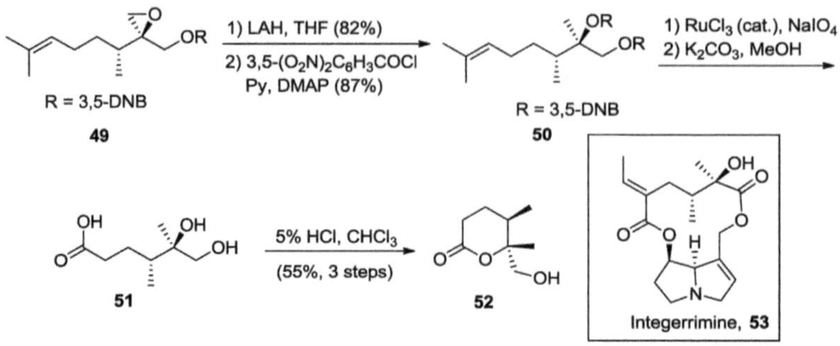

Scheme 9 White and Jayasinghe synthesis of integerrinecic acid lactone

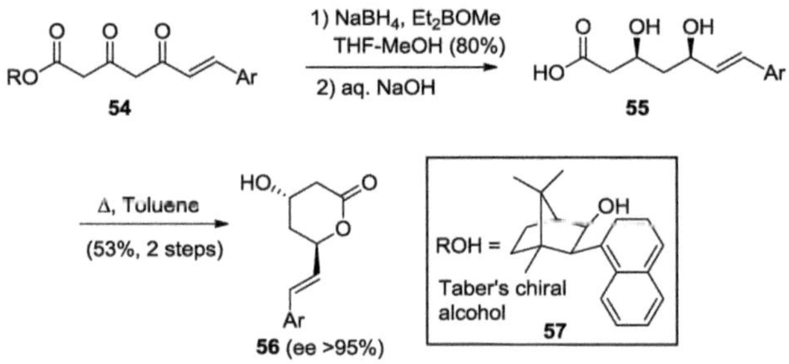

Scheme 10 Hiyama et al. synthesis of β-hydroxy δ-lactones

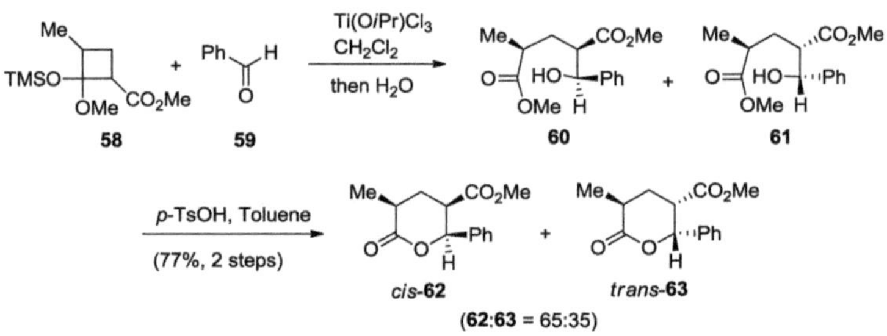

Scheme 11 Saigo et al. synthesis of 4,5-substituted δ-lactones

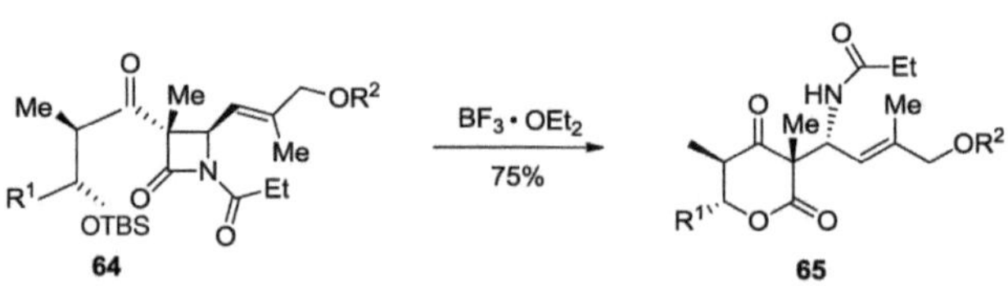

Scheme 12 Synthesis of δ-lactones from azetidinones

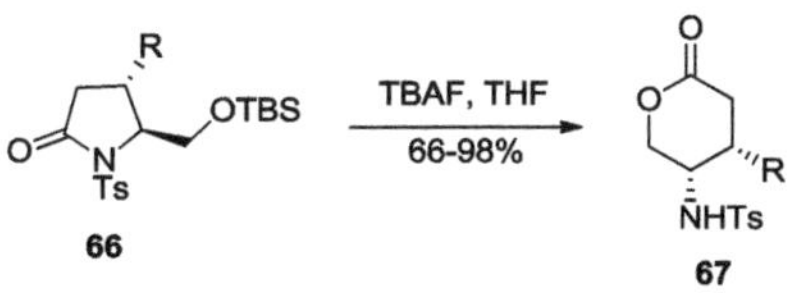

Scheme 13 Synthesis of 5-amino-δ-lactones from pyrrolidinones

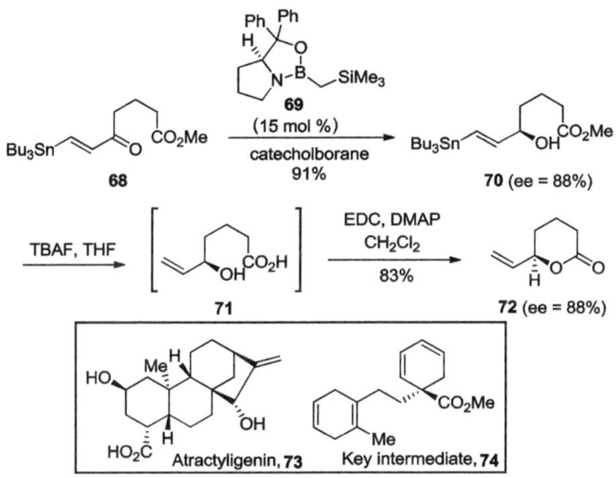

Scheme 14 Corey et al. synthesis of a δ-lactone for an enantioselective synthesis of atractyligenin

Cossy et al. have reported a simple and general one-step protocol for the synthesis of various 5-amino-δ-lactones from pyrrolidinones through the opening of a lactam ring [50] (Scheme 13). The fluoride-induced silyl deprotection of pyrrolidinones **66**, followed by opening of the lactam ring via transacylation, paved the way for the synthesis of a variety of 5-amino-δ-lactone **67** in one pot.

Corey et al. have reported the synthesis of a δ-lactone, an intermediate in their early synthesis of the key intermediate **74** of atractyligenin **73**, via lactonization of a δ-hydroxy acid [51] (Scheme 14). Corey–Bakshi–Shibata reduction of enone **68** under optimized conditions in the presence of freshly prepared catalyst **69** afforded alcohol **70** in 88 % enantiomeric excess. A fluoride-induced destannylation of **70** gave seco-acid **71**, which was lactonized using *N*-(3-dimethylaminopropyl)-*N*′-ethylcarbodiimide hydrochloride (EDC) and 4-(dimethylamino)pyridine (DMAP) to provide lactone **72**. This intermediate was then elaborated to the key intermediate **74** via alkylation and Ireland–Claisen rearrangement.

Honda et al. have utilized acid-catalyzed lactonization of a δ-hydroxy ester in an enantioselective synthesis of the lactone moiety of HMG CoA reductase inhibitor [52] (Scheme 15). An enantioselective deprotonation reaction of *meso*-ketone **75** with lithium (*S*,*S*)-α,α′-dimethyldibenzylamide as the chiral base in the presence of

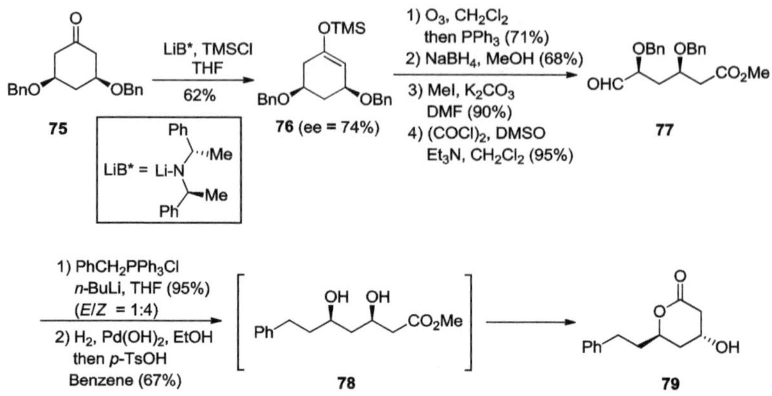

Scheme 15 Synthesis of the β-hydroxy-δ-lactone framework of mevinic acid

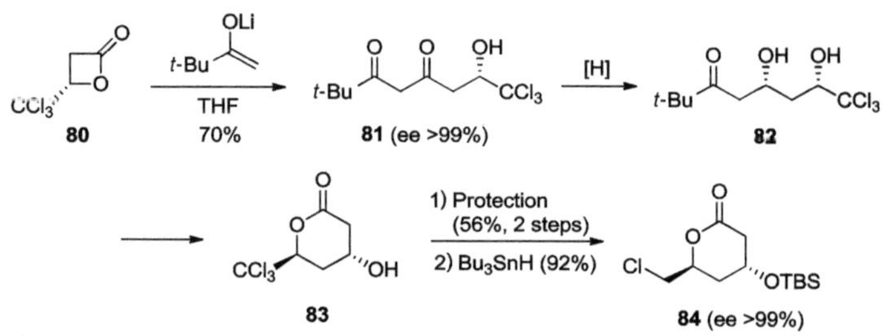

Scheme 16 Synthesis of the 4-hydroxy-δ-lactone component of mevinic acid

trimethylsilyl chloride gave the silyl ether **76** in 62 % yield. Ozonolysis of the enol ether **76** followed by a reduction step provided the corresponding hydroxy acid, which upon esterification and Swern oxidation furnished aldehyde **77**. A Wittig reaction applied to aldehyde **77** with benzyltriphenylphosphonium chloride and *n*-butyllithium, cleavage of the benzyl ethers, followed by acid-catalyzed lactonization generated the β-hydroxy-δ-lactone framework of mevinic acid.

Shimizu et al. have reported the synthesis of the 4-hydroxy-δ-lactone component of mevinic acid by lactonization of a δ-hydroxy ester [53] (Scheme 16). Reaction of lithio-*tert*-butylacetate with β-trichloromethyl-β-lactone **80** gave δ-hydroxy-β-keto ester **81**. A stereoselective *syn* reduction of the ketone, lactonization, and protection of the secondary alcohol provided the lactone. The trichloromethyl group was reduced with tri-*n*-butyltin hydride to furnish δ-chloromethyl lactone intermediate **84**.

An enantiospecific route to the synthesis of tetrasubstituted δ-lactones has been developed by Staunton et al. by lactonization of δ-hydroxy acids [54] (Scheme 17). The Evans aldolization of oxazolidinone **85** with optically active aldehyde **86** gave the Evans aldol product **87**. An oxidative removal of the chiral auxiliary followed by hydroxyl deprotection and cyclization furnished the tetrasubstituted δ-lactone **90**

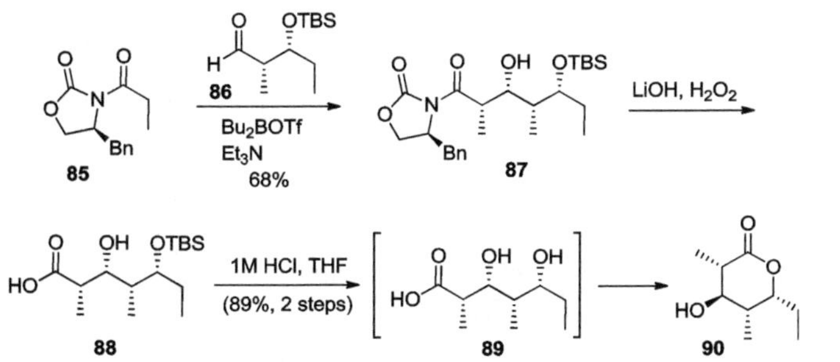

Scheme 17 Synthesis of tetrasubstituted δ-lactones

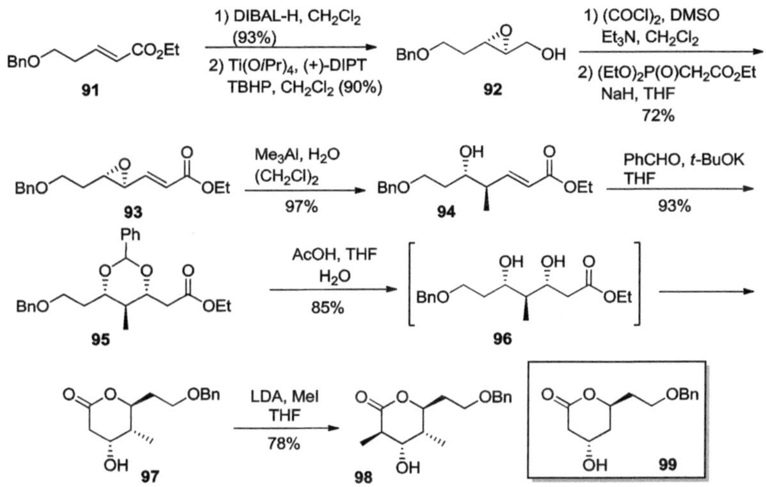

Scheme 18 Miyashita et al. synthesis of the δ-lactone segment of discodermolide

in good yield. It is worth pointing out that this protocol has been shown as a general enantiospecific route to synthesize all possible diastereomers of **90** starting with the appropriate enantiomer of the chiral auxiliary and the aldehyde.

Another example that entailed a stereoselective synthesis of the δ-lactone fragment of discodermolide has been demonstrated by Miyashita et al. [55] (Scheme 18). Reduction of the known ester **91** followed by the Katsuki–Sharpless asymmetric epoxidation of the resulting allylic alcohol gave the α-epoxy alcohol **92**. Transformation of **92** into intermediate **94** was achieved through a sequence of transformations involving a Swern oxidation, a Horner–Wadsworth–Emmons reaction, and a methylation of the epoxide with trimethylaluminium. Treatment of **94** with excess benzaldehyde under Evans conditions facilitated an oxa-Michael addition to give benzylidene acetal **95**, which upon acidic hydrolysis and enolate alkylation led to the δ-lactone backbone of discodermolide. In addition, epoxide **93** was also converted into a series of lactones like **99** through a palladium-catalyzed hydrogenolysis of the epoxide instead of opening with trimethylaluminium [56].

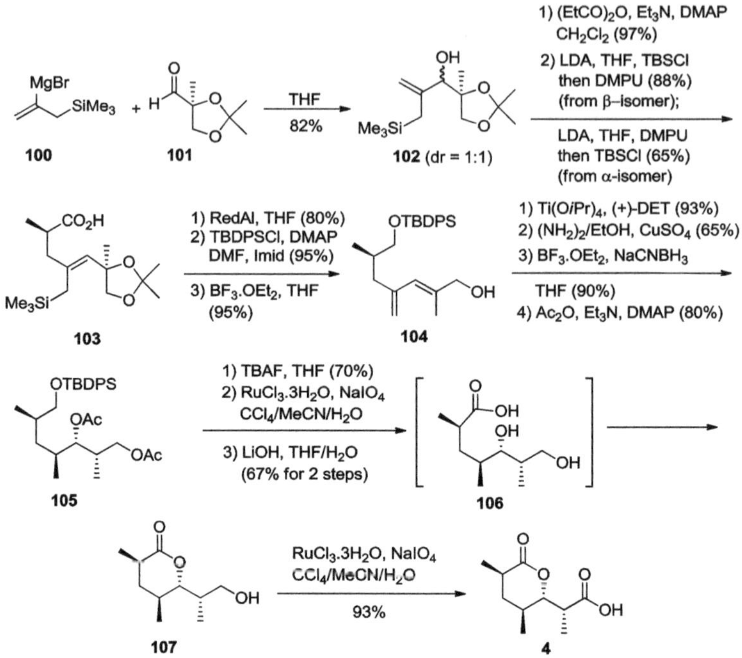

Scheme 19 Parsons et al. synthesis of (+)-Prelog–Djerassi lactonic acid

Another example has been shown by Parsons et al. in the synthesis of Prelog–Djerassi lactonic acid **4** [57] (Scheme 19). The reaction of Grignard reagent **100** with aldehyde **101** gave a separable 1:1 mixture of diastereomers **102**, which were independently converted to a single diastereomeric acetonide acid **103** via an Ireland–Claisen rearrangement under different conditions. Diene alcohol **104** was then transformed into diacetate **105** by Sharpless epoxidation, diimide reduction, and opening of the epoxide. Cleavage of the silyl ether and oxidation of obtained primary alcohol generated the carboxylic acid, which was elaborated to the target compound **4** by a concomitant saponification and lactonization followed by oxidation with ruthenium chloride.

Another example for the construction of δ-lactone from a δ-hydroxy ester is evident in the synthesis of (−)-iridolactone as reported by Ogasawara et al. [58] (Scheme 20). The chiral tricyclic diol **108** was protected and then converted to the tetracyclic intermediate **109** via enolate alkylation, reduction of a ketone, and bromoetherification. Oxidative cleavage of the double bond followed by oxidation of the resulting aldehyde gave the carboxylic acid **110** with concurrent removal of the ethoxyethyl protecting groups. Although seco-acid **110** failed to undergo direct lactonization, exposure of **110** to diazomethane facilitated spontaneous lactonization to reveal (−)-iridolactone **7**.

Another example on lactonization of a δ-hydroxy amide has been shown in the synthesis of 3-aryl-δ-lactones by Smitrovich et al. [59] (Scheme 21). Generation of

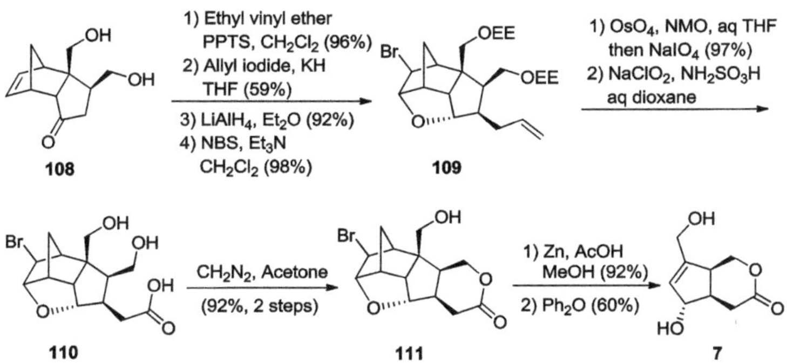

Scheme 20 Total synthesis of (−)-iridolactone

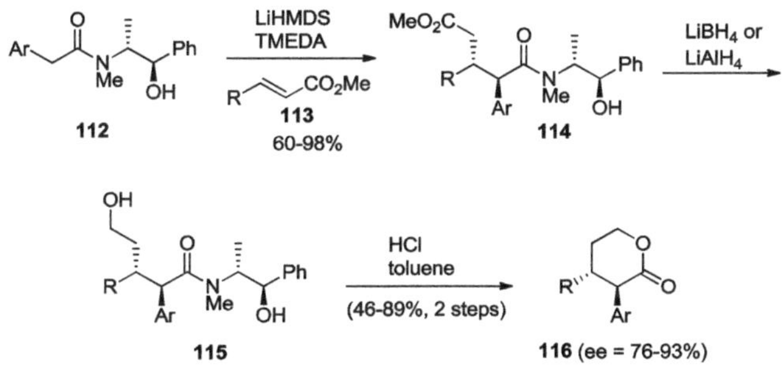

Scheme 21 Synthesis of 3-aryl-δ-lactones

the dianion of pseudoephedrine-derived amide **112** in the presence of LiHMDS and two equivalents of TMEDA at low temperature followed by addition of the unsaturated ester **113** gave the *anti*-isomer **114** with optimal selectivity. Selective reduction of the ester in the presence of the amide was accomplished using a variety of reducing agents, including $LiAlH_4$, $LiBH_4$, and $LiAlH(Ot\text{-}Bu)_3$, to give alcohol **115**, which underwent lactonization under acidic conditions to provide a variety of *trans*-3,4-disubstituted δ-lactone **116** in good yield and with good enantioselectivity. Use of HCl for the lactonization allowed recovery of the pseudoephedrine chiral auxiliary by filtration of the HCl salt.

Schulz et al. have reported the synthesis and absolute configuration of δ-lactones present as components in the scent glands, so-called hair pencils, of the giant white butterfly *Idea leuconoe* (Scheme 22) [60]. A highly enantioselective hydrogenation of dioxo ester **117** in the presence of commercially available (*R*)-Ru-BINAP catalyst afforded the dihydroxy ester **118**, which underwent lactonization under acidic conditions to provide (*R*,*R*)-δ-lactone **119** in excellent yield with excellent selectivity. Epimerization of lactone **119** was achieved by means of a Mitsunobu reaction providing the (*R*,*S*)-δ-lactone **13**. In a similar way, use of (*S*)-Ru-BINAP catalyst in the hydrogenation led to the enantiomers of **119** and **13**. Chiral-phase gas chromatographic analysis of the synthetic and natural products revealed that the

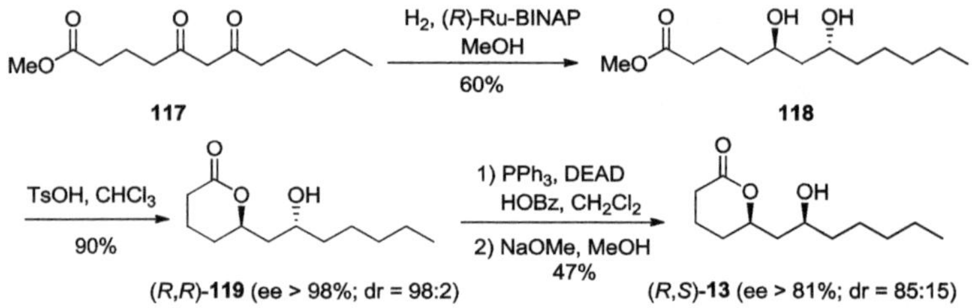

Scheme 22 Synthesis of δ-lactones present in the white butterfly *Idea leuconoe*

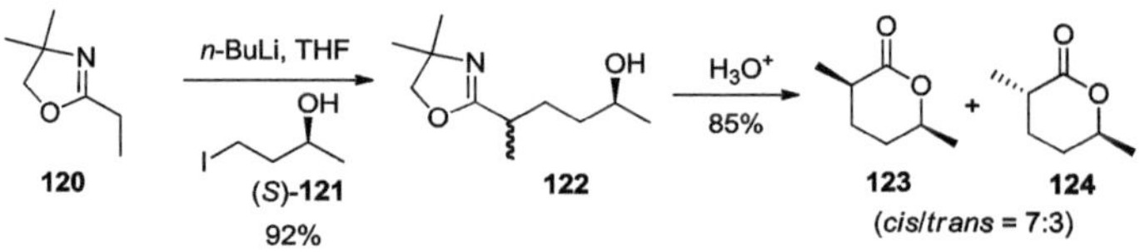

Scheme 23 Synthesis of chiral pheromone lactones

δ-lactones occur as mixtures of all enantiomers, the (*S*,*S*)-*anti*-enantiomer being the major part of this mixture.

Similarly, Oliveira et al. have reported the synthesis of all four possible diastereomers of a component of the pheromone blend of the carpenter bee *Xylocopa hirutissima* via alkylation of 4,4-dimethyl-2-oxazoline derivatives [61] (Scheme 23). In this case, the carboxylic acid, allowing the formation of the δ-lactone, is masked as an oxazoline. The reaction of the anion of 2-ethyl-4,4-dimethyl-2-oxazoline **120** with iodide **121** gave compound **122** in 92 % yield, which was hydrolyzed and cyclized in one-pot under acidic conditions to provide a mixture of stereoisomers **123** and **124**.

Jahn et al. have reported the synthesis of dihydronepetalactone **131** [62, 63] and its nor-analog **130** by lactonization of a δ-hydroxy diester derivative [64] (Scheme 24). Deprotonation of malonates **125** by LDA and subsequent treatment with two to three equivalents of ferrocenium hexafluorophosphate **126** in DMF gave access to a variety of 2-alkenyl-cyclopentan-1,1-dicarboxylates **127** in good yields as the sole products. No traces of compounds resulting from either the dimerization of malonyl radicals or a 1,5-hydrogen transfer were observed in any case. The scope of this method was further extended to the synthesis of cyclopentanoid monoterpenes, namely, dihydronepetalactone **131** and its nor-methyl analog **130**. Hydroboration/oxidation of alkene **127** followed by lactonization under acidic conditions afforded the six-membered lactones, which were transformed into dihydronepetalactone **131** and its nor-analog **130** by decarboxylation (Scheme 24).

Sibi and He have described the formation of δ-lactone from an in situ formed δ-hydroxy ester in the synthesis of ricciocarpins A and B [65] (Scheme 25). Enantioselective conjugate addition of the tertiary radical derived from **133** onto

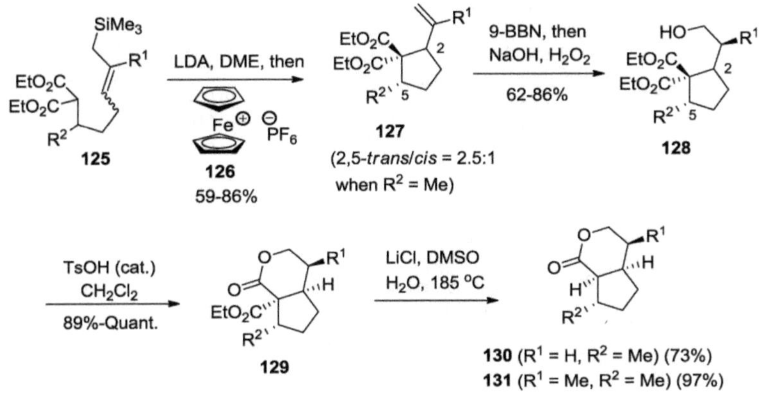

Scheme 24 Synthesis of dihydronepetalactone and its analog

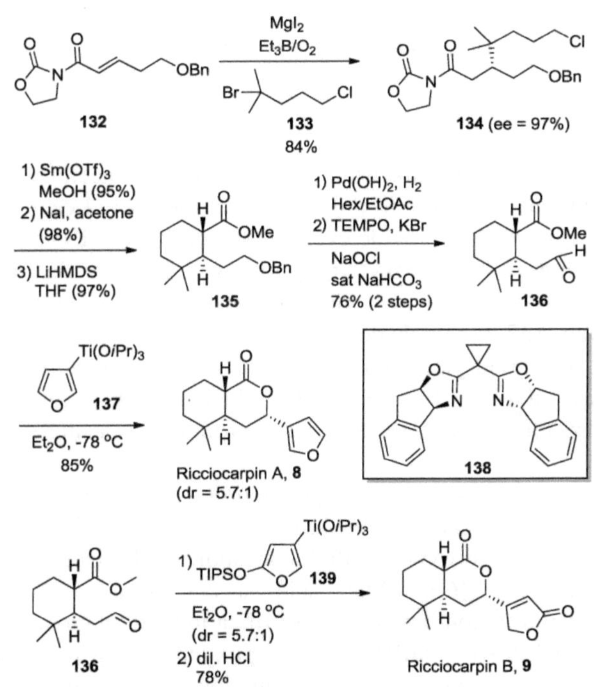

Scheme 25 Synthesis of ricciocarpins A and B

oxazolidinone-derived enoate **132** in the presence of 100 mol% of the chiral Lewis acid, derived from MgI_2 and chiral ligand **138**, afforded product **134** in excellent yield and with high enantioselectivity. Lowering the catalyst loading significantly reduced the yields but with similar levels of enantioselectivity. Conversion of **134** into the methyl ester using Otera's protocol ($Sm(OTf)_3$, MeOH), followed by

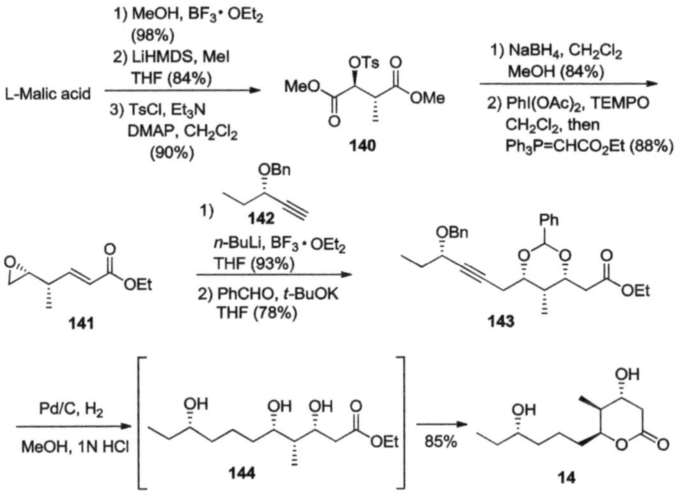

Scheme 26 Synthesis of δ-lactone **14**

halogen interchange under Finkelstein conditions (NaI, acetone), produced the iodoester, which was subjected to the crucial six-membered ring construction in the presence of LiHMDS to provide cyclohexane derivative **135** as a single *trans*-isomer. Unmasking of the hydroxyl group followed by immediate oxidation gave aldehyde **136** in good yield. After extensive optimization, addition of 3-titanyloxyfuran **137** to aldehyde **136** in ether at low temperature revealed ricciocarpin A **8** in good yield and with decent diastereoselectivity (dr = 5.7:1). In a similar way, addition of reagent **139** to aldehyde **136** followed by treatment with dilute hydrochloric acid furnished ricciocarpin B **9** as a single isomer in 78 % yield (Scheme 25).

Yadav et al. have reported a one-pot deprotection of hydroxyl group with concurrent lactonization of the resulting hydroxyl ester in the synthesis of (3*R*,4*S*,5*S*,9*S*)-3,5,9-trihydroxy-4-methylundecanoic acid δ-lactone **14** in a convergent fashion (Scheme 26) [66]. L-Malic acid was converted to tosylate **140** through a three-step transformation involving esterification, alkylation, and tosylation. Reduction of ester **140** with $NaBH_4$ gave an epoxy alcohol with concurrent formation of an epoxide. A one-pot oxidation/olefination of the resulting epoxy alcohol, following Vatale's protocol [$PhI(OAc)_2$, TEMPO, then $Ph_3P{=}CHCO_2Et$], afforded epoxy unsaturated ester **141**. Opening of epoxide **141** with the anion of alkyne **142** derived from D-mannitol gave the corresponding coupled product, which underwent oxy-anion assisted Michael addition with benzaldehyde to provide acetal **143**. Finally, deprotection of the benzylidene acetal, lactonization, reduction of the triple bond, and debenzylation were all achieved in one pot using Pd/C in acidic methanol to reveal lactone **14** in excellent yield.

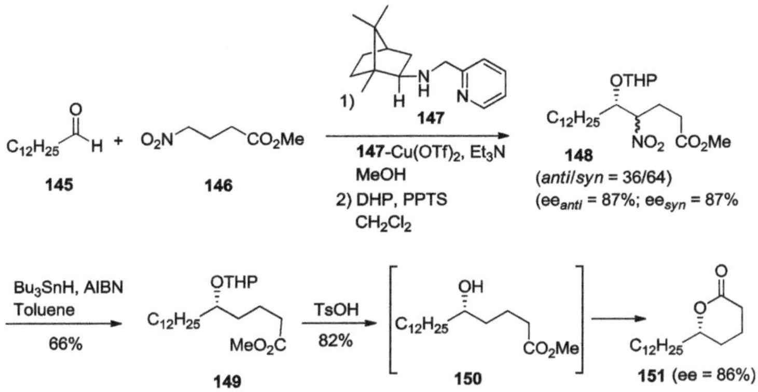

Scheme 27 Synthesis of δ-lactones via enantioselective Henry reaction

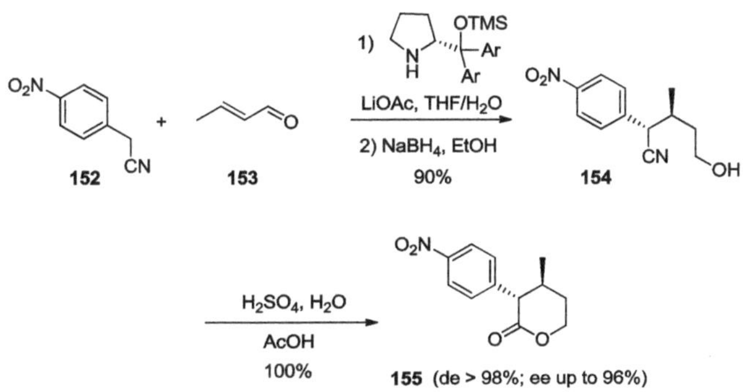

Scheme 28 Enantioselective synthesis of δ-lactones

Similarly, Blay, Pedro, et al. have reported the synthesis of δ-lactones via a catalytic highly enantioselective Henry addition of methyl 4-nitrobutyrate to aldehydes [67] (Scheme 27). The Henry reaction between aldehyde **145** and methyl 4-nitrobutyrate in the presence of $Cu(OTf)_2$-amine **147** complex and Et_3N afforded the addition product with moderate diastereoselectivity and with excellent enantioselectivity. Protection of the resulting hydroxyl group provided intermediate **148**. Removal of the nitro group present in **148** by treatment with Bu_3SnH/AIBN was followed by deprotection/lactonization under acidic conditions to give δ-lactone **151** with good enantioselectivity.

An interesting example on lactonization of a δ-hydroxy α-nitrophenyl nitrile has been reported by Cid, García Ruano et al. [68] (Scheme 28). The nitro group was deliberately incorporated in the phenyl ring to increase the acidity of the benzylic protons. Treatment of nitrile **152** with aldehyde **153** in the presence of proline-derived organocatalyst afforded the addition product with moderate diastereoselectivity and with excellent enantioselectivity. The resulting addition product was reduced with $NaBH_4$ and then cyclized under acidic conditions to provide δ-lactone **155**. In the absence of additive such as LiOAc, the reaction did not proceed.

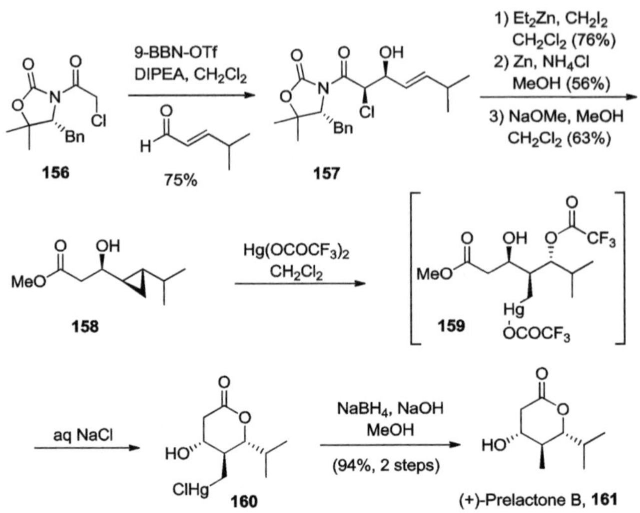

Scheme 29 Asymmetric synthesis of (+)-prelactone B

Bull et al. have reported an asymmetric synthesis of prelactone, which has been isolated as a shunt metabolite of polyketide metabolism from the bafilomycin-producing organism *Streptomyces griseus* [69, 70]. In this example, lactonization of an in situ formed δ-hydroxy ester has been used for the ring closure [71] (Scheme 29). Evans aldol reaction of oxazolidinone **156** with (*E*)-4-methylpent-2-enal afforded *syn*-aldol **157**, which was converted into cyclopropyl ester **158** via a series of cyclopropanation, dechlorination, and methanolysis reactions. Treatment of **158** with $Hg(OCOCF_3)_2$ facilitated the opening of the cyclopropane ring to result in intermediate **159**, which was hydrolyzed upon work-up to provide lactone **160**. Finally, reductive demercuration of **160** with alkaline $NaBH_4$ led to (+)-prelactone **161**.

3 Oxidation

3.1 Baeyer–Villiger Oxidation

Another interesting way to construct δ-lactones is the Baeyer–Villiger oxidation [72] of cyclopentanones. For example, Hacini and Santelli have reported an efficient synthesis of the Prelog–Djerassi lactone methyl ester using Baeyer–Villiger oxidation [73] (Scheme 30). Hydrolysis of enol ether **162** derived from (−)-*trans*-pulegenic acid furnished an inseparable diastereomeric mixture of the corresponding dimethylacetals. Ozonolysis of the olefin and subsequent

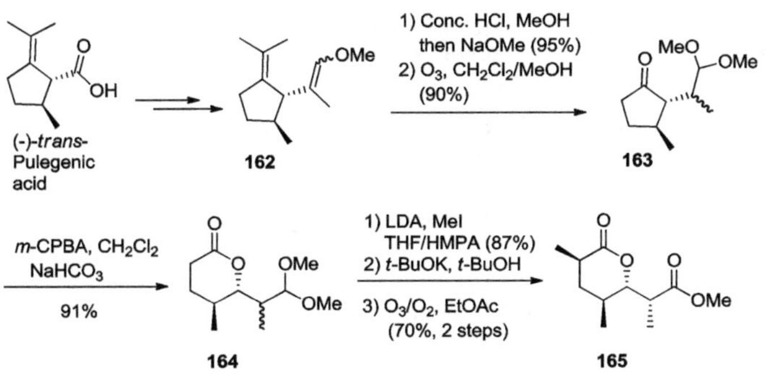

Scheme 30 Santelli and Hacini synthesis of the Prelog–Djerassi lactone

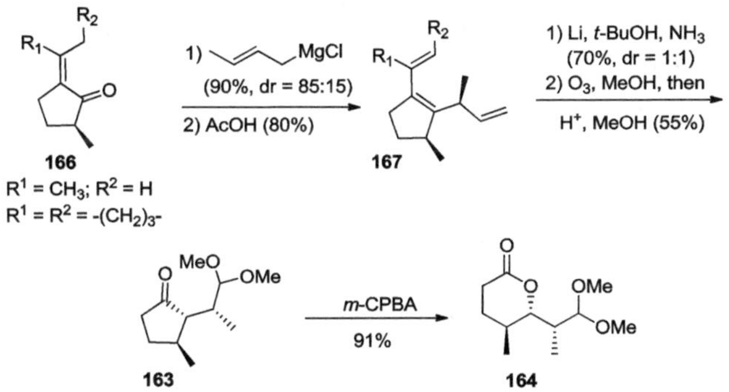

Scheme 31 Santelli et al. approach to the synthesis of Prelog–Djerassi lactone

Baeyer–Villiger oxidation of **163** afforded a separable mixture of lactone **164**. Alkylation of the lactone with LDA and MeI and equilibration of the resulting epimeric lactones gave the dimethylacetal, which upon ozonolysis under the Deslongchamps conditions (O_3/O_2) provided the Prelog–Djerassi lactone methyl ester **165**.

Santelli et al. have also reported another convenient synthesis of the Prelog–Djerassi lactone from a cyclopentanone derivative via Baeyer–Villiger oxidation [74] (Scheme 31). Addition of crotyl Grignard reagent to 2-alkylidene-5-methylcyclopentanones **166** followed by dehydration afforded triene **167**. A conjugate addition of the 1,3-diene moiety gave a 1:1 diastereomeric mixture of 1,5-dienes, which underwent ozonolysis to provide the corresponding cyclopentanone derivative **163** as a single diastereomer. Finally, Baeyer–Villiger oxidation of **163** furnished lactone **164**, a well-known precursor of the Prelog–Djerassi lactone.

Adger et al. have reported an enantioselective method for the synthesis of δ-lactones via application of enzymatic Baeyer–Villiger oxidation [75] (Scheme 32). Oxidation of cyclopentanone derivative **168** using a monooxygenase

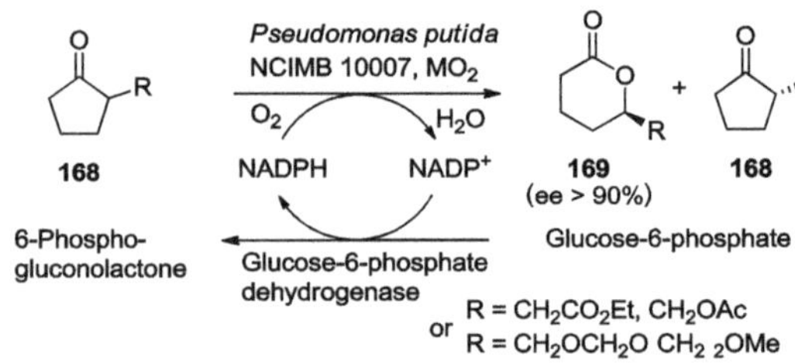

Scheme 32 Synthesis of δ-lactones via application of enzymic Baeyer–Villiger oxidation

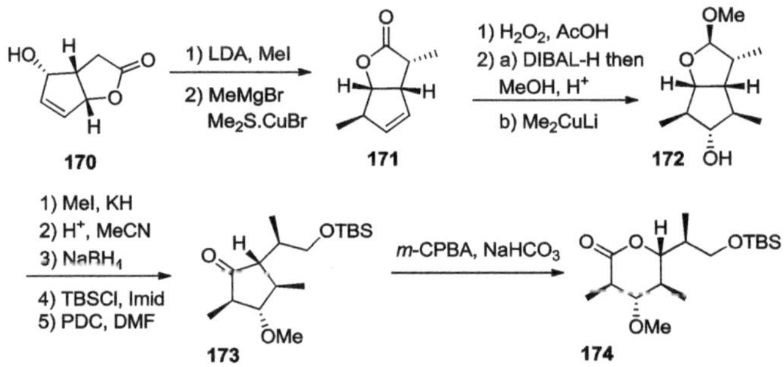

Scheme 33 Willis et al. synthesis of δ-lactones

from *Pseudomonas putida* NCIMB 10007 gave lactone **169** with good enantioselectivity. Recycling of NADPH was induced using glucose-6-phosphate plus glucose-6-phosphate dehydrogenase. Whereas cycloalkanones with polar and non-bulky side chains gave excellent enantiomeric excesses, ketones with a bulkier and less polar side chain were oxidized with very poor enantioselectivity.

A versatile approach to δ-lactones has been developed by Willis et al. via oxidation of tetrasubstituted cyclopentanones and Baeyer–Villiger oxidation as the key steps [76] (Scheme 33). Alkylation of bicyclic lactone **170** was followed by treatment with lithium dimethylcuprate to provide lactone **171**. A stereoselective epoxidation, acetal formation, and epoxide opening with lithium dimethylcuprate gave alcohol **172**. After protection of the free hydroxyl group, alcohol **172** was transformed into cyclopentanone **173** through a series of functional group transformations. Baeyer–Villiger oxidation of **173** occurred with complete regioselectivity, giving δ-lactone **174** as the sole product.

Yamamoto et al. have reported the synthesis of optically active δ-lactones by Baeyer–Villiger oxidation of chiral cyclopentanones [77] (Scheme 34). Asymmetric hydrogenation of enones **175** and **176** in the presence of 0.01 equivalent of $Ru_2Cl_4[(S)$-p-Tolyl-Binap$]_2NEt_3$ catalyst afforded chiral ketones **177** and **178** in good yield and with excellent enantioselectivity. These chiral ketones were found to show a fundamentally jasmine-like floral odor. Chiral ketones **177** and **178** were then transformed to δ-lactones **179** and **180** by Baeyer–Villiger oxidation with

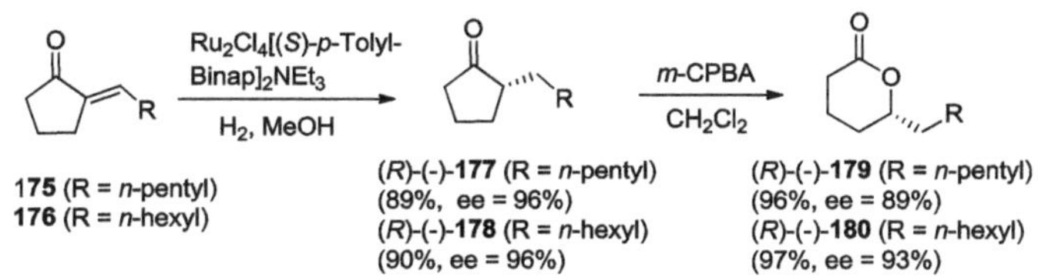

Scheme 34 Synthesis of optically active δ-lactones

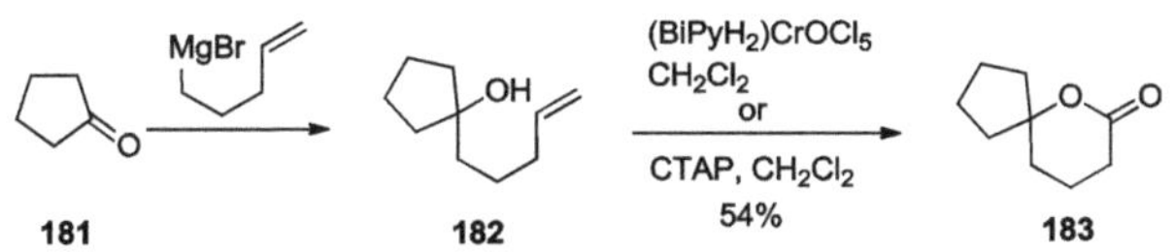

Scheme 35 Chandrasekaran et al. approach to the synthesis of δ-lactone

m-CPBA. Repeating the sequence with 0.01 equivalent of Ru_2Cl_4[(*R*)-*p*-Tolyl-BINAP]$_2$NEt$_3$ gave the antipodes of these lactones. All these δ-lactones were found to exhibit fruity and sweet odor properties.

3.2 *Oxidative Lactonization*

Synthesis of δ-lactones can also be achieved by oxidative lactonization of appropriately functionalized alkenols. For instance, Chandrasekaran et al. have reported a substituent-directed oxidative cyclization for the synthesis of δ-lactones using pentavalent chromium reagent, (BiPyH$_2$)CrOCl$_5$ [78], or cetyltrimethylammonium permanganate (CTAP) [79] to effect the oxidative cyclization of hydroxyolefin **182** derived from the corresponding ketone **181** (Scheme 35). A remarkable feature of this protocol is that cetyltrimethylammonium permanganate (CTAP) did not oxidize primary and secondary alcohols to the respective carbonyl compounds.

Similarly, Schlecht and Kim have reported a substituent-directed oxidation method for the synthesis of δ-lactones by oxidative cyclization of hydroxyalkenes [80] (Scheme 36). Addition of alkenyl Grignard reagent to ketones **184** afforded hydroxyalkene **185**, which upon treatment with chromium trioxide in acetic acid and acetic anhydride provided spiro-δ-lactone **186**.

A novel, high-yielding, and rapid oxidative cyclization of δ-stannyl carboxylic acids has been developed by Yamamoto et al. for the synthesis of 4-hydroxy-δ-lactones [81] (Scheme 37). Treatment of β-hydroxy-δ-stannyl carboxylic acid **187** with lead tetraacetate facilitated the oxidative cyclization to afford 4-hydroxy-δ-lactone **188** in good yield. The presence of the free hydroxyl group in the β-position was essential to coordinate the metal and deliver cyclization and to avoid the formation of an alkene through oxidative elimination.

Hiroi et al. have reported a novel approach to δ-lactones by an oxidative lactonization of 1,5-diols using an amino alcohol-based iridium bifunctional

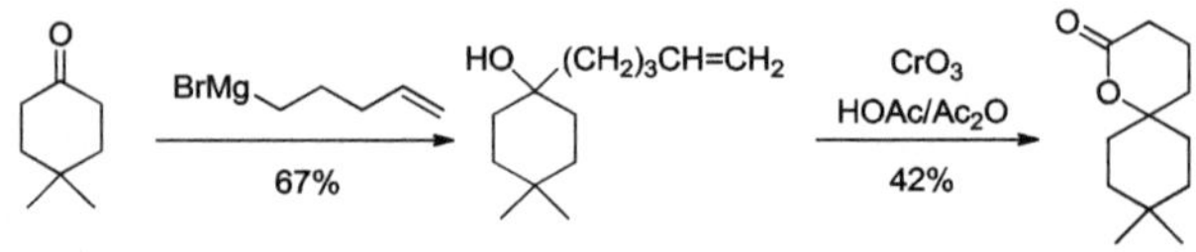

Scheme 36 Schlecht and Kim synthesis of δ-lactones

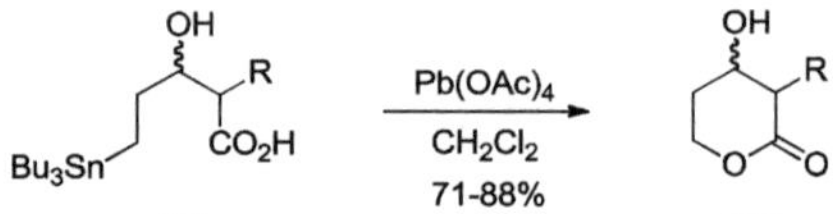

Scheme 37 Synthesis of 4-hydroxy-δ-lactones

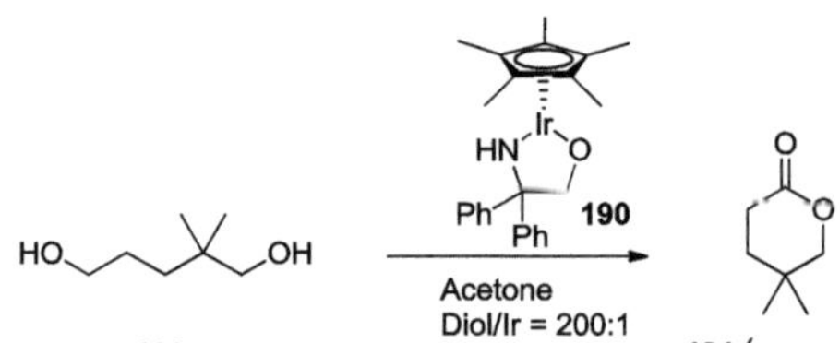

Scheme 38 An Ir-catalyzed oxidative lactonization of diols

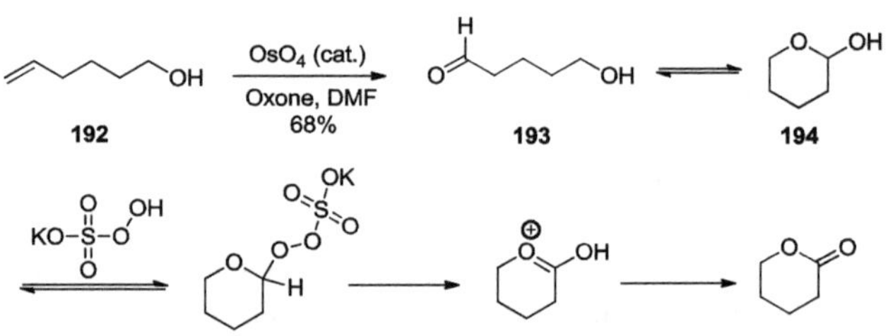

Scheme 39 Synthesis of δ-lactones via oxidative lactonization of alkenols

complex [82] (Scheme 38). Thus, treatment of a variety of 1,5-diols **189** with the iridium catalyst **190** derived from the corresponding amino alcohol in acetone afforded a range of δ-lactone **191** in excellent yields. The efficiency of the reaction was shown by the high substrate/catalyst molar ratio of 200/1,000 in acetone or butanone. In the case of unsymmetrical diols, the less hindered hydroxyl groups were oxidized selectively to give the corresponding δ-lactones.

A highly efficient, mild, and simple protocol for the synthesis of δ-lactones has been reported by Borhan et al. using tandem OsO_4-mediated oxidative cleavage/oxidative lactonization of alkenols [83] (Scheme 39). According to the mechanism, initial oxidative cleavage of the alkene **192** gives aldehyde **193** which is trapped

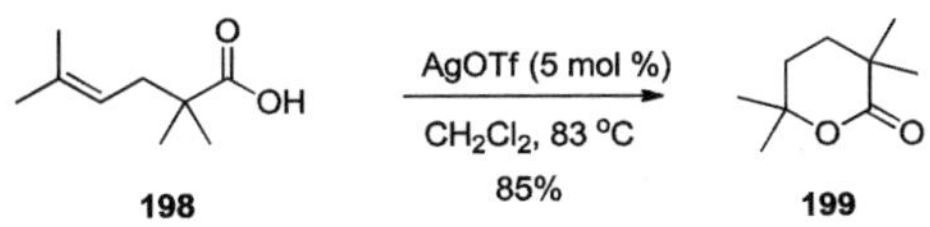

Scheme 40 Silver-catalyzed oxidative cyclization of olefinic acid to δ-lactones

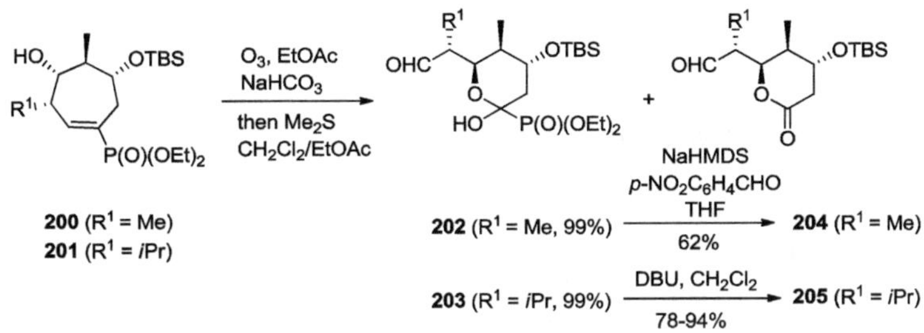

Scheme 41 Synthesis of δ-lactones by ozonolytic cleavage of vinyl phosphonates

intramolecularly by the alcohol moiety to form the hemiacetal **194**. Equilibration of the hemiacetal with oxone leads to the hemiperoxymonosulfate acetal **195**, which undergoes a Baeyer–Villiger-like rearrangement to provide δ-lactone **197**.

He et al. have reported a silver(I)-catalyzed intramolecular addition of carboxylic acid to inert olefins for the synthesis of δ-lactones [84] (Scheme 40). For example, treatment of olefinic acid **198** with 5 mol% of AgOTf afforded δ-lactone **199** in 85 % yield. In some cases, the δ-lactone was predominant along with small amounts of the γ-lactone, which could presumably arise from olefin migration catalyzed by silver(I).

Fuchs et al. have reported that their study on the ozonolytic reactivity of transposed cyclic vinyl phosphonates attempted to probe the anticancer SAR (structure–activity relationship) of a series of computer-designed (+)-discodermolide analogs [85] (Scheme 41). Ozonolytic cleavage of **200** and **201** in the presence of O_3 followed by quenching with Me_2S provided **202** and **203** along with the desired lactones **204** and **205** as minor products in a 6:1 ratio. Addition of catalytic DBU to **203** could drive lactonization to completion giving lactone **205** in excellent yield. As aldehyde **202** was less tolerant to DBU, it required a dropwise addition of NaHMDS in the presence of *p*-nitro-benzaldehyde to trap the expelled diethyl phosphate.

3.3 Oxidation of Lactol

The more conventional and straightforward way to obtain δ-lactone is through oxidation of the corresponding lactol. For example, a direct synthesis of δ-lactones from 2-(3-lithiopropyl)-1,3-dioxolane and carbonyl compounds has

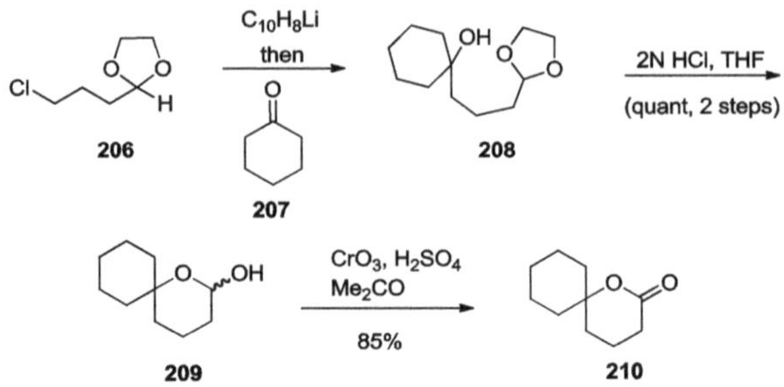

Scheme 42 Yus and Ramón synthesis of δ-lactones

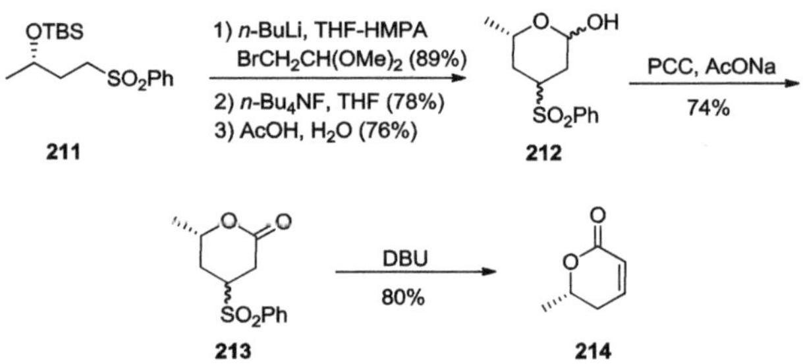

Scheme 43 Huet and Robin synthesis of δ-lactones

been reported by Ramón and Yus [86] (Scheme 42). The reaction of 2-(3-lithiopropyl)-1,3-dioxolane, prepared in situ by lithiation of the chloroacetal **206** with lithium naphthalenide, with aldehydes and ketones **207** followed by hydrolysis of the acetal yielded the lactol **209**, which underwent oxidation with PCC or Jones reagent to afford δ-lactone **210**.

Robin and Huet have reported a general method for the synthesis of δ-lactones as well as lactones with several ring sizes via oxidation of the corresponding lactol [87] (Scheme 43). Alkylation of sulfone **211** with bromoacetaldehyde dimethyl acetal, cleavage of the silyl ether, and treatment with acetic acid afforded lactol **212**. Oxidation of lactol **212** with PCC furnished the corresponding saturated δ-lactone **213**, which underwent elimination of sulfinic acid in the presence of DBU to provide lactone **214**.

De Brabander et al. have reported a very rapid enantioselective synthesis of the Prelog–Djerassi lactonic acid through an asymmetric aldol reaction [88] (Scheme 44). The Oppolzer sultam-derived *N*-propionyl derivative **215** was used to desymmetrize *meso*-dialdehyde **216**, and the diastereoselectivity was found to be 80 %. Oxidation of the resulting lactol **217** to lactone **218** was followed by oxidative removal of the chiral auxiliary. The unwanted diastereoisomer resulting from the aldol reaction was removed chromatographically after the oxidation step.

Synthesis of a synthetic equivalent of the δ-lactone in mevinic acids has been reported by Suemune et al. [89] (Scheme 45). Asymmetric hydrolysis of diacetate

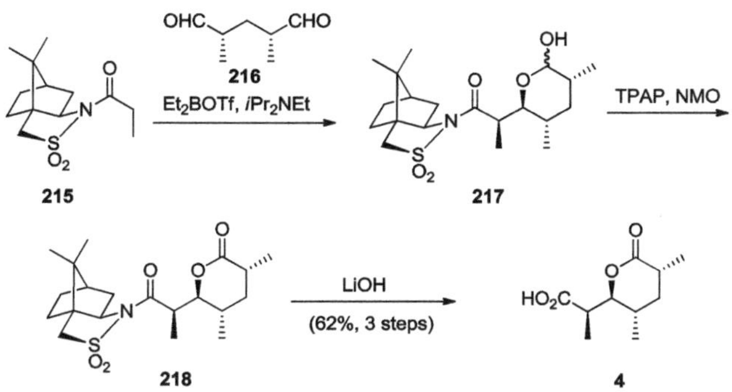

Scheme 44 Synthesis of Prelog–Djerassi lactonic acid

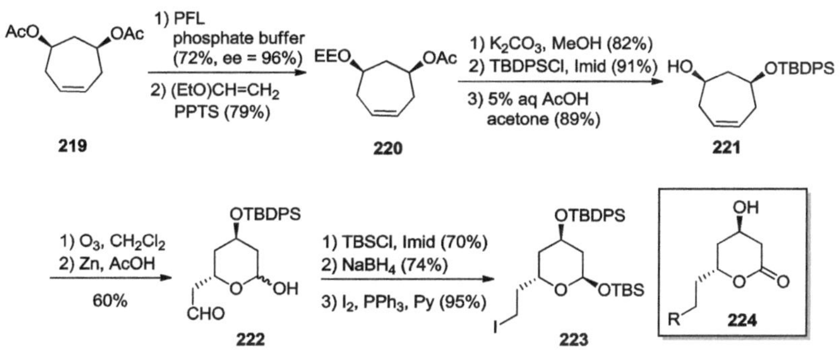

Scheme 45 Preparation of a synthetic equivalent for the δ-lactone in mevinic acids

219 catalyzed by *Pseudomonas fluorescence* lipase was followed by protection of the free alcohol to provide intermediate **220**. After a series of protective group transformations, alkene **221** was subjected to ozonolysis, and subsequent treatment with Zn/AcOH gave hemiacetal **222** as a 1:1 diastereomeric mixture at the C2 position. Protection of hemiacetal **222** gave the protected hemiacetal as the sole diastereomer with the defined stereochemistry. Reduction of the aldehyde and subsequent iodination of the resulting alcohol gave iodide **223** as a synthetic equivalent for the δ-lactone in mevinic acid analogues, **224**.

A couple of template-directed approaches have been developed for the synthesis of mevalonolactone [90]. In the first approach, the spiro-fused lactone **225** constructed from diacetone-D-glucose was subjected to a stereoselective epoxidation and then reduction to provide diol **226** (Scheme 46). Cleavage of the acetals followed by exhaustive oxidation allowed the cleavage of the carbon skeleton from the template, providing (*R*)-mevalonolactone **10** in moderate yield.

In another approach, ketone **228** was obtained in optically pure form by an enzymatic desymmetrization and a stereoselective epoxidation [91] (Scheme 47). A sequential reduction of epoxide and the carbonyl group gave diol **229**, which was

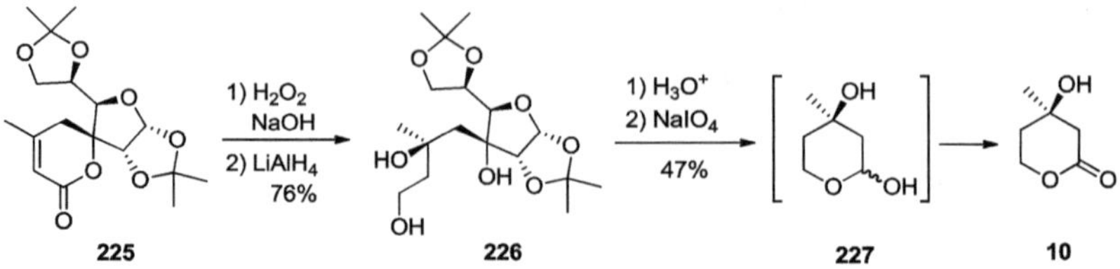

Scheme 46 Synthesis of (*R*)-mevalonolactone

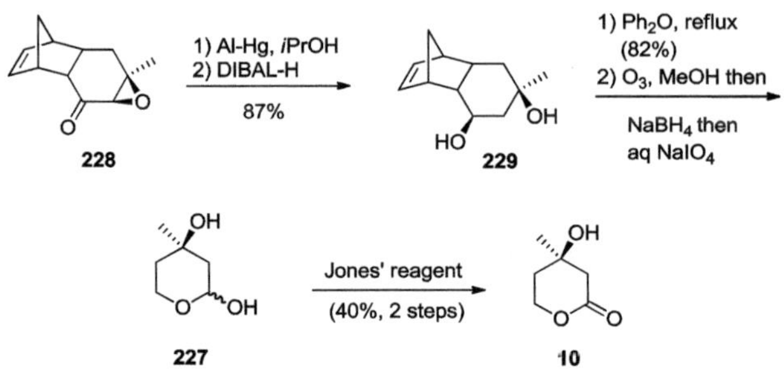

Scheme 47 Synthesis of (*R*)-mevalonolactone

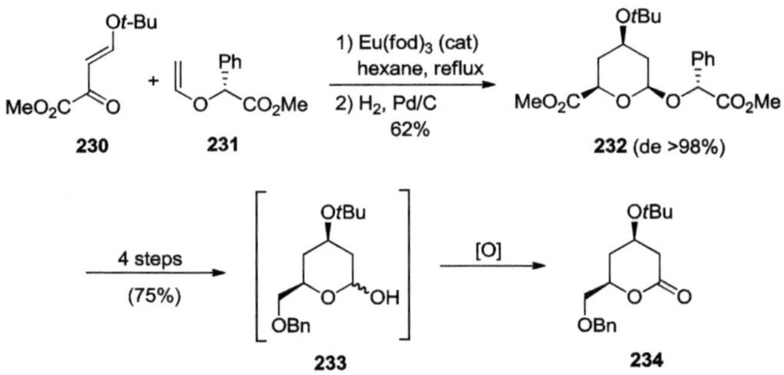

Scheme 48 Synthesis of a mevalonolactone derivative

liberated from the template by retro-Diels–Alder reaction, and converted into mevalonolactone via an exhaustive oxidation sequence.

Dujardin et al. have reported the synthesis of simple optically pure mevalonolactone derivatives via a hetero Diels–Alder cycloaddition [92] (Scheme 48). The lanthanide Lewis acid-catalyzed hetero Diels–Alder cycloaddition of *tert*-butyloxymethylenepyruvate **230** with a vinyl ether derived from mandelic acid **231** afforded a cycloadduct, which on hydrogenolysis gave the saturated adduct **232**. This intermediate was then transformed into the mevalonolactone derivative **234** in a few steps through oxidation of a lactol intermediate **233**.

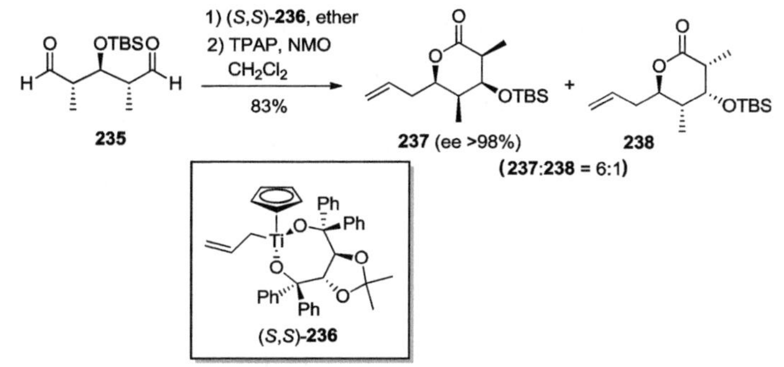

Scheme 49 Synthesis of δ-lactones by desymmetrization of *meso*-dialdehydes

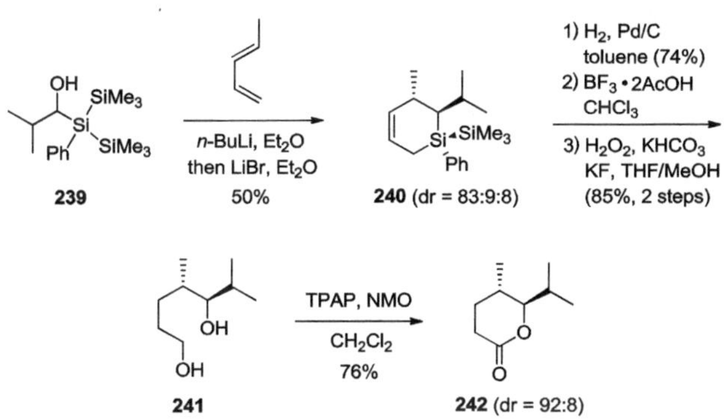

Scheme 50 Synthesis of δ-lactones from silences

Cossy et al. have developed a stereoselective approach for the synthesis of δ-lactones by desymmetrization of *meso*-dialdehydes with optically active cyclopentadienyldialkoxyallyltitanium complexes [93] (Scheme 49). Addition of the (*S*,*S*)-**236** TADDOL complex to aldehyde **235** gave the corresponding lactols, which on oxidation with TPAP/NMO led to lactones **237** and **238** in excellent yields and with good diastereoselectivity. Similarly, addition of the (*R*,*R*)-**236** TADDOL complex to aldehyde **235** provided the opposite two enantiomers in moderate yield and with similar diastereoselectivity.

Steel et al. have reported a novel approach for the synthesis of δ-lactones using silene chemistry [94] (Scheme 50). Treatment of silene precursor **239** with diene in the presence of *n*-BuLi followed by addition of LiBr gave the silacyclohexene **240** in moderate yield and with good diastereoselectivity via a highly stereoselective Diels–Alder reaction. Reduction of alkene followed by reaction with $BF_3{\cdot}2AcOH$ afforded the corresponding fluorosilane, which underwent a Tamao–Fleming-type oxidation to provide diol **241**. Further oxidation of diol **241** with TPAP and NMO furnished a diastereomeric mixture of δ-lactone **242** via the corresponding lactol.

Córdova et al. have developed a direct amino acid-catalyzed asymmetric synthesis of δ-lactones by oxidation of lactols [95] (Scheme 51). The strategy involves

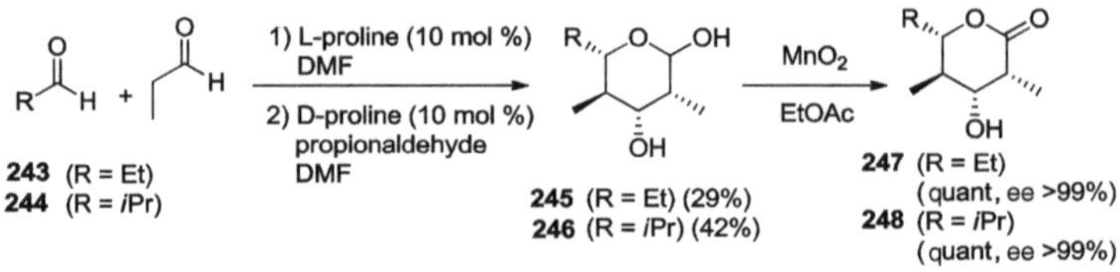

Scheme 51 Direct catalytic enantioselective synthesis of δ-lactones

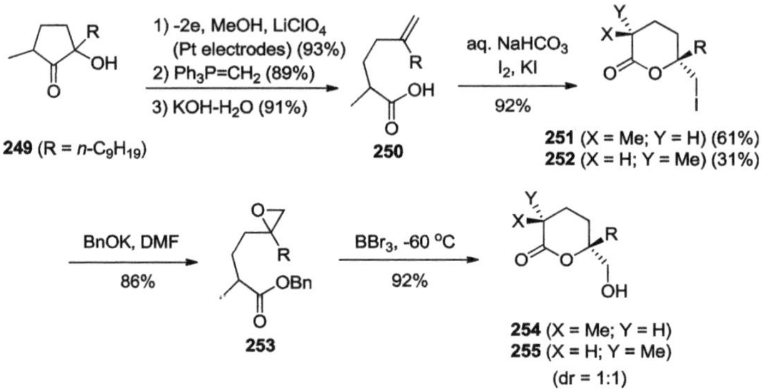

Scheme 52 Torii et al. synthesis of malyngolide

an iterative aldolization of aldehydes **243** and **244** with propionaldehyde to give hexoses **245** and **246** with excellent chemo-, diastereo-, and enantioselectivity. The resulting hexoses were quantitatively oxidized with MnO_2 to furnish δ-lactones **247** and **248**. Thus, this novel protocol allows access for the synthesis of δ-lactones with four contiguous stereocenters and with excellent stereocontrol.

4 Halolactonization and Selenolactonization

Lactone ring can also be constructed by either halolactonization or selenolactonization of olefinic acids. For instance, Torii et al. have reported the total synthesis of malyngolide, a marine antibiotic δ-lactone [96]. Electrooxidative cleavage of α-hydroxycyclopentanone **249** followed by Wittig olefination and saponification afforded carboxylic acid **250**, which underwent iodolactonization to provide a diastereomeric mixture of iodides **251** and **252** [97] (Scheme 52). Alcoholysis with potassium benzyl oxide gave benzyl ester **253**, which upon treatment with boron tribromide furnished malyngolide **254/255** through hydrolysis of **253** and subsequent intramolecular attack of the carboxylate on the epoxide.

In another example, Greeves et al. have reported the synthesis of δ-lactones from olefinic acids via iodolactonization and phenylselenolactonization [98] (Scheme 53). The precursors were synthesized from diallyl ether **256** by a tandem

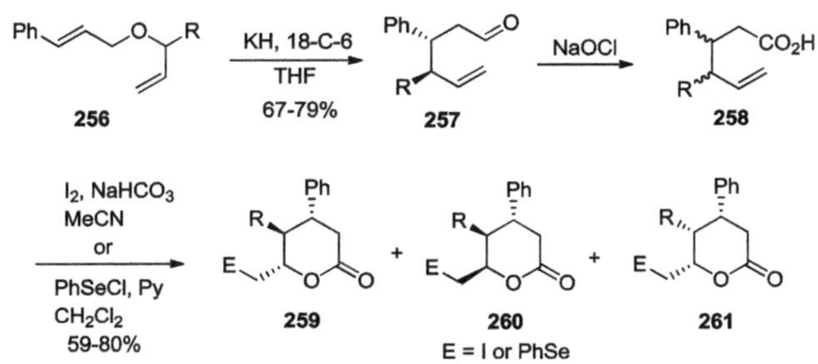

Scheme 53 Stereoselective synthesis of trisubstituted δ-lactones

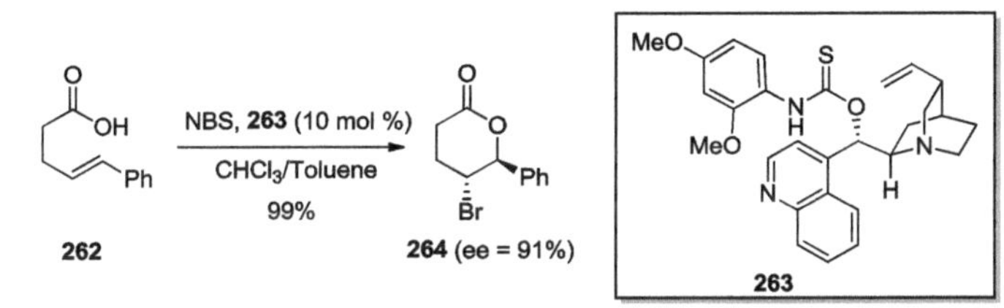

Scheme 54 Aminothiocarbonate-catalyzed asymmetric bromolactonization

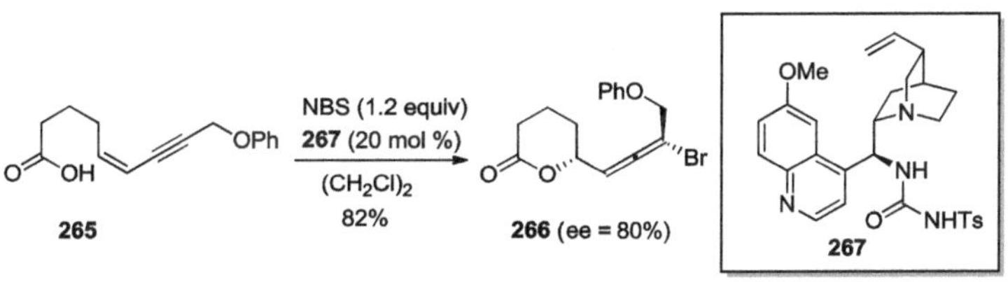

Scheme 55 Catalytic enantioselective bromolactonization of enynes

[2,3]-Wittig and anionic oxy-Cope rearrangements using potassium hydride. The anionic oxy-Cope rearrangement went through a chair-like transition state to yield the *syn*-substituted aldehydes **257**, which were oxidized to acid **258**. Treatment of δ,ε-unsaturated carboxylic acids **258** with either iodine or phenylselenium chloride afforded the cyclized iodo-/phenylselenolactones **259–261** as a mixture of diastereomers. In most cases, lactone **259** was obtained as the major or sole product.

An efficient and enantioselective bromolactonization of 1,2-disubstituted olefinic acids has been developed by Yeung et al. [99] (Scheme 54). Reaction of olefinic acid **262** with NBS in the presence of 10 mol% of quinidine-derived aminothiocarbamate catalyst **263** afforded δ-lactone **264** in excellent yield and with excellent enantioselectivity.

A similar catalytic enantioselective bromolactonization of enyne carboxylic acids has been reported by Tang et al. [100] (Scheme 55). Treatment of **265** with NBS in the presence of 20 mol % of catalyst **267** obtained by tethering a cinchona alkaloid skeleton to an urea group afforded bromoallenyl-δ-lactone **266** in excellent yield and with good enantioselectivity.

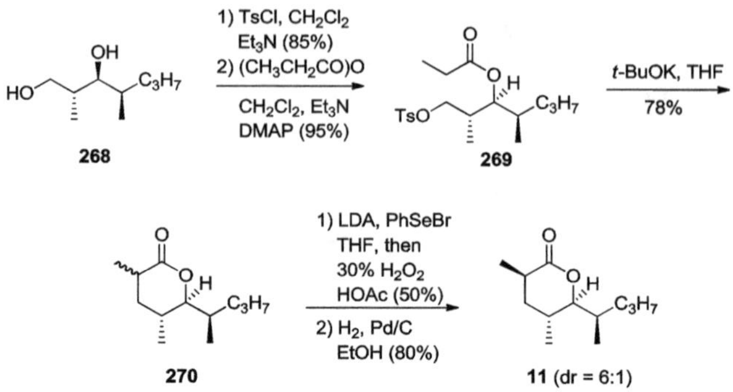

Scheme 56 Pilli and Murta total synthesis of invictolide

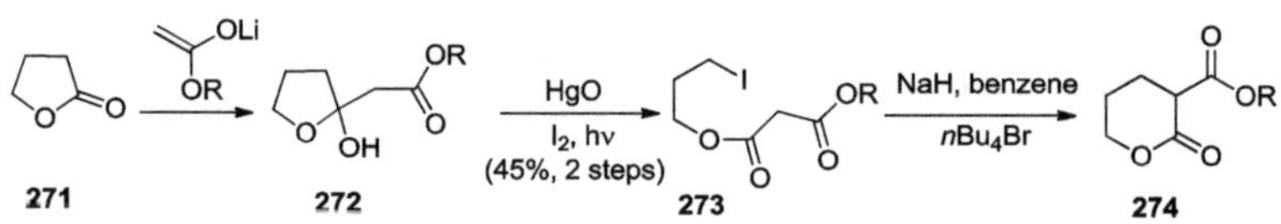

Scheme 57 Suginome et al. one-carbon intercalation approach to the synthesis of δ-lactones

5 Intramolecular Nucleophilic Displacement

There are a few reports on the synthesis of δ-lactones wherein an intramolecular nucleophilic displacement has been used for the ring closure. For example, Pilli and Murta have reported a stereoselective total synthesis of invictolide **11**, a component of the queen recognition pheromone of *Solenopsis invicta* [101] (Scheme 56). Selective tosylation of diol **268** followed by acylation of the secondary alcohol gave the propionate ester, which underwent an intramolecular nucleophilic displacement to afford a 2:3 diastereomeric mixture of lactone **270** in favor of the undesired isomer. The correct stereochemistry at C3 of the target molecule was gained through the catalytic hydrogenation of the corresponding unsaturated lactone obtained from **270** through α-selenylation and oxidative selenoxide elimination.

In another instance, Suginome et al. have reported a general method for the synthesis of δ-lactones using a radical-mediated endocyclic cleavage of a tetrahydrofuranyl derivative [102] (Scheme 57). Addition of lithium enolate of acetate ester to lactone **271** afforded tetrahydrofuranyl hydroxyl ester **272**, which underwent a radical-mediated endocyclic cleavage in the presence of mercuric oxide, iodine, and light, to give iodo ester **273**. Treatment of iodo ester **273** with sodium hydride facilitated an intramolecular alkylation providing lactone **274**.

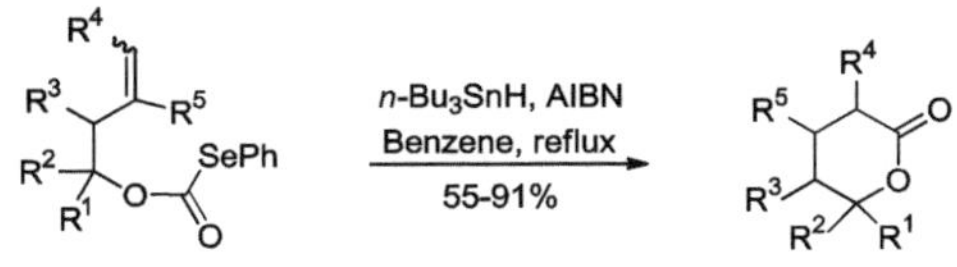

275 **276**

Scheme 58 Bachi and Bosch free-radical annelation for the synthesis of δ-lactones

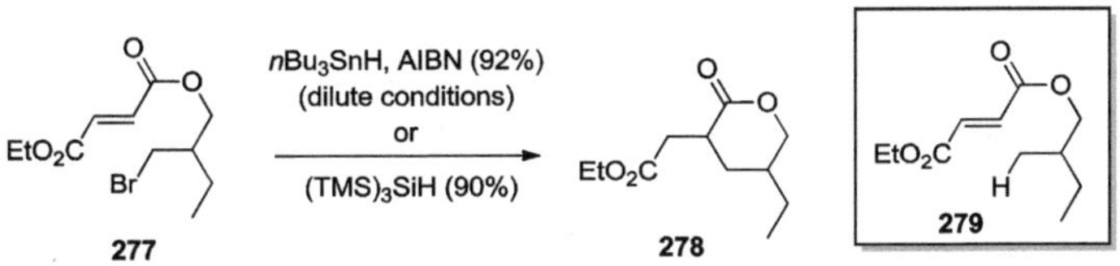

277 **278** **279**

Scheme 59 A radical cyclization approach to δ-lactones

6 Radical Cyclization

Described in this section are a couple of reports on the synthesis of δ-lactones wherein the ring closure has been achieved through a radical cyclization. Bachi and Bosch have reported the synthesis of δ-lactones by free-radical annelation of phenylselenyl carbonates [103] (Scheme 58). An intramolecular addition of an alkoxycarbonyl radical, formed by reaction of phenylselenyl carbonates **275** with tri-*n*-butyltin hydride in the presence of AIBN, to carbon–carbon multiple bonds provided the highly substituted lactone **276**. The salient features of this free-radical cyclization include high regioselectivity favoring *exo* addition and a high ratio of cyclization to reduction products.

A radical cyclization approach for the synthesis of δ-lactones has also been reported by Ihara et al. [104] (Scheme 59). Treatment of bromo ester **277** with tri-*n*-butyltin hydride in the presence of AIBN under dilute conditions facilitated a 6-*exo-trig* radical cyclization to provide diastereomeric lactone **278** in 92 % yield. However, the reaction was found to be dramatically concentration dependent as it exclusively gave the reduced product **279** under concentrated conditions. Nevertheless, this problem was overcome by using tris(trimethylsilyl)silane, which exclusively led to the desired δ-lactone irrespective of the concentration of the reaction.

7 Samarium Iodide-Mediated Reductive Cyclization

There are few reports in which the δ-lactone ring is constructed through a samarium iodide-mediated reductive cyclization of 1,5-dicarbonyl compounds. In the first example, Fang, Tsai et al. have reported the synthesis of δ-lactones using samarium

Scheme 60 Synthesis of δ-lactones via reductive cyclization

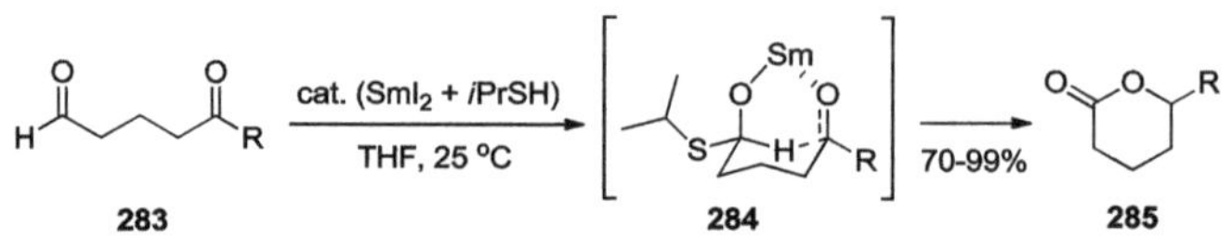

Scheme 61 Synthesis of δ-lactones from 5-oxopentanals

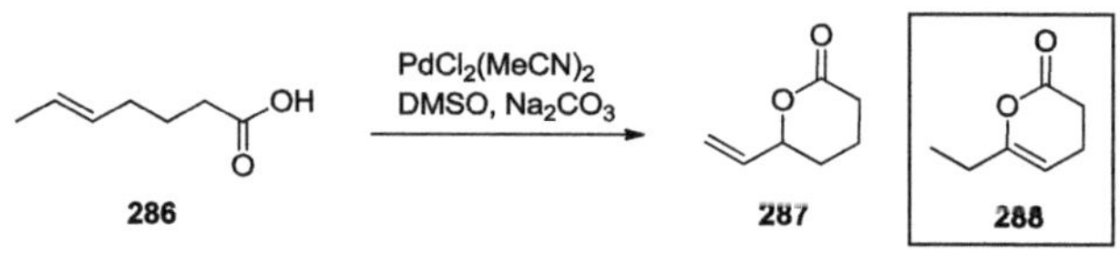

Scheme 62 Synthesis of δ-lactones through Pd(II)-assisted oxidative lactonization

iodide-mediated reductive cyclization of acyl silanes [105] (Scheme 60). Treatment of 1-(trimethylsilyl)-1,5-pentanedione **280** with SmI_2 in the presence of MeOH gave δ-silyl-δ-lactone **282**. The reaction is presumably initiated by addition of MeOH to the aldehyde group to form hemiacetal **281**, followed by Tishchenko reaction through a hydride transfer to the acyl silane, and subsequent cyclization to afford lactone **282**.

In the second example, a general method for the conversion of various 5-oxopentanals to substituted δ-lactones by the synergistic catalysis of samarium diiodide and 2-propanethiol has been demonstrated by Fang et al. [106] (Scheme 61). A series of 5-alkyl- and 5-phenyl-5-oxopentanals **283** were successfully converted to the corresponding δ-substituted-δ-lactone **285** by the catalysis of SmI_2/*i*PrSH (10–50 mol%) via the Tishchenko reaction. The reaction is believed to go through the transition state **284**, and notably, no aldol or pinacol products were observed under the reaction conditions. In addition, the deliberate use of 2-propanethiol is beneficial to facilitate the catalytic cycle.

8 Palladium-Catalyzed Lactonization

Annby and Andersson have reported their studies on Pd(II)-assisted lactonization for the formation of six-membered lactones [107] (Scheme 62). Treatment of carboxylic acid **286** with $PdCl_2(MeCN)_2$ in DMSO in the presence of sodium

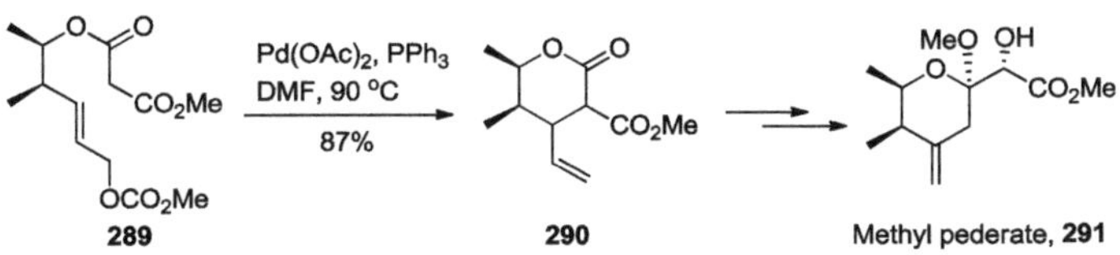

Scheme 63 Synthesis of a δ-lactone intermediate in the synthesis of methyl pederate

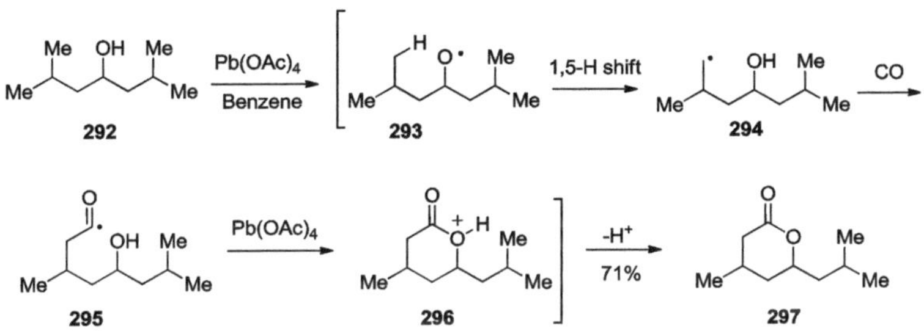

Scheme 64 A remote carbonylation approach to the synthesis of δ-lactones

carbonate gave δ-lactone **287**. However, the same reaction in acetonitrile in the presence of sodium acetate resulted in the formation of the endocyclic lactone **288**.

In another example, a palladium-catalyzed intramolecular allylic alkylation protocol has been developed by Toyota et al. to construct the δ-lactone intermediate **290** in a synthesis of (+)-methyl pederate **291** (Scheme 63) [108, 109]. Treatment of allyl carbonate with palladium acetate in the presence of PPh_3 in DMF gave the cyclized product **290** in excellent yield via a regioselective 6-*exo-trig* cyclization. The efficiency of the reaction was highly dependent on solvent, catalyst, and temperature. While preformed palladium(0) complexes gave very poor yields, the reaction was favored in highly polar solvents at elevated temperatures. The lactone **291** was then transformed into methyl pederate, an intermediate in the synthesis of mycalamides.

9 Carbonylation

δ-Lactone rings have also been constructed using carbonylation of appropriately functionalized compounds by the insertion of carbon monoxide. Ryu, Sonoda et al. have reported the synthesis of δ-lactones from saturated alcohols and carbon monoxide via remote carbonylation [110, 111] (Scheme 64). Treatment of saturated alcohol **292** with lead tetraacetate led to oxygen-centered radical **293**, which underwent a 1,5-hydrogen transfer reaction to produce carbon-centered radical **294**. Trapping of this radical with carbon monoxide and oxidation followed by cyclization gave lactone **297**.

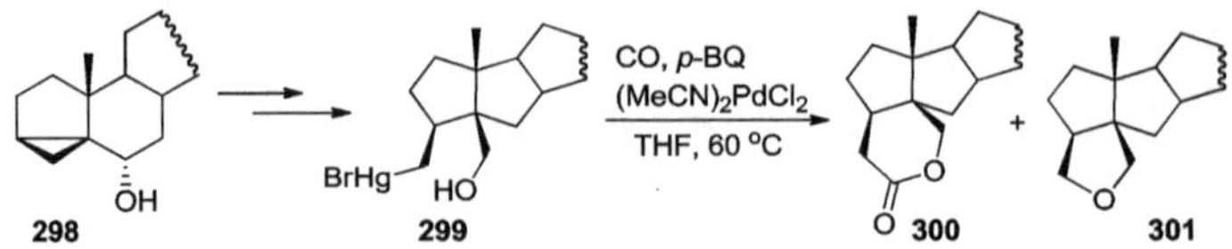

Scheme 65 Synthesis of Pd(II)-catalyzed carbonylation of halomercurio alcohols

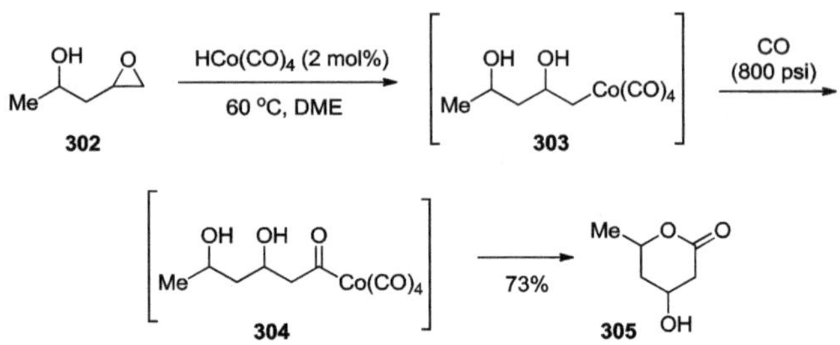

Scheme 66 Synthesis of 3-hydroxy-δ-lactones via carbonylation of homoglycidols

In another instance, Kočovský et al. have reported a Pd(II)-catalyzed carbonylation of halomercurio alcohols for the synthesis of δ-lactones [112] (Scheme 65). Bromomercurio alcohol **299** derived from cyclopropyl derivative was subjected to a Pd(II)-catalyzed carbonylation in the presence of *p*-benzoquinone. Whereas the use of stoichiometric amount of palladium catalyst led to lactone **300** in 55 % yield along with the tetrahydrofuran derivative **301** (11 %), the catalytic version gave rise to a mixture of **300** (14 %) and **301** (44 %).

In addition, Coates et al. have developed a cobalt-catalyzed carbonylation of epoxides for the synthesis of substituted 3-hydroxy-δ-lactones [113] (Scheme 66). After screening for several catalysts, $HCo(CO)_4$ was identified as the best catalyst to effect this transformation. The proposed mechanism of the carbonylation involves protonation and ring opening of the epoxide **302** by the catalyst to form cobalt alkyl complex **303**, followed by insertion of CO and subsequent cyclization to generate the 3-hydroxy-δ-lactone framework **305**.

10 Carboxylation

Carboxylation has also been seldom used especially for the construction of various spiro-δ-lactones. Rieke et al. have developed a direct synthesis of spiro-δ-lactones from conjugated dienes and epoxides [114, 115] (Scheme 67). Treatment of 1,2-bis (methylene)-cyclohexane-magnesium reagent derived from diene **306**, with an excess of ethylene oxide, gave intermediate **307**, which upon reaction with carbon dioxide and hydrolysis afforded the spiro-δ-lactone **308**.

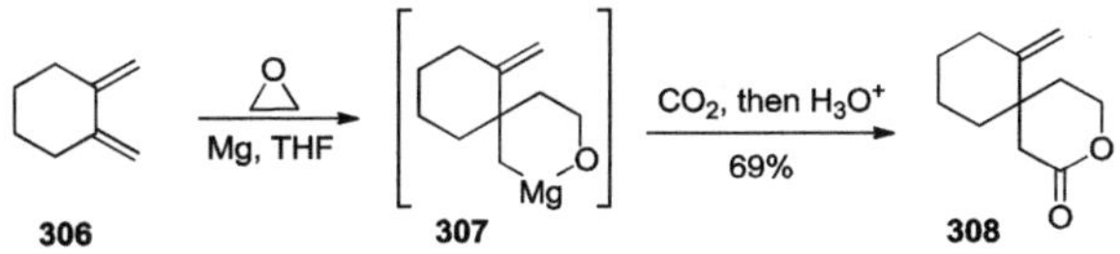

Scheme 67 Rieke et al. synthesis of spiro-δ-lactones

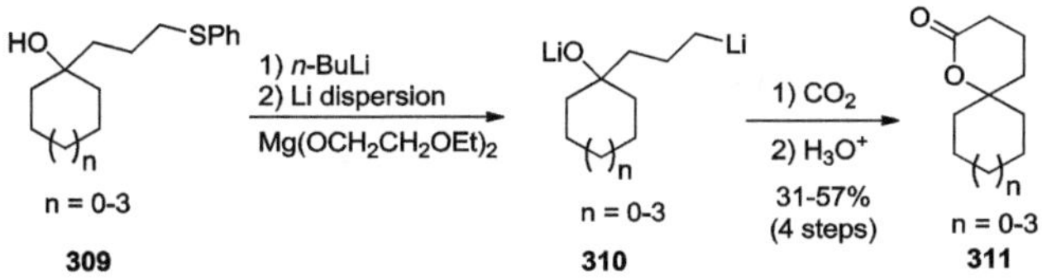

Scheme 68 Synthesis of spiro-δ-lactones

A general method for the synthesis of spiro-δ-lactones has been reported by Kostas and Screttas, exploiting the chemistry of (lithioalkoxy)lithiums [116] (Scheme 68). The cleavage of sulfide **309** using an excess of lithium dispersion in the presence of magnesium 2-ethoxyethoxide generated the organometallic reagent **310**. The presence of the magnesium alkoxide dramatically reduces the propensity of the initially formed metallated species to react with ethereal solvents, thereby increasing their stability in tetrahydrofuran. Carboxylation of **310** and subsequent acidic hydrolysis yielded a range of δ-lactones and spiro-δ-lactones.

11 Miscellaneous

In this category a collection of some unsorted methods for the synthesis of δ-lactones are described. Canonne et al. have reported a one-step spiroannelation for the synthesis of spiro-δ-lactones from cyclic anhydrides [117]. Addition of 1,4-bis(bromo-magnesio)butane to spirocyclic anhydride **312** led to intermediate **314** via the formation of **313**, and subsequent treatment with HCl provided spiro-δ-lactone **315** (Scheme 69). The scope of this method was further demonstrated with several anhydrides to synthesize a variety of spiro-δ-lactones.

A similar intramolecular approach for the synthesis of δ-lactones was reported by Watt et al. [118] (Scheme 70). After screening of several conditions, iodotrimethylsilane was found to effect the desired cyclization. Thus, treatment of α-iodoacetate **316** with iodotrimethylsilane led to the formation of lactone **317**, probably through the formation of *O*-trimethylsilyl ketene acetal **318**.

White and Jayasinghe have reported a synthesis of integerrinecic acid lactone via anchimerically assisted opening of an epoxide [46] (Scheme 71). Ozonolysis of olefin **319**, followed by oxidative work-up with Jones' reagent, and subsequent

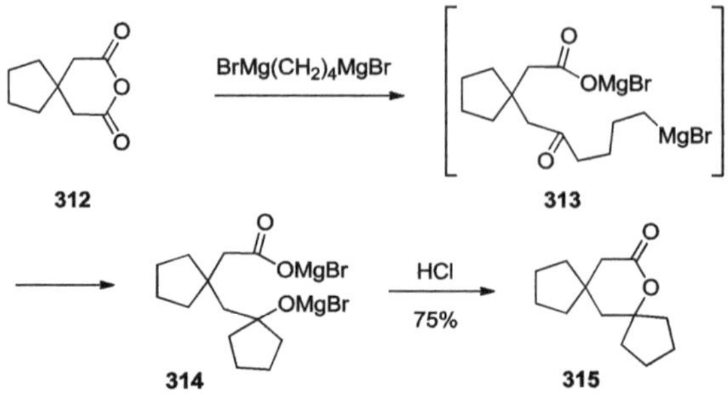

Scheme 69 Canonne et al. one-step spiroannelation to the synthesis of spiro-δ-lactones

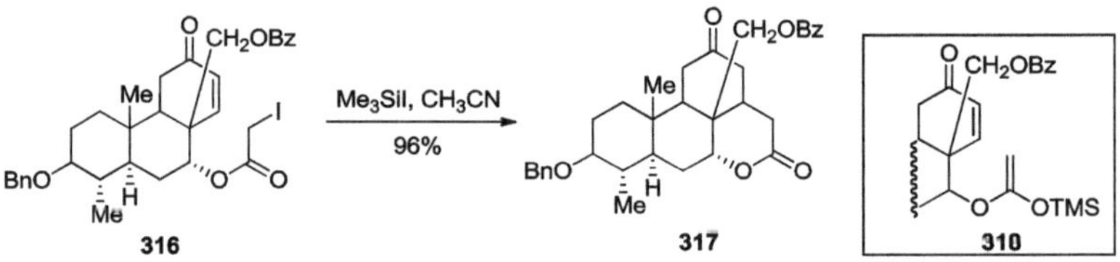

Scheme 70 Watt et al. synthesis of δ-lactones

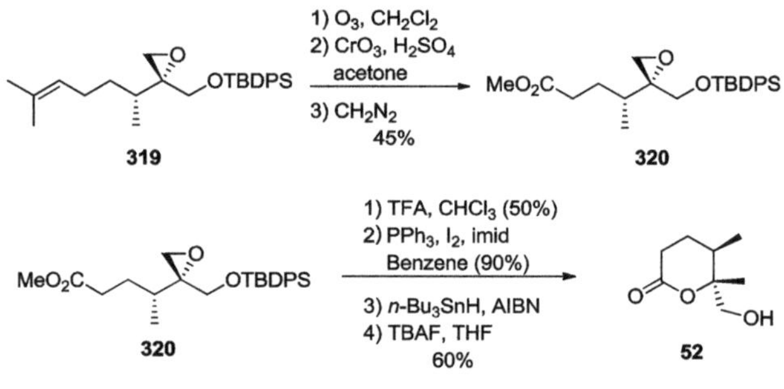

Scheme 71 Synthesis of integerrinecic acid lactone

treatment with diazomethane furnished ester **320**. Acid-catalyzed opening of the epoxide was anchimerically assisted by the ester function to give the lactone. The primary alcohol of the resulting lactones was reduced to a methyl group via the iodide, and removal of the silyl group yielded lactone **52**, which was elaborated integerrinecic acid lactone.

The synthesis of δ-lactones by an intramolecular Claisen-type condensation of β-acetoxy amides and imides has been reported by Brandänge and Leijonmarck [119] (Scheme 72). The *syn*-aldol acetate ester **321** derived from Oppolzer chiral auxiliary was subjected to an intramolecular Claisen-type condensation with LiHMDS to afford β-keto-δ-lactone **322**. Conversion of enol **322** into the corresponding enol tosylate was followed by hydrogenation with H_2 and Pd/C using $MgHPO_4{\cdot}7H_2O$ as acid scavenger, providing δ-lactone **323**.

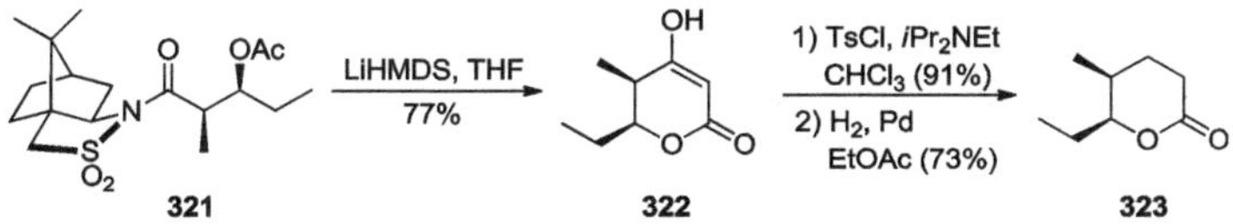

Scheme 72 Brandänge and Leijonmarck Claisen-type condensation to the synthesis of δ-lactones

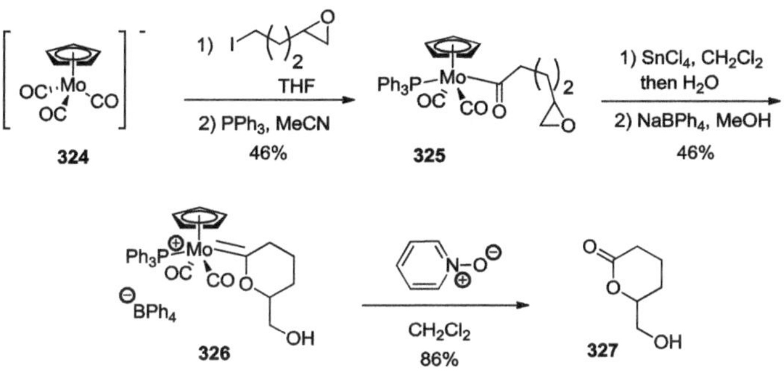

Scheme 73 Synthesis of hydroxymethyl δ-lactone via cyclization of epoxy oxacarbene complexes

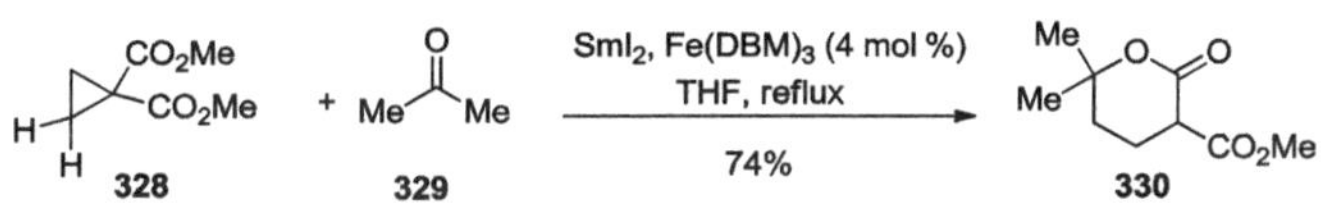

Scheme 74 Imamoto et al. synthesis of δ-lactones

Marson et al. have reported a general method for the synthesis of hydroxymethyl lactones via cyclization epoxy oxacarbene complexes [120] (Scheme 73). Reaction of 3-iodo-propyloxirane with molybdenum complex **324** gave the corresponding molybdenum alkyl, which underwent an alkyl to carbonyl migration in the presence of triphenylphosphine to give molybdenum acyl **325**. A cationic cyclization of molybdenum acyl with $SnCl_4$ followed by anion exchange with $NaBPh_4$ afforded the oxacyclohexylidene **326**. Oxidative cleavage of the molybdenum carbene bond was best achieved using pyridine *N*-oxide to provide hydroxymethyl lactone **327**.

Imamoto et al. have reported a samarium(II) iodide promoted reductive ring opening of cyclopropane-1,1-dicarboxylic esters for the synthesis of δ-lactones [121] (Scheme 74). Treatment of various cyclopropane-1,1-dicarboxylate esters **328** with a variety of ketones **329** in the presence of samarium iodide and a catalytic amount of tris(dibenzoylmethiodo)iron(III) gave a diverse array of 2-methoxycarbonyl-5,5-disubstituted δ-lactone **330** in good yields. The use of a catalytic amount of the iron complex was imperative to accelerate the reductive ring opening of the cyclopropane.

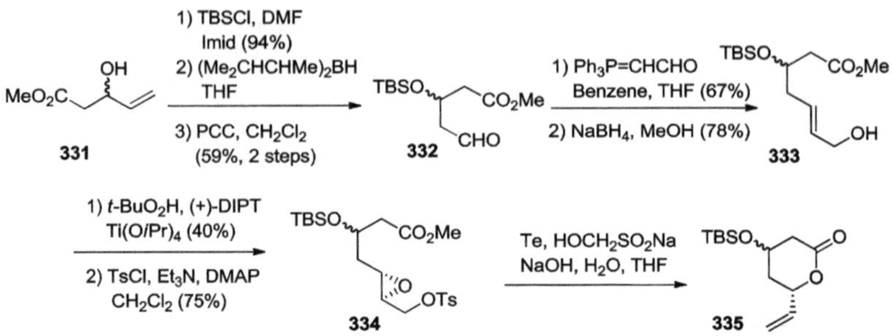

Scheme 75 Dittmer and Kumar synthesis of the lactone moiety of mevinic acid

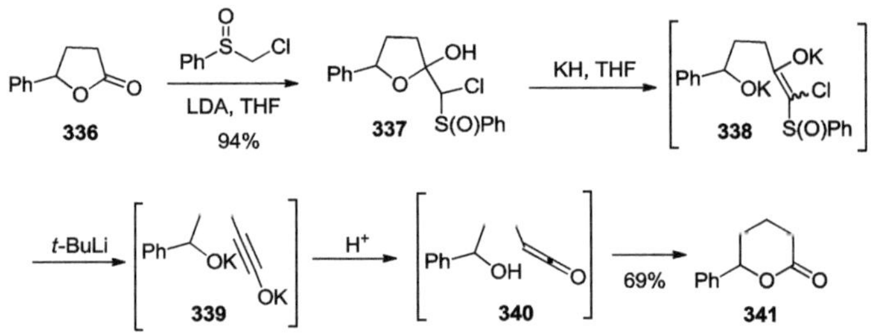

Scheme 76 Synthesis of δ-lactones via one-carbon ring enlargement

Kumar and Dittmer have reported the synthesis of intermediates for the lactone moiety of mevinic acids using tellurium-induced nucleophilic reduction developed by their own group as the key step [122] (Scheme 75). Alcohol **331** was protected and converted into aldehyde **332**, which upon Wittig reaction and reduction gave allylic alcohol **333**. Sharpless epoxidation of **333** and tosylation of the primary alcohol afforded tosylate **334**, which underwent the tellurium-induced transposition providing lactone **335** through spontaneous lactonization. However, in the case of the corresponding *tert*-butyl ester, there was no spontaneous lactonization observed.

A one-carbon ring enlargement approach for the synthesis of δ-lactones has been developed by Satoh and Kurihara [123] (Scheme 76). Reaction of the lithium carbanion of chloromethyl phenyl sulfoxide with lactone **336** afforded a diastereomeric mixture of the hemiacetal **337**. Treatment of this adduct with three equivalents of KH in THF generated potassium enolate **338**, which on addition of four equivalents of *t*-BuLi underwent an alkylidene carbenoid rearrangement to give alkynolate **339**. Protonation of **339** by the addition of sulfuric acid facilitated the formation of ω-hydroxy ketene **340**, which underwent an intramolecular cyclization to afford δ-lactone **341**.

Suárez et al. have reported a general method for the synthesis of alduronic acid lactones via fragmentation of carbohydrate anomeric alkoxy radicals [124] (Scheme 77). Treatment of hexopyranose derivative of the galacturonic acid, **342**, with (diacetoxyiodo)benzene and iodine under mild conditions or with

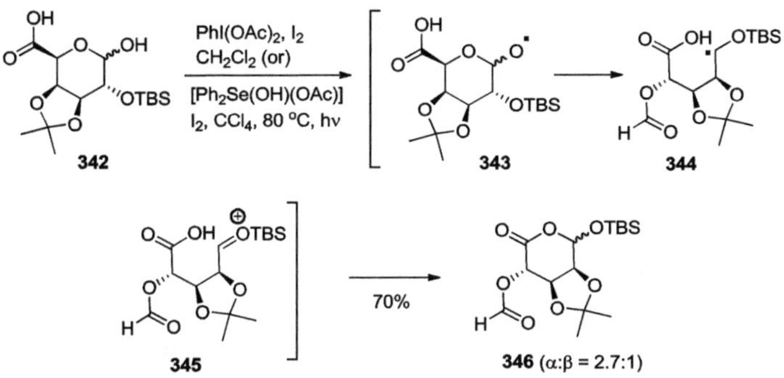

Scheme 77 Synthesis of alduronic acid lactones

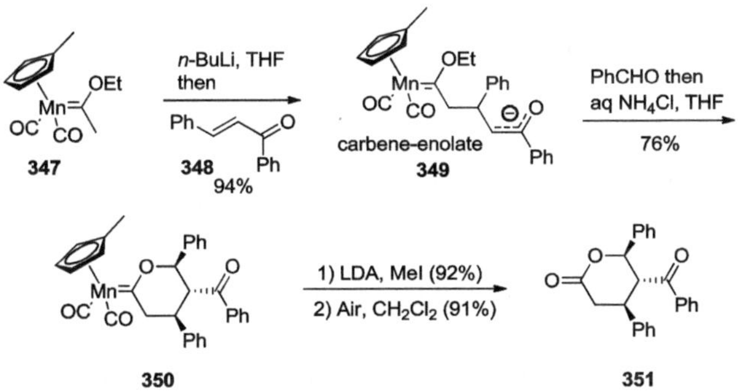

Scheme 78 Synthesis of γ-acyl-δ-lactones

diphenylhydroxyselenium acetate and iodine under visible light irradiation generated the anomeric radical **343**, which underwent a β-fragmentation of the C1–C2 bond to afford the C2 radical **344**. Oxidation of intermediate radical **344** with the reagent gave oxonium ion **345**, which reacted intramolecularly with the nucleophilic carboxyl group to furnish a separable anomeric mixture of 5,1-lyxuronic acid lactone **346**.

A general method for the synthesis of γ-acyl-δ-lactones has been developed by Lugan et al. by vicinal di-functionalization of α,β-unsaturated ketones via manganese carbene intermediates [125] (Scheme 78). Treatment of manganese carbene **347** with *n*-BuLi generated the carbene anion, which underwent Michael addition with benzylidene acetone and chalcone to give carbene-enolate intermediate **349**. A cascade aldolization/transesterification of this intermediate with benzaldehyde afforded the oxacyclocarbene complexes **350**, which were readily oxidized by air to release γ-acyl-δ-lactone **351**.

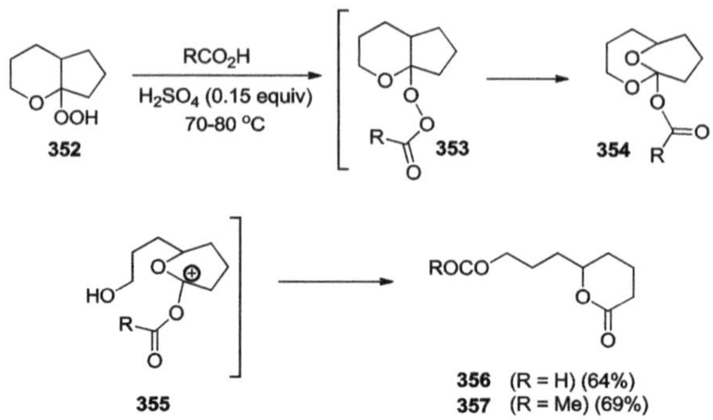

Scheme 79 Synthesis of δ-lactones via rearrangement of 1-hydroperoxy-2-oxabicycloalkanes

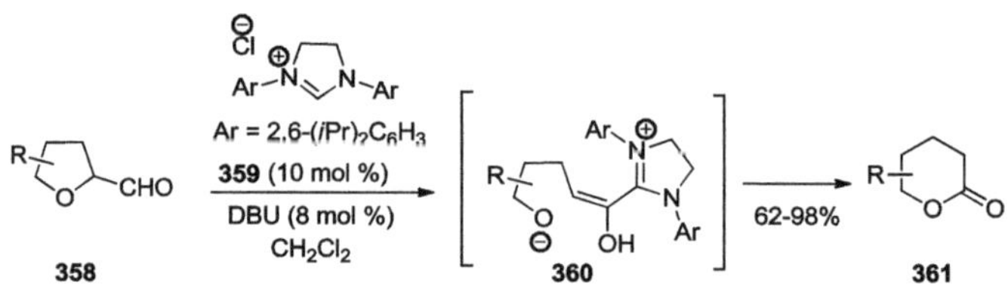

Scheme 80 *N*-heterocyclic carbene-catalyzed synthesis of δ-lactones

Ogibin et al. have described a rearrangement of 1-hydroperoxy-2-oxabicycloalkanes into ω-acyloxy-δ-lactones [126] (Scheme 79). Thus, 1-hydroperoxy-2-oxabicycloalkanes **352** on heating with formic or acetic acid containing a catalytic amount of sulfuric acid resulted in the formation of ω-acyloxy-δ-lactones **356** and **357**. The rearrangement is proposed to go through a mechanism related to the Criegee reaction. In the first step, sulfuric acid catalyzes the acylation of hydrogen peroxide with carboxylic acid to generate bicyclic peroxy ester **353**, which undergoes a 1,2-shift followed by rearrangement to provide the δ-lactones **356** and **357**. Furthermore, the fact that there was no reaction in the absence of sulfuric acid further underscores the proposed mechanism.

Gravel et al. have disclosed a general method for the versatile synthesis of δ-lactones by *N*-heterocyclic carbene (NHC)-catalyzed ring expansion of oxacycloalkane-2-carboxaldehydes [127] (Scheme 80). Treatment of aldehyde **358** with 10 mol% of NHC catalyst **359** and 8 mol% of DBU afforded a variety of δ-lactone **361** in good to excellent yield via intermediate **360**. Screening of various NHC catalyst for this transformation revealed that the electronic factors play a major role in this reaction.

12 Conclusions

In view of the biological importance of the δ-lactone moiety, extensive efforts have been devoted for the development of various methods for the synthesis of saturated δ-lactones. Among the various methods, the more classical methods include lactonization of the δ-hydroxy acid derivatives, Baeyer–Villiger oxidation of cyclopentanones, and oxidation of lactols. Besides, more challenging and attractive methods such as oxidative lactonization, radical cyclization, and carbonylation have also been used efficiently for the synthesis of δ-lactones. The past two decades have witnessed remarkable growth in the development of catalytic and asymmetric methods for the synthesis of δ-lactones in optically pure form. In the next decade, new and more exciting advances in the development of efficient and catalytic enantioselective methods and their application in the synthesis of complex δ-lactone natural products can be expected.

Acknowledgments K.P.K. thanks DST, New Delhi, for the award of the Swarnajayanti fellowship, and K.P. thanks CSIR, New Delhi, for a fellowship. Further, we thank Sankar R. Suravarapu for his valuable initial work on this program.

References

1. Rasmussen JK (1977) Synthesis: 91
2. Reetz MT (1982) Angew Chem Int Ed 21:96
3. Rocca JR, Tumlinson JH, Glancey BM, Lofren CS (1983) Tetrahedron Lett 24:1889
4. Gunasekara SP, Longley RE, Schulte GK (1991) J Org Chem 56:1346
5. Endo A, Kuroda M, Tsujita M (1976) J Antibiot 29:1346
6. Barua NC, Schmidt RR (1986) Synthesis: 1067
7. Thompson CM (1987) Tetrahedron Lett 28:4243
8. Mori K (1989) Tetrahedron 45:3233
9. Boucard V, Broustal G, Campagne JM (2007) Eur J Org Chem: 225
10. Florence GJ, Gardner NM, Paterson I (2008) Nat Prod Rep 25:342
11. Chiu P, Leung LT, Ko CBB (2010) Nat Prod Rep 27:1066
12. Schreiber SL (1991) Science 251:283
13. Rosen MK, Schreiber SL (1992) Angew Chem Int Ed 31:384
14. Endo A (1979) J Antibiot 32:854
15. Alberts AW, Chen J, Kuron G, Hunt V, Huff J, Hoffman CH, Rothrock J, Lopez M, Joshua H, Harris E, Patchett AA, Monaghan R, Currie S, Stapley E, Albers-Schönberg G, Hensens O, Hirschfield J, Hoogsteen K, Liesch J (1980) Proc Natl Acad Sci USA 77:3957
16. Endo A, Kuroda M, Tsujita Y (1976) J Antibiot 26:1346
17. Brown AG, Smale TC, King TJ, Hasenkamp R, Thompson RH (1976) J Chem Soc Perkin 1: 1165
18. Anliker R, Dvornik D, Gubler K, Heusser H, Prelog V (1956) Helv Chim Acta 39:1785
19. Djerassi C, Zderic JA (1956) J Am Chem Soc 78:6390
20. Li ZL, Pan DJ, Hu CQ, Wu QL, Yang SS, Xu GY (1982) Acta Chimi Sin 40:447
21. Li ZL, Pan DJ, Wu QL, Xu GY (1982) Acta Chimi Sin 40:757
22. Zhou BN, Ying B, Song GQ, Chen ZX, Han J, Yan YF (1983) Planta Med 47:35
23. Hamburger M, Shieh H, Pezzuto J, Cordell GA, Zhou BN (1988) Planta Med 54:579

24. Hamburger MO, Shieh HL, Zhou BN, Pezzuto JM, Cordell GA (1989) Magn Reson Chem 27:1025
25. Akdemir Z, Calis I, Junior P (1991) Planta Med 57:584
26. Ono M, Ito Y, Kubo S, Nohara T (1997) Chem Pharm Bull 45:1094
27. Wurzel G, Becker H, Eicher T, Tiefensee K (1990) Planta Med 56:444
28. Wurzel G, Becker H (1990) Phytochemistry 29:2565
29. Zinsmeister HD, Becker H, Eicher T (1991) Angew Chem Int Ed 30:130
30. Thiel R, Adam KP, Zapp J, Becker H (1997) Pharm Pharmacol Lett 7:103
31. Lakner FJ, Hager LP (1996) J Org Chem 61:3923
32. Takano S, Shimazaki Y, Iwabuchi Y, Ogasawara K (1990) Tetrahedron Lett 31:3619
33. Ikan R, Gottlieb R, Bergmann ED, Ishay J (1969) J Insect Physiol 15:1709
34. Schulz S, Nishida R (1996) Bioorg Med Chem 4:341
35. Crouse GD, Sparks TC (1998) Rev Toxicol 2:133
36. Martin CJ, Timoney MC, Sheridan RM, Kendrew SG, Wilkinson B, Staunton J, Leadlay PF (2003) Org Biomol Chem 1:4144
37. Pappo R, Jung CJ (1962) Tetrahedron Lett 3:365
38. Chavdarian CG, Heathcock CH (1975) J Org Chem 40:2970
39. Coke JL, Richon AB (1976) J Org Chem 41:3516
40. Khan HA, Paterson I (1982) Tetrahedron Lett 23:5083
41. Yamaguchi M, Shibato K, Hirao I (1984) Tetrahedron Lett 25:1159
42. Gerth DB, Giese B (1986) J Org Chem 51:3726
43. Sakamoto A, Yamamoto Y, Oda J (1987) J Am Chem Soc 109:7188
44. Priestap HA, Bonafede JD, Rúveda EA (1977) Phytochemistry 16:1579
45. Carretero JC, Ghosez L (1988) Tetrahedron Lett 29:2059
46. White JD, Jayasinghe LR (1988) Tetrahedron Lett 29:2139
47. Reddy GB, Minami T, Hanamoto T, Hiyama T (1991) J Org Chem 56:5752
48. Shimada S, Tohno I, Hashimoto Y, Saigo K (1993) Chem Lett: 1117
49. Roe JM, Thomas EJ (1995) J Chem Soc Perkin 1: 359
50. Cossy J, Cases M, Pardo DG (1996) Tetrahedron Lett 37:8173
51. Corey EJ, Guzman-Perez A, Lazerwith SE (1997) J Am Chem Soc 119:11769
52. Honda T, Ono S, Mizutani H, Hallinan KO (1997) Tetrahedron Asymmetry 8:181
53. Shimizu M, Ishi K, Fujisawa T (1997) Chem Lett: 765
54. Wilkinson AL, Hanefeld U, Wilkinson B, Leadlay PF, Staunton J (1998) Tetrahedron Lett 39:9827
55. Miyazawa M, Oonuma S, Maruyama K, Miyashita M (1997) Chem Lett: 1191
56. Miyazawa M, Matsuoka E, Sasaki S, Oonuma S, Maruyama K, Miyashita M (1998) Chem Lett: 109
57. Hiscock SD, Hitchcock PB, Parsons PJ (1998) Tetrahedron 54:11567
58. Tanaka H, Kamikubo T, Yoshida N, Sakagami H, Taniguchi T, Ogasawara K (2001) Org Lett 3:679
59. Smitrovich JH, Boice GN, Qu C, DiMichele L, Nelson TD, Huffman MA, Murry J, McNamara J, Reider PJ (2002) Org Lett 4:1963
60. Stritzke K, Schulz S, Nishida R (2002) Eur J Org Chem: 3884
61. Zarbin PHG, Oliveira ARM, Simonelli F, Villar JAFP, Delay O Jr (2004) Tetrahedron Lett 45:7399
62. Fleming I, Terrett NK (1998) J Chem Soc Perkin 1: 2645
63. Uyehara T, Shida N, Yamamoto Y (1992) J Org Chem 57:3139
64. Jahn U, Hartmann P, Kaasalainen E (2004) Org Lett 6:257
65. Sibi MP, He L (2004) Org Lett 6:1749
66. Yadav JS, Reddy PMK, Reddy PV (2007) Tetrahedron Lett 48:1037
67. Blay G, Hernández-Olmos V, Pedro JR (2010) Org Lett 12:3058
68. Cid MB, Duce S, Morales S, Rodrigo E, García Ruano JL (2010) Org Lett 12:3586
69. Bindseil KU, Zeeck A (1993) Helv Chim Acta 76:150

70. Boddien C, Gorber-Nolte J, Zeeck A (1996) Liebigs Ann: 1381
71. Peed J, Domínguez IP, Davies IR, Cheeseman M, Taylor JE, Kociok-Köhn G, Bull SD (2011) Org Lett 13:3592
72. Renz M, Meunier B (1999) Eur J Org Chem: 737
73. Hacini S, Santelli M (1990) Tetrahedron 46:7787
74. Santelli-Rouvier C, Lefrère S, Santelli M (1994) Tetrahedron Lett 35:6101
75. Adger B, Teresa Bes M, Grogan G, McCague R, Pedragosa-Moreau S, Roberts SM, Villa R, Wan PWH, Willetts AJ (1995) J Chem Soc Chem Commun: 1563
76. Clissold C, Kelly CL, Lawrie KWM, Willis CL (1997) Tetrahedron Lett 38:8105
77. Yamamoto T, Ogura M, Amano A, Adachi K, Hagiwara T, Kanisawa T (2002) Tetrahedron Lett 43:9081
78. Chakraborty TK, Chandrasekaran S (1980) Tetrahedron Lett 21:1583
79. Rathore R, Vankar PS, Chandrasekaran S (1986) Tetrahedron Lett 27:4079
80. Schlecht MF, Kim H-J (1985) Tetrahedron Lett 26:127
81. Yamamoto M, Hirota M, Kishikawa K, Kohmoto S, Yamada K (1997) Synthesis: 937
82. Suzuki T, Morita K, Tsuchida M, Hiroi K (2002) Org Lett 4:2361
83. Schomaker JM, Travis BR, Borhan B (2003) Org Lett 5:3089
84. Yang C-G, Reich NW, Shi Z, He C (2005) Org Lett 7:4553
85. Mollat du Jourdin X, Noshi M, Fuchs PL (2009) Org Lett 11:543
86. Ramón DJ, Yus M (1990) Tetrahedron Lett 31:3767
87. Robin S, Huet F (1993) Tetrahedron Lett 34:2945
88. Oppolzer W, Walther E, Balado CP, De Brabander J (1997) Tetrahedron Lett 38:809
89. Kaku H, Tanaka M, Norimine Y, Miyashita Y, Suemune H, Sakai K (1997) Tetrahedron Asymmetry 8:195
90. Kishida M, Yamauchi N, Sawada K, Ohashi Y, Eguchi T, Kakinuma K (1997) J Chem Soc Perkin 1: 891
91. Shimizu M, Kamikubo T, Ogasawara K (1997) Tetrahedron Asymmetry 8:2519
92. Dujardin G, Rossignol S, Brown E (1998) Synthesis: 763
93. BouzBouz S, Popkin ME, Cossy J (2000) Org Lett 2:3449
94. Berry MB, Griffiths RJ, Sanganee MJ, Steel PG, Whelligan DK (2003) Tetrahedron Lett 44:9135
95. Casas J, Engqvist M, Ibrahem I, Kaynak B, Córdova A (2005) Angew Chem Int Ed 44:1343
96. Cardllina JH II, Moore RE, Arnold EV, Clardy J (1979) J Org Chem 44:4039
97. Torii S, Inokuchi T, Yoritaka K (1981) J Org Chem 46:5030
98. Greeves N, Lee W-M, Barkley JV (1997) Tetrahedron Lett 38:6453
99. Tan CK, Zhou L, Yeung Y-Y (2011) Org Lett 13:2738
100. Zhang W, Liu N, Schienebeck CM, Decloux K, Zheng S, Werness JB, Tang W (2012) Chem Eur J 18:7296
101. Pilli RA, Murta MM (1993) J Org Chem 58:338
102. Kobayashi K, Minakawa H, Sakurai H, Kujime S, Suginome H (1993) J Chem Soc Perkin 1: 3007
103. Bachi MD, Bosch E (1992) J Org Chem 57:4696
104. Ihara M, Setsu F, Shohda M, Taniguchi N, Fukumoto K (1994) Heterocycles 37:289
105. Chuang T-H, Fang J-M, Jiaang W-T, Tsai Y-M (1996) J Org Chem 61:1794
106. Hsu J-L, Chen C-T, Fang J-M (1999) Org Lett 1:1989
107. Annby U, Stenkula M (1993) Tetrahedron Lett 34:8545
108. Toyota M, Hirota M, Nishikawa Y, Fukumoto K, Ihara M (1998) J Org Chem 63:5895
109. Toyota M, Nishikawa Y, Fukumoto K (1998) Heterocycles 47:675
110. Tsunoi S, Ryu I, Sonoda N (1994) J Am Chem Soc 116:5473
111. Tsunoi S, Ryu I, Okuda T, Tanaka M, Komatsu M, Sonoda N (1998) J Am Chem Soc 120:8692
112. Kočovský P, Dunn V, Grech JM, Šrogl J, Mitchell WL (1996) Tetrahedron Lett 37:5585
113. Kramer JW, Joh DY, Coates GW (2007) Org Lett 9:5581

114. Sell MS, Xiong H, Rieke RD (1993) Tetrahedron Lett 34:6007
115. Rieke RD, Sell MS, Xiong H (1995) J Org Chem 60:5143
116. Kostas ID, Screttas CG (1997) J Org Chem 62:5575
117. Canonne P, Bélanger D, Lemay G, Foscolos GB (1981) J Org Chem 46:3091
118. Demir AS, Gross RS, Dunlap NK, Bashir-Hashemi A, Watt DS (1986) Tetrahedron Lett 27:5567
119. Brandänge S, Leijonmarck H (1992) Tetrahedron Lett 33:3025
120. Marson CM, Randall L, Winter MJ (1994) Tetrahedron Lett 35:6717
121. Imamoto T, Hatajima T, Yoshizawa T (1994) Tetrahedron Lett 35:7805
122. Kumar A, Dittmer DC (1994) J Org Chem 59:4760
123. Satoh T, Kurihara T (1998) Tetrahedron Lett 39:9215
124. Francisco CG, Martin CG, Suárez E (1998) J Org Chem 63:2099
125. Mongin C, Gruet K, Lugan N, Mathieu R (2000) Tetrahedron Lett 41:7341
126. Ogibin YN, Terent'ev AO, Kutkin AV, Nikishin GI (2002) Tetrahedron Lett 43:1321
127. Wang L, Thai K, Gravel M (2009) Org Lett 11:891

Synthesis of 7-Oxabicyclo[2.2.1]heptane and Derivatives

Antonio J. Moreno-Vargas and Pierre Vogel

Contents

Abstract Nature provides a number of 7-oxanorbornanes (7-oxabicyclo[2.2.1] heptanes), many of them with interesting biological activity. There are several routes to their synthesis; the most common remains the Diels–Alder reaction of furans with olefinic or acetylenic dienophiles. Several 7-oxanorbornane derivatives

A.J. Moreno-Vargas
Facultad de Química, Departamento de Química Orgánica, Universidad de Sevilla,
C/Prof. García González, 1, 41012 Sevilla, Spain

P. Vogel (✉)
LGSA, EPFL, 1015 Lausanne, Switzerland
e-mail: pierre.vogel@epfl.ch

J. Cossy (ed.), *Synthesis of Saturated Oxygenated Heterocycles I*, Topics in Heterocyclic Chemistry 35, DOI 10.1007/978-3-642-41473-2_4,

can be prepared enantiomerically enriched readily. They are extremely useful chirons for the total asymmetric synthesis of all kinds of natural products and bioactive compounds such as rare sugars and analogues, monosaccharides, and disaccharide mimetics. There are several methods for the C–O and C–C bond cleavage of 7-oxanorbornanes. Because of their bicyclic structure, 7-oxanorbornanes permit to generate a wide chemodiversity in a highly stereoselective manner.

Keywords 1,4-Cineole • Acid promoted • Aldol reaction • Alkaloid • Antibiotic • Antitumor • Asymmetric total synthesis • Aza-*C*-disaccharide • Banyaside • Cantharidin • Carba-pyranose • Carotenoid pigment • C–C bond cleavage • *C*-disaccharide • *C*-glycoside • Chiral auxiliary • Conduramine • Conduritol • Cyclization • Cyclophellitol • Diels–Alder reaction • Diterpenoid • Elimination • Ether cleavage • Etherification • Ethisolide • Fragrance • Furan • Glycinoeclepin • Heterolysis • Hexoses • Illudin • Iminodideoxyalditol • Isoavenaciolide • Ketyl radical anion • Maneonene • Methyl nonactate • Molecular device • "Naked sugars" • Natural 7-oxanorbornanes • Nonactin • Norcantharidin • Palasonin • Peduncularine • Phosphatase inhibitors • Photo-induced reduction • Pinacolic rearrangement • Prostaglandin • Radical reaction • Reduction • Sesquiterpenoid • Single-electron transfer • Solanoeclepin • Strain of 7-oxanorbornane • Sylvan • Template • Triterpenoid • Uronolactone

1 Introduction

As other cyclic ethers, unsubstituted 7-oxabicyclo[2.2.1]heptane (1,4-epoxycyclohexane, or 7-oxanorbornane, **1** should be called 7-oxatrinorbornane as three methyl groups have been removed from the corresponding bornane) and alkyl-substituted derivatives generate useful polymers upon oxa ring openings [1–5]. On its side, 2-methylidene- 7-oxanorbornane **2** has been used in radical-induced alkene polymerizations [6, 7]. 7-Oxanorbornane derivatives are found in Nature, some of these have interesting biological properties, and analogues of these compounds have also been found to be bioactive [8], for instance, as herbicides [9, 10]. In the laboratory, 7-oxanorbornanes are readily available through Diels–Alder reactions of furans or by other methods that will be illustrated. A large number of these bicyclic templates are available enantiomerically enriched, either through classical resolution of diastereomers or through asymmetric catalysis [11, 12]. The various reactions of 7-oxanorbornanes permit a high chemodiversity in organic chemistry in general and for the synthesis of compounds of biological interest. In addition, the 7-oxanorbornane skeleton allows the construction of unusual templates and molecular devices of interest for biology [13] and material sciences that profit of its shape and rigidity [14] (Fig. 1).

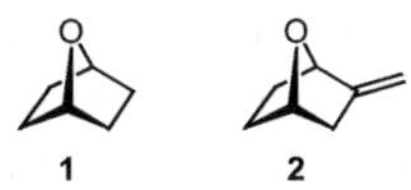

Fig. 1 7-Oxanorbornane and 2-methylidene-7-oxanorbornane

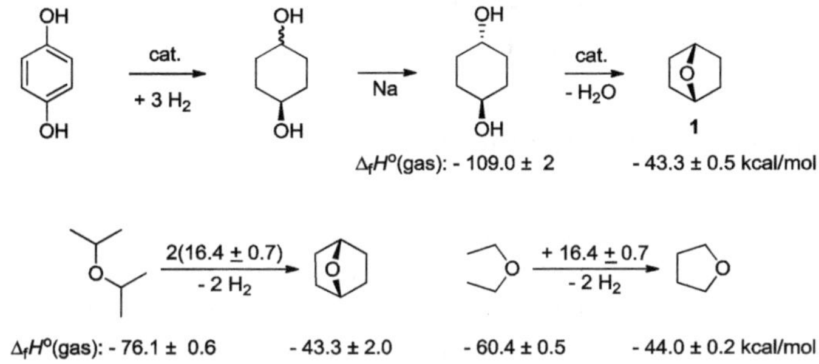

Scheme 1 Preparation of 7-oxabicyclo[2.2.1]heptane and gas phase thermochemical data

2 Preparation of 7-Oxabicyclo[2.2.1]heptane

The parent 7-oxanorbornane, **1**, is commercially available. Its preparation starts with the catalytic hydrogenation of hydroquinone to generate a mixture of *trans*- and *cis*-cyclohexane-1,4-diol [15–18]. *cis*-Cyclohexane-1,4-diol can be isomerized into the more stable *trans* isomer with metallic sodium [17]. Dehydratation of the latter on A4 zeolites, on alumina [19], or over nickel–kieselguhr catalyst [20] provides **1**. This reaction is exergonic ($\Delta_rG^o = \Delta_rH^o - T\Delta_rS^o = -3.1 \pm 2.5$ kcal/mol) at room temperature as its standard gas phase heat of reaction amounts to $\Delta_rH^o = +7.3 \pm 2.5$ kcal/mol. A variation of entropy of reaction of ca. +35 eu is assumed for this fragmentation, what leads to $-T\Delta_rS^o = 298(0.035) = -10.4$ kcal/mol. The standard gas phase heat of formation of *trans*-cyclohexane-1,4-diol (Scheme 1) is estimated from that of cyclohexanol (-69.0 ± 2.0 kcal/mol) and the standard heat of oxidation of cyclohexane into cyclohexanol (-39.5 kcal/mol) [21]. Chickos and Acree [22] give $\Delta_fH^o(\mathbf{1}) = -43.4 \pm 0.5$ kcal/mol (see also [23–25]).

Ring strain of **1** corresponds to the ring strain of two tetrahydrofuran moieties. Their annulation into the bicyclo[2.2.1]heptane system does not introduce extra ring strain as demonstrated by the thermochemical analysis reported in Scheme 1. Double dehydrogenation (twice $+16.4 \pm 0.7$ kcal/mol, the heat of dehydrogenation of diethyl ether into tetrahydrofuran) of diisopropyl ether into **1** gives a standard gas phase heat of formation for **1** that amounts to -43.3 ± 2.0 kcal/mol, practically the same value as experimental $\Delta_fH^o(\mathbf{1}) = -43.4 \pm 0.5$ kcal/mol. The same analysis (double dehydrogenation of 2,3-dimethylpentane) for bicyclo[2.2.1]heptane (norbornane) gives an estimated gas phase standard heat of formation of -14.4

$\pm$ 0.9 kcal/mol to be compared with the experimental Δ_fH^o (norbornane, gas) = $-$13.13 $\pm$ 0.25 kcal/mol [21]. Considering experimental mean deviations, it can be stated that 7-oxanorbornane is not more or less strained than norbornane itself.

3 Natural 7-Oxabicyclo[2.2.1]heptanes and Bioactive Analogues

3.1 Cantharidin and Analogues

The Meloidae family of Coleoptera (beetles) has been known since antiquity to produce a defensive agent, a vesicant principle called cantharidin, **3** (*exo*,*exo*-2,3-dimethyl 7-oxabicyclo[2.2.1]hepta-2,3-dicarboxylic acid anhydride). This substance was first obtained in the crystalline form by Robiquet in 1810 [26]. During mating the male beetle deposits a spermatophore containing μg of **3** in the female spermatophoral receptacle, a copulatory gift that is then used to protect the fertilized eggs from predation [27]. Cantharidin is found in over 1,500 species of flies including *Lytta vesicatoria* (L.), which is found in the Mediterranean area (Spanish fly), *Lytta tenuicollis* (Pallas) in India, *Mylabris* sp. in India and China, and *Epicauta* sp. in Asia and North America [28]. Cantharidin is listed as a drug under the name Mylabris in the medical monograph *Materia Medica* published in 77 A.D. [29]. In recent times, **3** has been used topically in the treatment of warts [30]. The anticancer activity of **3** was known already in the thirteenth century; although **3** is cytotoxic to cancer cells and stimulatory on the bone marrow, its toxicity prevents its use in mainstream oncology [31].

Palasonin, (−)-**4**, was first isolated by Raj and Kurup [32] from the seeds of *Butea frondosa*. Its structure was established in 1968 by Bochis and Fischer [33] (see also [34, 35]). Norcantharidin, **5** (should be called dinorcantharidin), the demethylated analogue of **3**, also possesses antitumor activity [36]. It does not show the nephrotoxicity associated with **3** [37]. Compound **5** results from the catalytic hydrogenation of the Diels–Alder cycloadduct of maleic anhydride to furan (see Sect. 4.2 for the Diels–Alder reactions of furans). Both **3** and **4** are protein phosphatase 1 (PP1) and protein phosphatase 2A (PP2A) inhibitors [38, 39]. Norcantharidin, **5**, also inhibits these enzymes and calcineurin (protein phosphatase PP2B) [40]. Mono methyl, ethyl, and *n*-propyl ester derived from **5** are also good inhibitors of these enzymes and show anticancer activities [41]. PP1 and PP2A, via reversible phosphorylation of serine and threonine residues, modulate cellular transduction events such as T-cell activation and cell proliferation [42–45]. These discoveries have stimulated the search for further derivatives of cantharidin such as the trimethylene anhydride **6** [46], monoamide **7** [47], and esters **8** [48]. The studies suggested that both the 7-oxa ethereal bridge and the *endo*-dicarboxylic anhydride unit are necessary for a good inhibition of protein phosphatases. Compounds **7** and **8** are good inhibitors of calcineurin [protein phosphatase

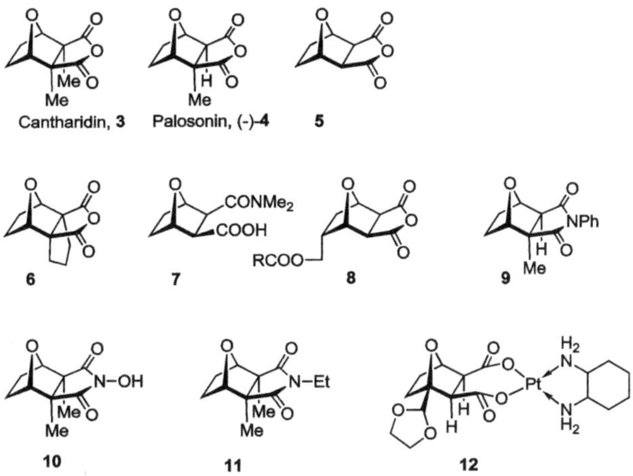

Fig. 2 Examples of bioactive 7-oxanorbornane derivatives

2B (PP2B)] which is a calcium- and calmodulin-regulated enzyme composed of a 59-kDa catalytic subunit (CnA) and a 19-kDa calcium-binding subunit (CnB) [49, 50]. This enzyme is a key signaling enzyme in T-lymphocyte activation. Its inhibition in T-lymphocytes prevents the formation of active transcription factors such as NF-AT and NF-IL2A, which are essential for interleukin-2 (IL2) gene expression [51]. Inhibition of calcineurin leads to the disruption of the cellular immune response since IL2 is necessary for the T-cell proliferation. Monoamide analogues of **7** have shown enhanced antiplasmodial activity compared with **5**; for instance, (1*S*,4*R*)-3-(allylcarbamoyl)-7-oxabicyclo[2.2.1]heptane-2-carbocylic acid is 20 times more active than norcantharidin [52].

Imide **9** has been isolated from the pod of *Butea monosperma* [53]. *N*-Hydroxycantharidinimide **10**, an ingredient of *Mylabris phalerata*, shows antitumor activity [54, 55]. The synthetic analogue **11** is active against mouse sarcoma 180 [56, 57]. Further imides have shown antitumor activities [58, 59]. *N*-Alkyl, *N*-aryl, and *N*-heteroaryl imides derived from anhydride **5** present anti-convulsant activity [60, 61]; other derivatives have shown nematocidal activity in *Haemonchus contortus* [62]. The diamine platinum complex **12** has a good anti-neoplastic activity against leukemia cells (P388) in mice [63, 64]. This compound derives from the Diels–Alder cycloadduct of maleic anhydride and ethylene glycol acetal of furfural (Fig. 2).

3.2 *Monoterpenoid 7-Oxabicyclo[2.2.1]heptanes*

In 1907 Wallach [65] identified 1,4-cineole (or cineole: 1-isopropyl-4-methyl-7-oxabicyclo[2.2.1]heptane, **13**) that forms by acid-promoted dehydratation of 1,8-terpin (*p*-menthan-1,8-diol) [66]. A two-step protocol preparation of **13** from α-terpineol (*p*-menth-1-en-8-ol) has been presented [67]; cineole is present in many

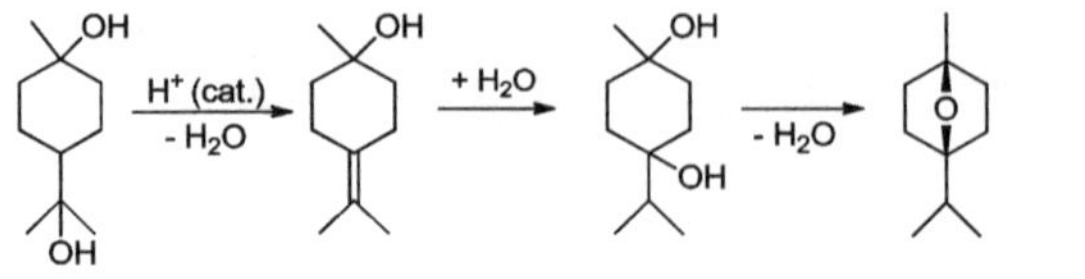

Fig. 3 Formation of 1,4-cineole in plants

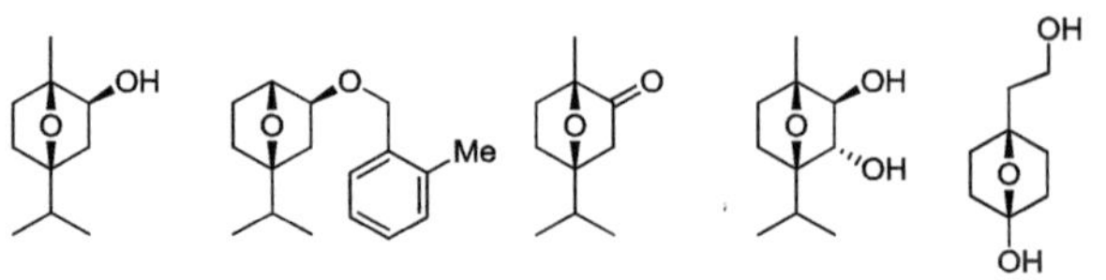

Fig. 4 7-Oxanorbornanols and derivatives from plants

plants and in perfumes [68]. It is found in Tequila [69], in leaves and flowers from *Bellis perennis* (the common daisy [70]), and in essential oils of various lemon tree leaves [71]. Cineole is formed together with myrtenol and *trans*-pinocarveol by fermentation of β-pinene with basidiomycetes [72] (Fig. 3).

Cineole, **13**, is a natural herbicide [73]. Its hydroxy derivative **14** (2-hydroxy-1,4-cineole: 1,4-epoxy-*p*-menthane-2-ol) is a constituent of oil from rhizomes of *Ferula jaeschkeana* [74]. Its 2-methylbenzyl ether **15** (cinmethylin) is a preemergence grass herbicide [75, 76]. Alcohol **14** can be prepared by microbial hydroxylation of **13** [77]. This also produces ketone **16** and its enantiomer [78]. The fragrance of ketone **16** and isomeric 1-isopropyl-4-methyl- 7-oxabicyclo[2.2.1] heptan-2-one is very similar to that of **14** and menthone [79]. Mullilam diol **17**, a dihydroxy derivative of **13**, has been isolated from *Zanthoxylum rhetsa*, a plant that exhibits antibiotic activity which is prescribed in dyspepsia and diarrhea. The eight-carbon system rengyoxide has been found in *Forsythia suspensa* fruits [80] (Fig. 4).

Ether (−)-**23** (3′,6′-epoxyaurapten) [81] has been isolated from various plants. Its total asymmetric synthesis has been realized by Aziz and Rouessac [82] and is outlined in Scheme 2. Aurapten **18** is obtained by displacement of geranyl bromide with 7-hydroxycoumarin. Its allylic oxidation with SeO_2 and *t*-BuOOH generates allylic alcohol **19** which undergoes asymmetric Katsuki–Sharpless epoxidation with (−)-diethyl D-tartrate/*t*-BuOOH/Ti/(*i*PrO)$_4$ into epoxide **20**. The hydroxymethyl group of **20** is converted into a methyl group by a sequence of alcohol tosylation and displacement of the tosylate by NaI/acetone and hydride reduction ($NaBH_3CN$). This sequence of reactions gives **21** that is isomerized into (−)-**23** upon treatment with $SnCl_4$ in CH_2Cl_2, probably through the zwitterionic intermediate **22**. This rearrangement was preceded in the literature by the isomerization of epoxide **24** into **25** [83, 84] and then into **26** (yields not given) [85].

An alternative approach to the synthesis of 1,3,3-trimethyl-7-oxabicylo[2.2.1] heptane derivatives has been presented by Sneden (Scheme 3) [86]. The Diels–Alder

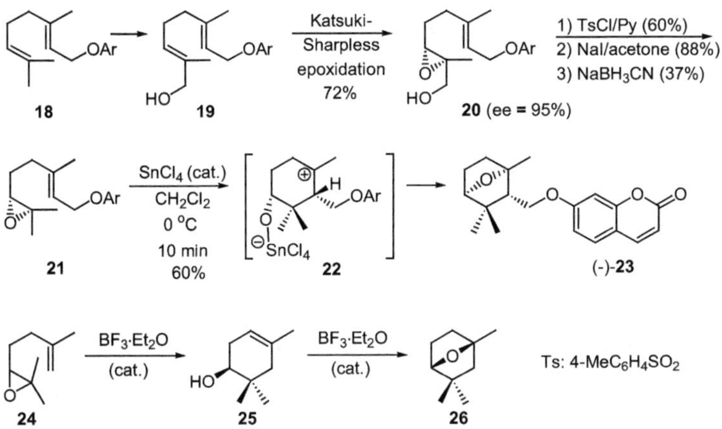

Scheme 2 Total asymmetric synthesis of 3′6′-epoxyauraptene

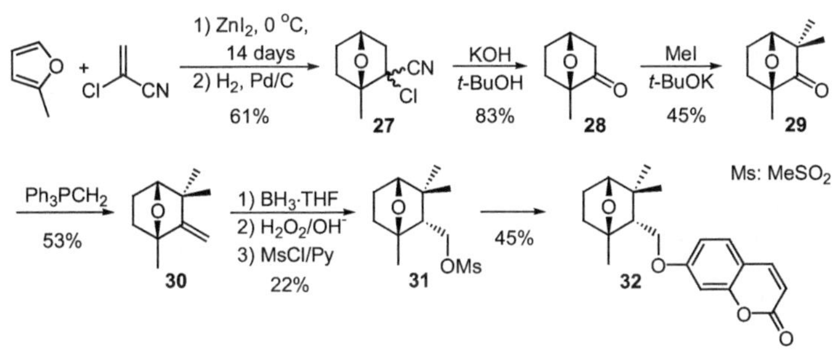

Scheme 3 Synthesis of the 2-epimer of racemic 3′,6′-epoxyauraptene

reaction of 2-methylfurane (sylvan) with 2-chloroacrylonitrile catalyzed by ZnI_2 gives a mixture of 7-oxanorbornene cycloadducts that are hydrogenated into 7-oxanorbornanes **27**. Their alkaline hydrolysis (KOH/*t*-BuOH/H_2O) generates ketone **28**. Double α-dimethylation with MeI and *t*-BuOK produces **29**, the Wittig methylenation of which gives alkene **30**. Hydroboration of **30** followed by oxidative work-up and subsequent transformation to a mesylate furnishes **31** that is then reacted with 7-hydroxycoumarin to yield racemic *endo*-3′6′-epoxyauraptene, *rac*-**32**.

3.3 Sesquiterpenoid 7-Oxabicyclo[2.2.1]heptanes

Creticacumarin **34**, an oxidized form of farnesiferol-C **33** [87, 88], has been isolated from Turkish species of the genus *Anthemis* [89]. Sesquiterpenol **35** has been found in *Artemisia barrelieri* [90] (Fig. 5).

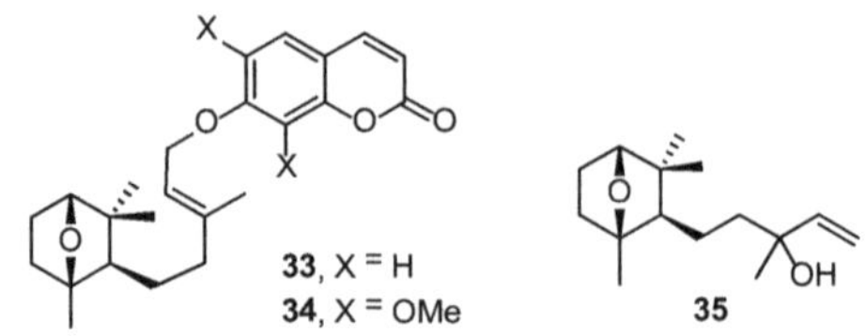

Fig. 5 Sesquiterpenoid 7-oxanorbornanes from plants

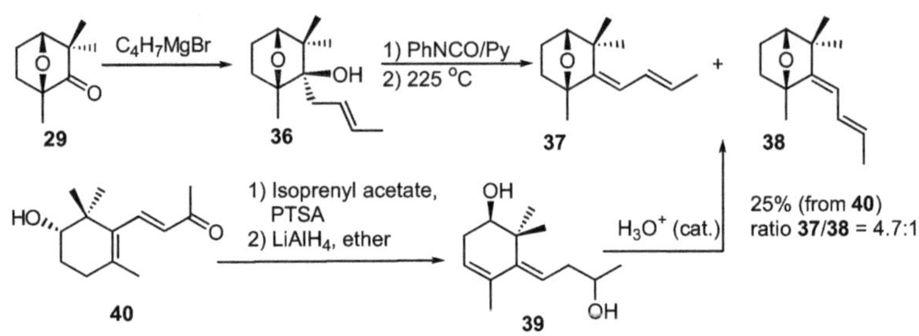

Scheme 4 Synthesis of *rac*-2,5-epoxy-6(*E*,*Z*),8(*E*)-megastigmadiene

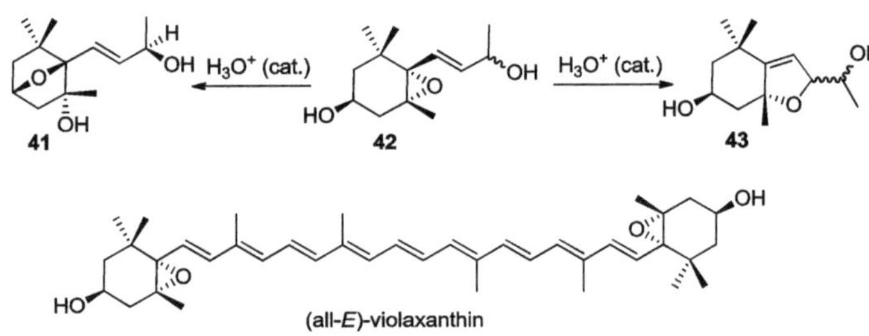

Scheme 5 Acid-induced isomerization of a 3,4-epoxycyclohexanol into a 7-oxanorbornanol

Bicyclic ketone **29** has been converted into *rac*-2,5-epoxy-6(*E*),8(*E*)-megastigmadiene **37** in three steps (Scheme 4) [86]. Compound **37** and its (6*Z*),(8*E*)-stereoisomer **38** have been found (0.02 %) in the extract of *Osmanthus* [91]. They can be prepared also by acidic treatment of diol **39**, which can be obtained from alcohol **40** [92].

The 7-oxanorbornanol derivative **41** has been isolated from sun-cured Greek tobacco [93, 94]. It is formed together with **43** upon acidic treatment of allylic epoxide **42**, a product of degradation of violaxanthin (Scheme 5) [95].

A synthesis of *rac*-farnesiferol-C, *rac*-**33**, is presented in Scheme 6. Baeyer–Villiger oxidation of ketone **44** resulted in the formation of the 7-oxabicyclo[2.2.1]heptane **46**, not in the expected hydroxyl lactone **45**. Standard Barbier–Wieland degradation of **46** (CH_2N_2, then PhMgBr followed by acidic

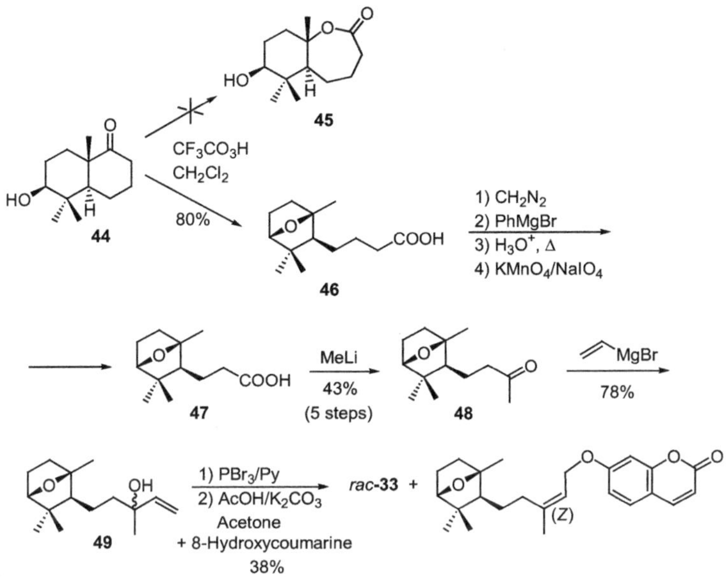

Scheme 6 Synthesis of *rac*-farnesiferol-C

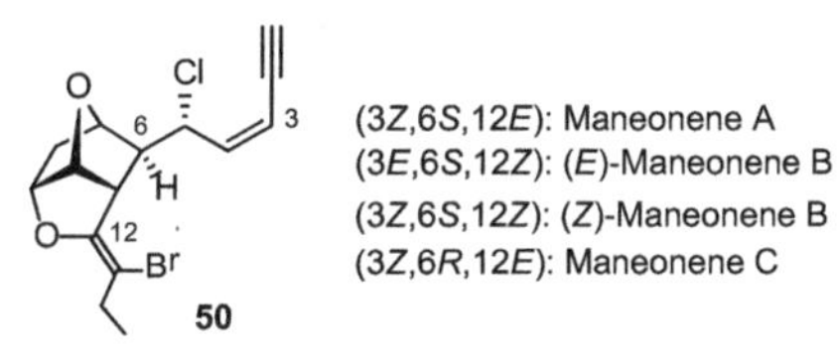

Fig. 6 Maneonenes

workup, and final oxidation with $KMnO_4/NaIO_4$) gives the lower homologue **47** which reacts with two equivalents of MeLi to produce methyl ketone **48**. Addition of vinyl MgBr to this ketone provides a mixture of stereoisomeric allylic alcohols **49** that are converted into the corresponding bromides upon treatment with PBr_3 and pyridine. Their displacement with 7-hydroxycoumarin furnishes a 3:1 mixture of *rac*-**33** and its (*Z*)-stereoisomer [96].

The maneonenes **50** are tricyclic diethers found in the marine alga *Laurencia nidifica* [97–99] (Fig. 6). Synthetic analogues like **54** have been prepared by Renaud and Vionnet through radical addition to ketene acetal **51** (Scheme 7) [100]. This generates radical **52** which isomerizes into 7-oxa-2-norbornyl radical **53**. Allylic quenching of radical **53** with allyltributyltin produces tricyclic compound **54**.

Vetiver oil contains small amounts of ether **55** [101]. The Mediterranean marine alga *Laurencia obtusa* has yielded 7-oxanorbornane **56** [102] and 7-oxanorbornene **57** [103]. The terpenoid 1,4-epoxy-6-eudesmanol has been isolated from *Sideris*

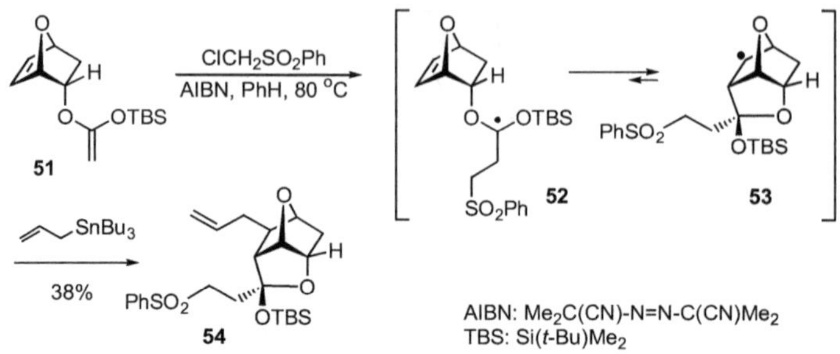

Scheme 7 Renaud and Vionnet's three-component synthesis of a polysubstituted 7-oxanorbornane derivative

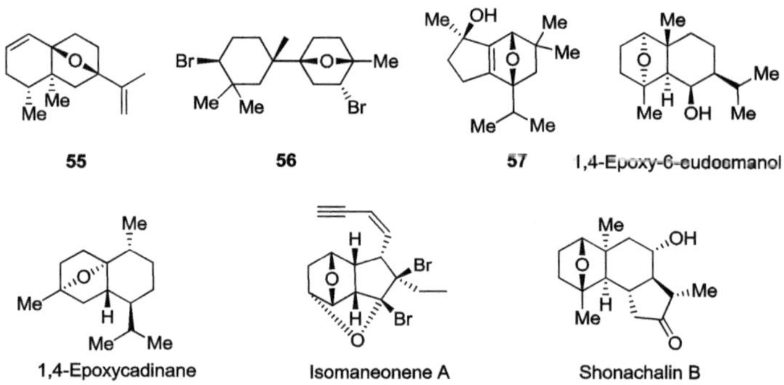

Fig. 7 Examples of natural polycyclic 7-oxanornornane derivatives

varoi [104] and from *Ambrosia artemisioides* [105]. 1,4-Epoxycadinane is a constituent of *Dilophus fasciola* [106, 107]. Isomaneonene A has been found in extracts of *Laurencia nidifica* [97]. Shonachalin B is an example of eudesmanolide isolated from the aerial parts of *Artemisia caerulescens* [108] (Fig. 7).

3.4 *Diterpenoid 7-Oxabicyclo[2.2.1]heptanes and Prostaglandin Analogues*

The bicyclic diterpene dactylomelol has been isolated from the shell-less mollusc *Aplysia dactylomela* [109]. Teupestalin B is a constituent of *Teucrium pestalozzae* [110]. 1,4-Epoxy-13-dolastene (Fig. 8) has been isolated from a *Dictyota* sp. brown alga [111].

The adversary relationship between prostacyclin (PGI_2) and thromboxane-A_2 (TXA_2), which modulates coronary blood vessel caliber [112] and platelet aggregation [113], presents opportunities for therapeutic intervention in cardiovascular diseases. Substances that inhibit TXA_2 synthetase or interfere at the TXA_2 receptor

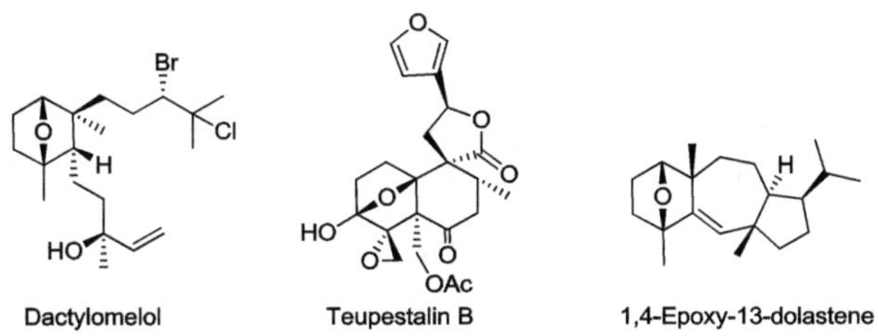

Fig. 8 Diterpenoid 7-oxanorbornanes

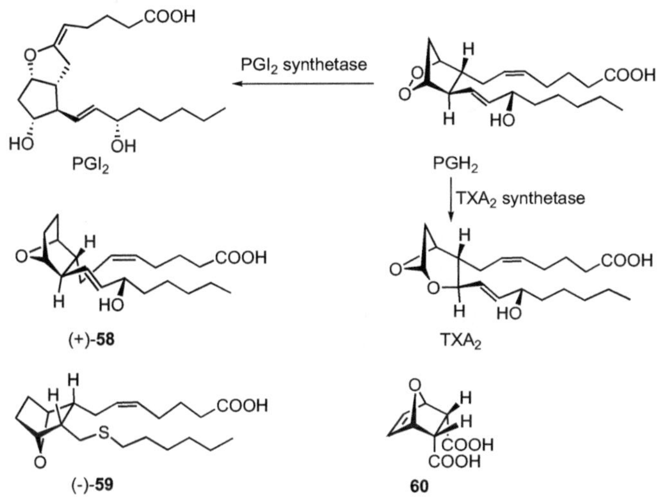

Scheme 8 Some TXA_2/PGH_2 agonists

have been sought because they are expected to normalize pathological events caused by over-biosynthesis of TXA_2 [114–117]. For instance, Hall and coworkers have shown that the 7-oxanorbornane derivative (+)-**58** is a TXA_2/PGH_2 agonist [118]. It was found also that (−)-**59** is a potent ligand for the PGH_2/TXA_2 receptor ($K_d = 1.6 \pm 0.4$ nM). All seven other stereoisomers are not active [119]. Compound (+)-**58** has been prepared enantiomerically pure (Scheme 7) starting from the Diels–Alder cycloadduct **60** of furan and maleic acid obtained at 20 °C [120]. The Alder *endo* rule is followed in this case [121] (Scheme 8).

Reduction of the mixed anhydride obtained by the reaction of **60** with Ac_2O in the presence of pyridine, with $NaBH_4$, generates the racemic lactone **61**. Its reduction with diisobutylaluminum hydride (DIBAL-H) produces lactol **62**, the optical resolution of which is realized through esters derived from (+)-ketopinic acid. The ring opening of lactol (+)-**62** with dimethylhydrazine, followed by acetylation of the primary alcohol so-obtained and subsequent hydrogenation of the 7-oxanorbornene moiety, provides **63**. Cu(II)-promoted hydrolysis of the hydrazine liberates an aldehyde that reacts (Wittig–Horner–Emmons) with $(MeO)_2P(O)$

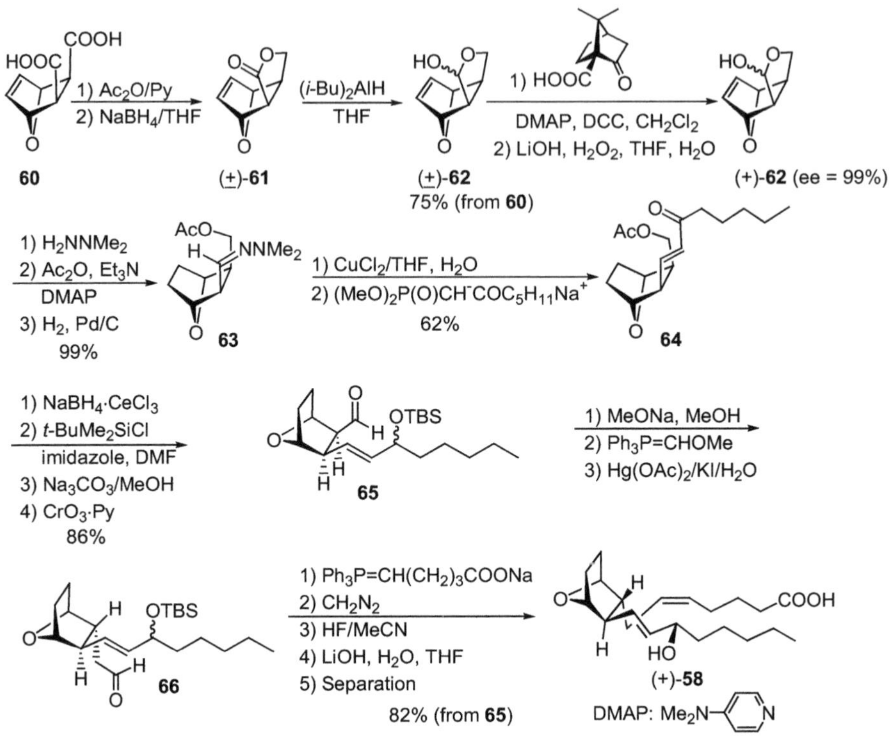

Scheme 9 Synthesis of a thromboxane mimetic incorporating a 7-oxabicyclo[2.2.1]heptane moiety

$CHCOCOC_5H_{11}Na$ furnishing alkene **64**. Under Luche's conditions, **64** is reduced into a mixture of allylic alcohols that are then converted into their silyl ethers. Methanolysis of the acetate liberates an alcohol that is oxidized (Collins) into aldehyde **65**. On treatment with MeONa in MeOH **65** epimerizes its *endo*-carbaldehyde moiety into the more stable *exo*-carbaldehyde. Subsequent homologation by Wittig olefination with Ph_3P=CH–OMe and subsequent alkenyl methyl ether hydrolysis provides **66**. Another Wittig olefination of **66**, followed by esterification with CH_2N_2, then desilylation, saponification, and chromatographic separation, produces (+)-**58** [122] (Scheme 9). Compound (−)-**59** has been derived from *exo*-anhydride **5** in a similar way.

The prostaglandin analogue (+)-**67** was also derived from **5**. It is a potent inhibitor of fatty acid cyclooxygenase [118], the enzyme catalyzing the formation of PGH_2 from arachidonic acid. Among the numerous 7-oxanorbornane-like prostaglandin analogues made, (+)-**68** which incorporates an oxazole carboxamide moiety was found to be potent, selective, and orally active TXA_2 antagonist with a long duration of action. In human platelet-rich plasma, (+)-**68** inhibits arachidonic acid-induced aggregation with an IC_{50} value of 7 nM [123–128] (for further analogues incorporating an *exo*-5,6-epoxy-7-oxanorbornane moiety, see [129]) (Fig. 9).

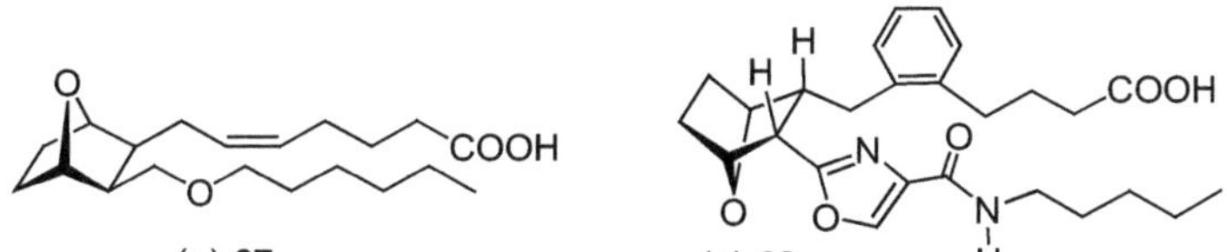

(+)-**67** (+)-**68**

Fig. 9 Bioactive prostaglandin analogues

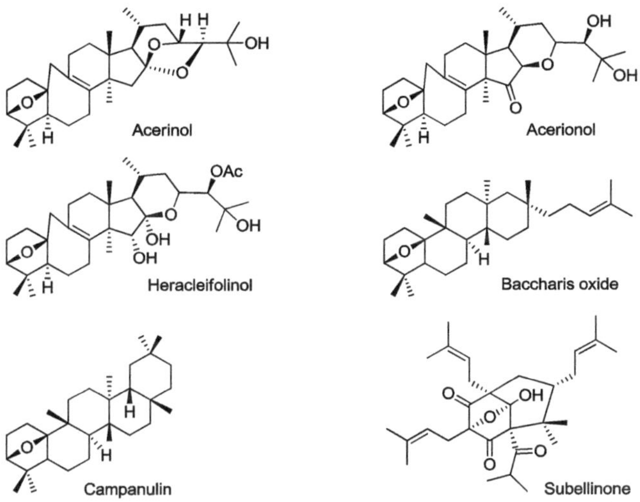

Fig. 10 Examples of natural triterpenoid 7-oxanorbornanes

3.5 *Triterpenoid 7-Oxabicyclo[2.2.1]heptanes*

Acerinol and acerionol are constituents of *Cimicifuga* sp. [130, 131]. Heracleifolinol has been isolated from *Cimicifuga heracleifolia* [132]. Baccharis oxide is a constituent of *Baccharis halimifolia* [133]. Campanulin, a 3,10-epoxyglutinane, is a constituent of *Rhododendron* sp. and *Dendropanax bifidus* [134]. Subellinone, a polyisoprenylated phloroglucinol derivative, has been isolated from the wood of *Garcinia subelliptica*, a biologically active plant growing in the Yaeyama Islands [135] (Fig. 10).

Glycinoeclepin A (Scheme 10) was isolated from the dried root of the kidney bean [136, 137]. It is a potent hatch-stimulating agent of the soybean cyst nematode, a devastating pest of host plants including soybean, kidney bean, and adzuki [138]. Four total syntheses of these compounds have been presented. In the first approach of Murai et al. [139, 140] (Scheme 10), the key synthetic intermediate is the 7-oxanorbornane derivative **73** derived from 2,2-dimethylcyclohexa-1,3-dione via baker's yeast reduction into aldol **69**. After protection as an ethoxyethyl ether,

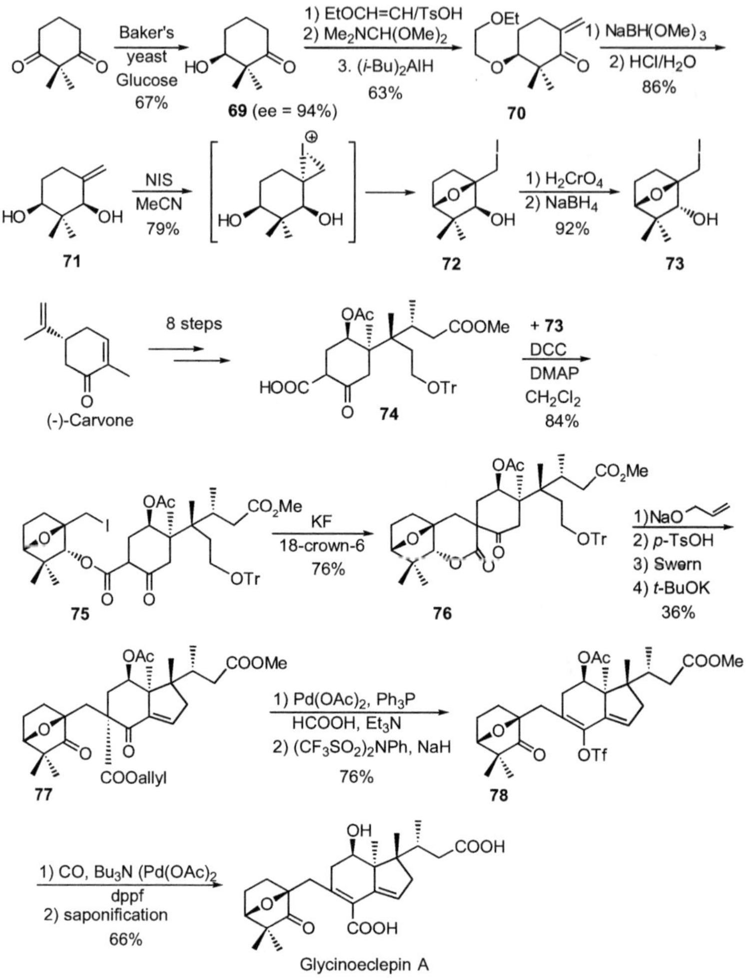

Scheme 10 Murai et al. asymmetric total synthesis of glycinoeclepin A

subsequent treatment with Bredereck's reagent [141] ($Me_2NCH(OMe)_2$) and with $(i\text{-}Bu)_2AlH$ enone **70** is obtained. Its 1,2-reduction and ethoxyethyl ether acidic hydrolysis delivers diol **71**. It is converted into 1-(iodomethyl)-3,3-dimethyl-7-oxabicyclo[2.2.1]heptan-2-*exo*-ol, **72**, by iodoetherification using *N*-iodosuccinimide (NIS) in acetonitrile. Epimerization of the *exo*-alcohol **72** into its *endo*-epimer **73** requires Jones' oxidation into the corresponding ketone and subsequent reduction with $NaBH_4$ (eight steps, 27.6 % yield). Carboxylic acid **74** was derived from (−)-carvone in several steps. In the presence of DCC and a base as catalyst, esterification of **73** by **74** furnishes **75**. Treatment of **75** by KF and 18-crown-6 ether gives spiro compound **76**, a lactone that is ring opened by reaction with sodium allyl alcoholate. This generates an oxanornornanol. The side chain of

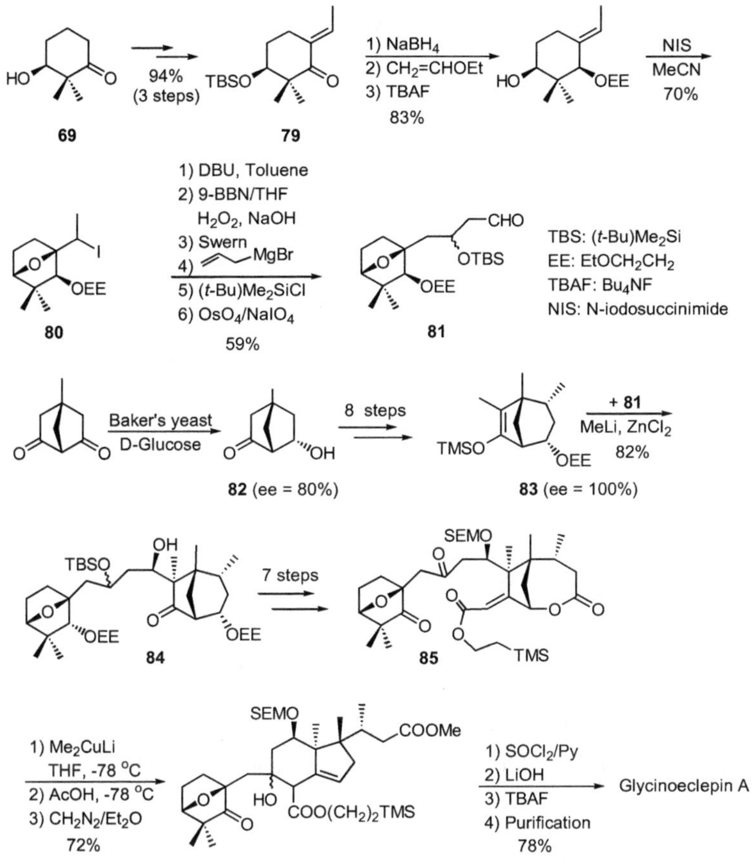

Scheme 11 Mori and Watanabe approach

this intermediate contains a trityl ether moiety. Its acidic hydrolysis generates a diol that is oxidized under Swern's conditions into the corresponding diketo-aldehyde. The latter undergoes intramolecular crotonalization with the cyclohexanone moiety providing the bicyclo[4.3.0]hept-6-ene system **77**. Palladium-catalyzed hydrolysis of the allyl ester of **77** and subsequent decarboxylation gives a cyclohexanone intermediate that is converted into enol triflate **78**. The Pd-catalyzed carbonylation of **78** and final saponification provides glycinoeclepin A.

The second total synthesis (Scheme 11) of glycinoeclepin A was reported by Mori and Watanabe [142, 143]. The 7-oxanorbornane unit **80** is also derived from 2,2-dimethylcyclohexa-1,3-dione. Following the method of Murai et al. (Scheme 10), aldol **69** is converted into enone **79** and then into 7-oxanorbornanone **80**. Further transformations convert **80** into the key intermediate **81**. The other key intermediate **83** is derived from 4-methylbicyclo[2.2.1]hepta-2,6-dione. The enantioselectivity is introduced by yeast-catalyzed reduction of this diketone into aldol **82**. Several steps convert **82** into the silyl enol ether **83** that is

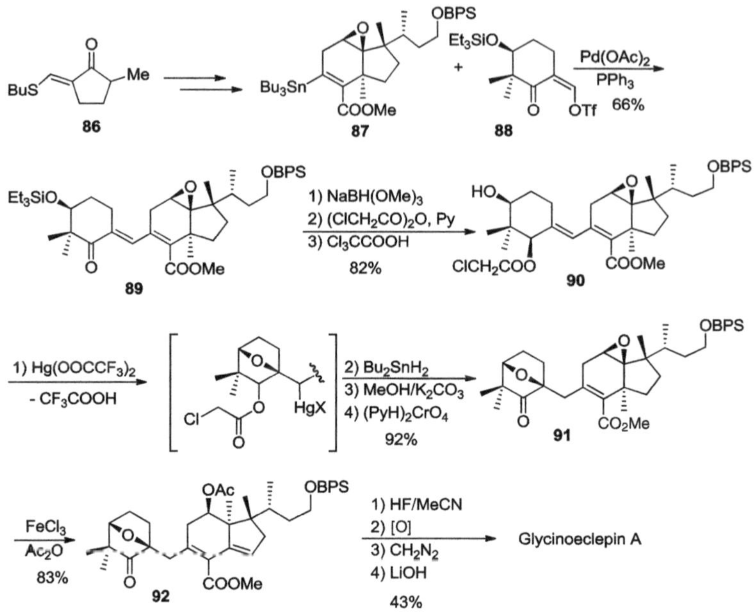

Scheme 12 Corey and Hong asymmetric total synthesis of glycinoeclepin A

condensed with aldehyde **81** through a Mukaiyama cross-aldol reaction giving **84**. Several steps then convert aldol **84** into lactone **85** and finally into glycinoeclepin A (Scheme 11).

Corey and Houpis [144] have reported a relatively short synthesis of glycinoeclepin A starting from cyclopentanone **86** (Scheme 12). Alkenyltin compound **87** was prepared from **86** in several steps. The Stille cross-coupling of **87** with alkenyl triflate **88** generates dienone **89**, the 1,2-hydride reduction of which followed by esterification as trichloroacetate furnishes **90**. The formation of the 7-oxanorbornane moiety is induced by mercuric trifluoroacetate that gives an intermediate mercurial product that undergoes demercuration with dibutyltin dihydride. Subsequent deprotection of the trichloroacetate liberates a 7-oxanorbornanol derivative that is oxidized into ketone **91**. Epoxide ring opening of **91** is induced by $FeCl_3$ in Ac_2O. This generates a tertiary carbenium ion intermediate that undergoes a 1,2-methyl shift and a proton elimination with formation of the acetate **92**. After deprotection, the primary alcohol is oxidized into the corresponding carboxylic acid that is esterified with CH_2N_2. Saponification with LiOH in aqueous dimethoxyethane provides glycinoeclepin A. Corey and Hong [145] have prepared the dimethyl ester of 12-deoxyglycinoeclepin A starting from cycloartenol.

In the most recent approach proposed by Tanino et al. [146], the C–C cross-coupling between the 7-oxanorbornane and bicyclo[4.3.0]heptane moieties is realized by alkylation of a 3,3-trimethyl-2-oxo-7-oxabicyclo[2.2.1]hept-1-yl anion

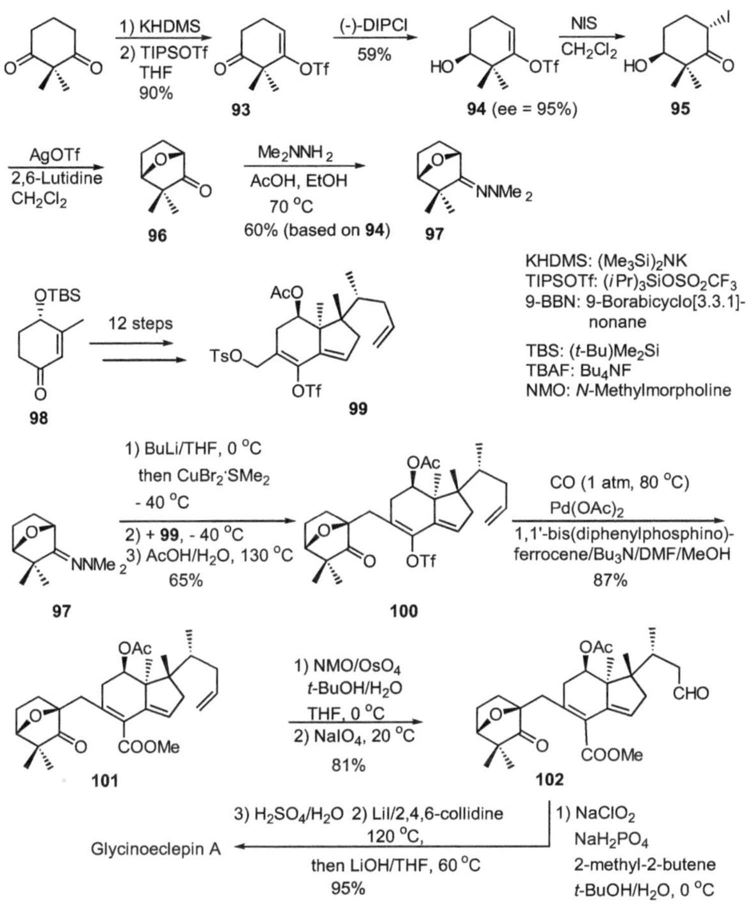

Scheme 13 Tanino et al. asymmetric total synthesis of glycinoeclepin A

equivalent (Scheme 13). The latter was derived from 2,2-dimethylcyclohexa-1,3-dione which is enolized into the corresponding triethylsilyl monoenol ether **93**. Asymmetric reduction of ketone **93** applies the Brown's protocol [147]. Thus, reaction of **93** with (−)-B-chlorodiisopinocampheylborane ((−)-DIPC) gives alcohol **94** with 95 % ee. Its iodination with *N*-iodosuccinimide (NIS) and intramolecular cyclization of the iodoalcohol intermediate **95** mediated by silver triflate generates 7-oxanorbornanone **96**. Its dimethylhydrazone **97** (mixture of geometric stereoisomers) is used for the coupling with allyl tosylate **99**. This key intermediate is constructed in 12 steps starting from cyclohexenone **98**. Hydrogen/metal exchange of 7-oxanorbornanone hydrazone **97** with an excess of butyllithium generates the bridgehead alkyllithium species that is converted into the corresponding cuprate by reaction with $CuBr/SMe_2$. The latter displaces the allylic tosylate **99** giving **100** after hydrolysis of the dimethylhydrazone. The vinyl triflate

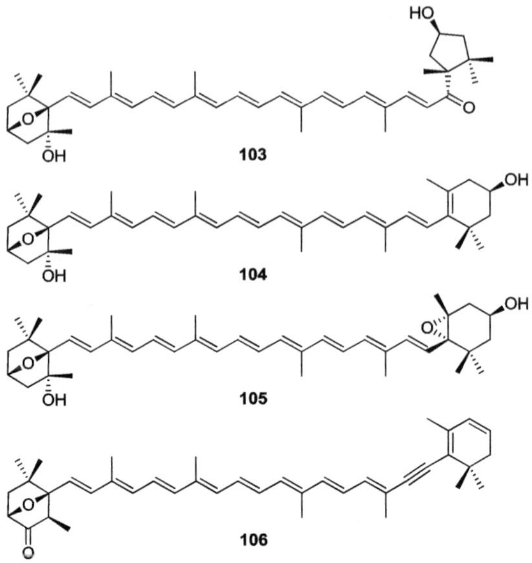

Fig. 11 Examples of carotenoids with terminal 7-oxanorbornane moieties

moiety of **100** is carbonylated in MeOH/DMF/Bu_3N, what produces methyl ester **101**. Chemoselective dihydroxylation of the acyclic terminal alkene unit of **101**, followed by Malaprade oxidation, provides aldehyde **102** that is then oxidized into the corresponding carboxylic acid. Hydrolysis of the acetate and methyl ester moieties finally liberates glycinoeclepin A.

3.6 Carotenoids with 7-Oxabicyclo[2.2.1]heptyl End Groups

Red paprika, *Capsicum annum*, contains a variety of carotenoid pigments among them capsanthin-5,6-epoxide, **103**, and cucurbitaxanthin A, **104**, that contain one 2-*endo*-hydroxy-3-*exo*,6,6-trimethyl-7-oxabicyclo[2.2.1]hept-1-yl end group [148]. Carotenoids **104** and **105** (cucurbitaxanthin B) are found in pumpkin, *Cucurbita maxima* [149]. Oxanorbornanone **106** (eutreptiellanone) has been isolated from the alga *Eutreptiella gymnastica* [150–152] (Fig. 11).

Synthetic analogues have been prepared as exemplified in Scheme 14 [95]. A first Wittig–Horner olefination of aldehyde **107**, followed by acidic treatment, generates cyclohexenone **108**. Chemo- and stereoselective reduction of the latter with 9-BBN followed by *syn*-selective epoxidation of the allylic alcohol (lateral hydroxyl group control) by *m*-chloroperbenzoic acid gives **109**. Selective tosylation of its secondary alcohol moiety (steric hindrance makes the tosylation of the tertiary alcoholic moiety difficult) and subsequent deprotonation of the tertiary alcohol with NaH provide an alcoholate that undergoes an intramolecular displacement reaction,

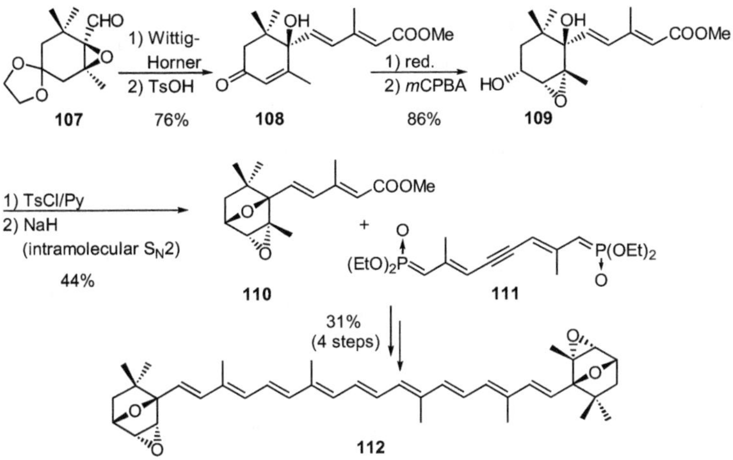

Scheme 14 Synthesis of a 3,6:3′,6′-diepoxy-5,6,5′,6′-tetrahydro-β,β-carotene analogue

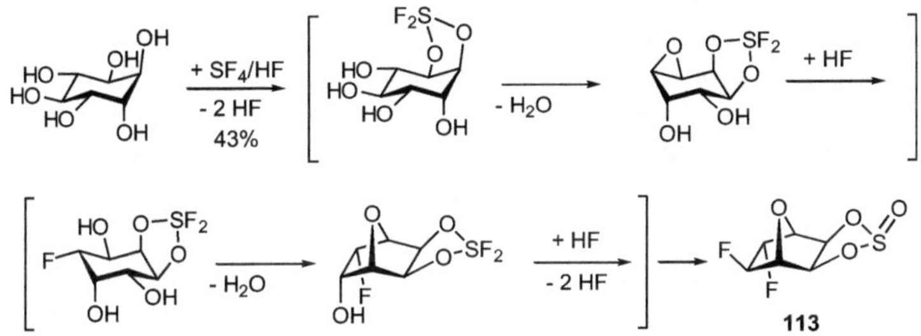

Scheme 15 Synthesis of a 7-oxanorbornane derivative from (+)-*chiro*-inositol

furnishing 7-oxanorbonane derivative **110**. Conversion of the methyl ester of **110** into a carbaldehyde group, the double Wittig–Horner condensation of the latter with diphosphonate **111**, and hydrogenation over Lindlar catalyst produce the carotenoid analogue **112**.

4 Preparation of 7-Oxabicyclo[2.2.1]heptane Derivatives

4.1 Non-Diels–Alder Approaches

With the synthesis of parent 7-oxanorbornane, **1**, (Scheme 1) and of 1,4-cineole, **13**, examples of 7-oxanorbornane synthesis through water elimination from cyclohexa-1,4-diols have been presented. We can add to these reactions the conversion of (+)-*chiro*-inositol with SF_4/HF to give 2α,3β-difluoro- 7-oxabicyclo[2.2.1]heptane-5α,6α-sulfite, **113** [153] (Scheme 15).

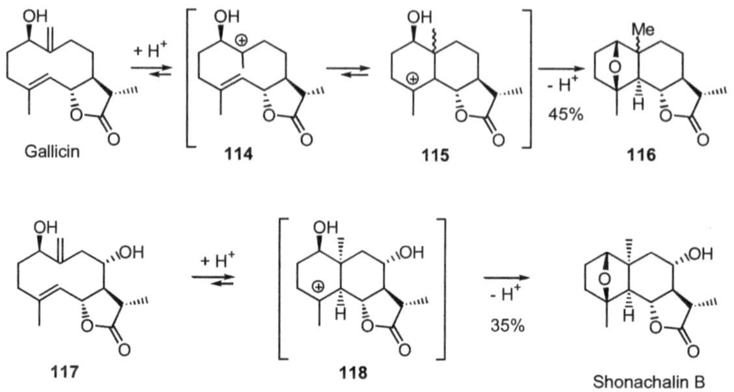

Scheme 16 Protic acid-promoted cyclizations of germacranolides

Related to the water elimination from cyclohexa-1,4-diols, one can cite the intramolecular nucleophilic displacement of the 4-hydroxycylohex1-yl tosylate derived from diol **109** into 7-oxanorbornane **110** (Scheme 14) and the synthesis of *rac*-farnesiferol C that uses an intramolecular displacement of a cyclohexa-1,4-diol-derived lactone (conversion of **45** into oxanorbornane **46**, Scheme 6).

With the total syntheses of (−)-3′,6′-epoxyauraptene, (−)-**23**, Lewis acid-promoted isomerization of alk-3-enyloxiranes into 7-oxanorbornane systems has been illustrated (Scheme 2). The protic acid-promoted isomerization of 7-oxabicyclo[4.1.0]heptan-3-ol into 7-oxanorbornanol derivatives has been illustrated with the conversion of **42** into **41**. The iodoetherification of 4-alkylidenecyclohexanols into 1-(1-iodoalkyl)-7-oxanorbornanes has been exemplified in the total synthesis of glycinoeclepin A by Murai et al. (intermediate **72**, **73**; Scheme 10), by Mori et al. (intermediate **80**; Scheme 11), and by Tanino et al. (**96**, **97**; Scheme 13). Alternatively, the mercurioetherification of 4-alkylidenecyclohexanols can be used to generate 7-oxanorbornane derivatives as shown by Corey (conversion of **90** into **91**, Scheme 12) in his total synthesis of glycinoeclepin A.

We can add to these methods the acid-catalyzed cyclization of the natural germacranolide gallicin that generates, among several products, the 1,4-epoxyeudesmanolide **116** that contains a *trans*-fused decalin moiety (Scheme 13). Under the same conditions, the closely related 8α-hydroxygallicin, **117**, is cyclized into shonachalin B with a *cis*-fused decalin system. These reactions proceed through cationic intermediates **114**, **115**, and **118** [154] (Scheme 16).

Padwa et al. have applied the 1,3-dipolar cycloaddition of carbonyl ylide **120** to cylopent-2-en-1-one to generate oxanorbornanone **121** as a 4:1 mixture of *exo*- and *endo*-isomers. The process starts with the reaction of diazoketone **119** with $Rh_2(OAc)_4$ [155, 156] (Scheme 17).

An intramolecular version of this cycloaddition has been developed by the same authors for the preparation of polycyclic systems **122** [157] and **123** (single diastereomers) [158] (Scheme 18). The latter compound has been converted into

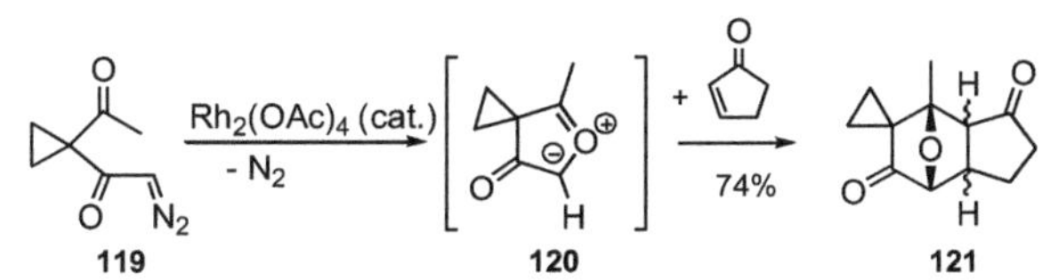

Scheme 17 Synthesis of 7-oxanorobornanes through 1,3-dicarbonyl cycloaddition of carbonyl ylides

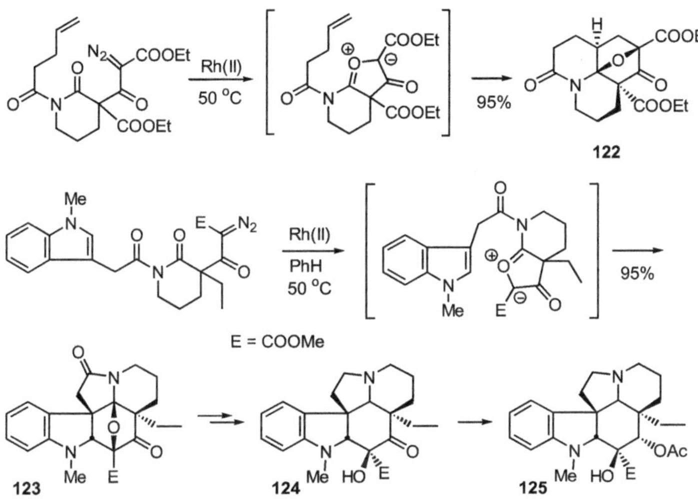

Scheme 18 Padwa et al.'s preparation of polycyclic systems containing 7-oxabicyclo[2.2.1] heptane units

vindoline analogues **124** and **125**, what realizes an efficient synthesis of aspidosperma alkaloids.

Warrener et al. have shown that cyclobutene epoxides **126** undergo C–C ring opening into cyclic carbonyl ylides **127** that can be quenched by alkenes in 1,3-dipolar cycloadditions, generating 7-oxanorbornane derivatives **128** (Scheme 19) [159, 160]. The method has been applied to construct molecular devices such as **131** by double cycloadditions of norbornene derivatives **129** to the bis-epoxide **130**.

4.2 *Diels–Alder Reaction of Furans*

The shortest and most general method to prepare 7-oxabicyclo[2.2.1]heptanes is the Diels–Alder reaction of furans with alkene and alkyne dienophiles that generate 7-oxabicyclo[2.2.1]hept-2-enes and 7-oxabicyclo[2.2.1]hepta-2,5-dienes, respectively

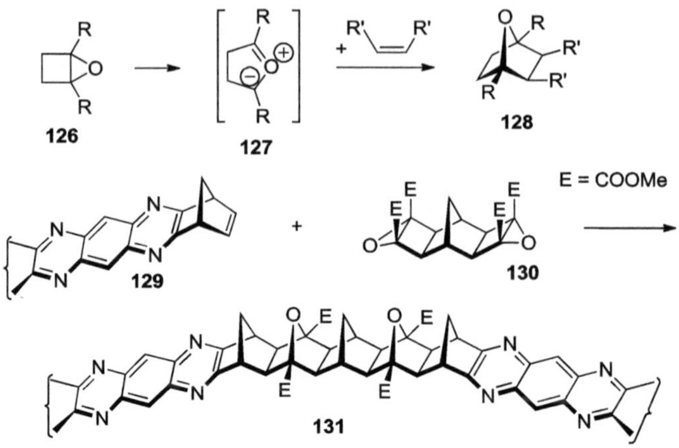

Scheme 19 Warrener et al.'s synthesis of 7-oxabicyclo[2.2.1]heptane-derived molecular devices

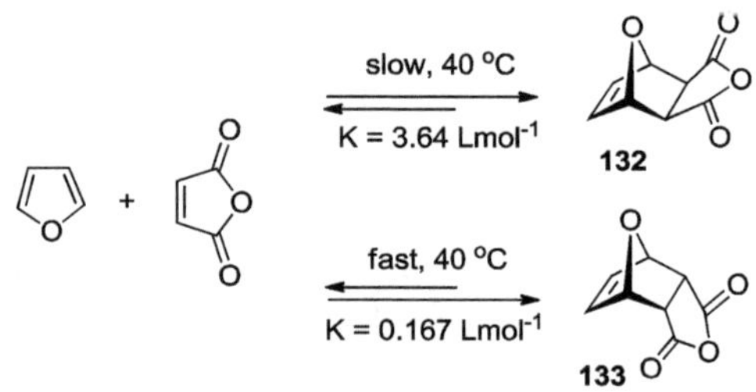

Scheme 20 Diels–Alder reaction of furan with maleic anhydride: kinetic (*endo* Alder rule) and thermodynamic control

[11, 161, 162]. The reaction of furans with ethyl 2,2-difluoro-1-diethylamino-carboxyacrylate generates fluorinated 7-oxanorbornenes [163]. These cycloadducts undergo alkene catalytic hydrogenation (or other reactions) into the corresponding 7-oxabicyclo[2.2.1]heptanes. Examples have been presented with the synthesis of norcantharidin, **5**, with the synthesis of the 2-epimer of *rac*-3′,6′-epoxyauraptene, **32** (Scheme 3), and with the synthesis of thromboxane mimetics (Scheme 8). In 1929, Diels and Alder first reported the reaction of furan with maleic anhydride that produces at room temperature the *exo*-adduct **132** [164]. In 1962, Anet found that at low temperature, the reaction produces first the *endo*-adduct **133** in agreement with the *endo* Alder rule [165]. At 25 °C and in MeCN, the *exo*-adduct **132** is more stable by 1.9 kcal/mol compared with **133** [166] (Scheme 20).

Berson and Swidler have shown that the Diels–Alder reaction of furan with maleic acid in water gives first the *endo*-adduct **60** that reacts with bromine to give

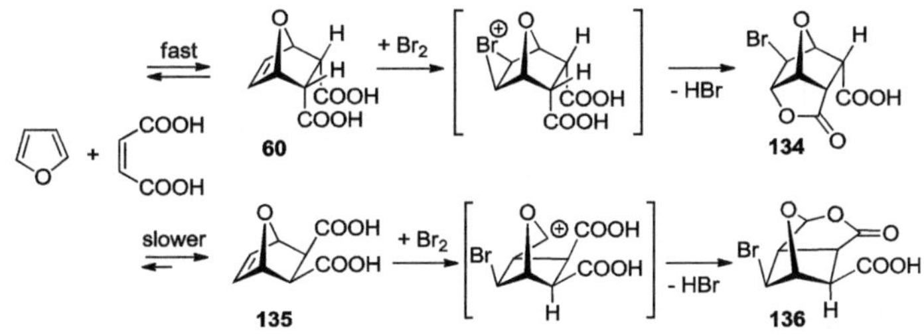

Scheme 21 Diels–Alder reaction of furan with maleic acid: kinetic (*endo* Alder rule) and thermodynamic control

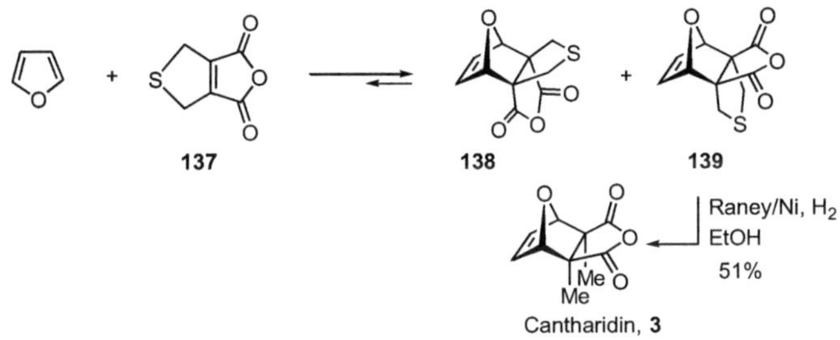

Scheme 22 Dauben's synthesis of cantharidin

the 7-oxanorbornane derivative **134** [167]. Adduct **60** gradually equilibrates with the more stable isomeric *exo*-cycloadduct **135**. Its reaction with bromine produces the rearranged (pinacolic rearrangement) product **136** (Scheme 21).

In an autoclave at 428 K, furan and ethylene produce 7-oxabicyclo[2.2.1]hep-2-ene in 5–8 % yield [168]. A minimal equilibrium constant of 0.02 L mol^{-1} was evaluated for this equilibrium at 428 K. Because of the aromaticity of furans (ca. −14 kcal/mol), 7-oxabicyclo[2.2.1]hept-2-enes undergo retro-Diels–Alder reaction on heating. Reluctant Diels–Alder reactions of furans can be accelerated and displaced in favor of the cycloadducts by applying very high pressures (5–20 kbar) [169].

In 1980 Dauben et al. developed a two-step synthesis of cantharidin, **3**, that involves the Diels–Alder reaction of furan with 2,5-dihydrothiophene-3,4-dicarboxylic anhydride, **137**, at 20 °C under 7 kbar of pressure [170, 171]. This leads to a 1:4 mixture of cycloadducts **138** and **139**. After desulfurization and alkene hydrogenation, a mixture of **3** and *epi*-cantharidin was obtained from which pure **3** could be isolated in 51 % yield after selective crystallization and recrystallization from EtOAc. More recently, using Griego's medium (5 M $LiClO_4$ in Et_2O), Dauben et al. found that the addition of furan to **137** could be carried out at 20 °C under one atmosphere. In this medium the equilibrium constant ($K = 3$ L mol^{-1}) is about 300 times larger than in pure furan [172] (Scheme 22).

The double furan **140** [173] [obtained in one step from 2,4-dimethylfuran (prepared in three steps from acetone) by reaction with acetaldehyde] adds to

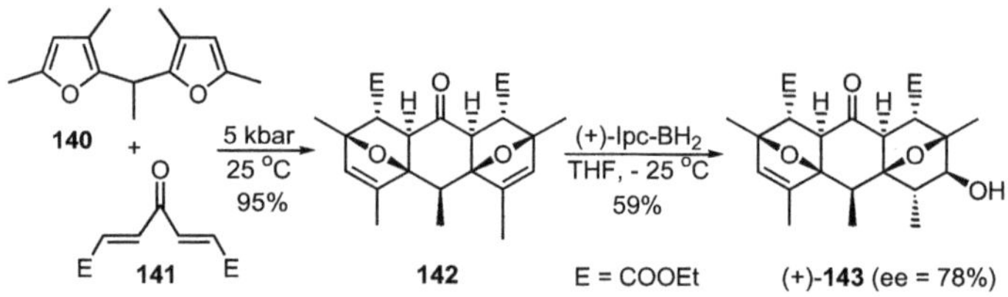

Scheme 23 Eleven stereogenic centers in two steps from two symmetric compounds

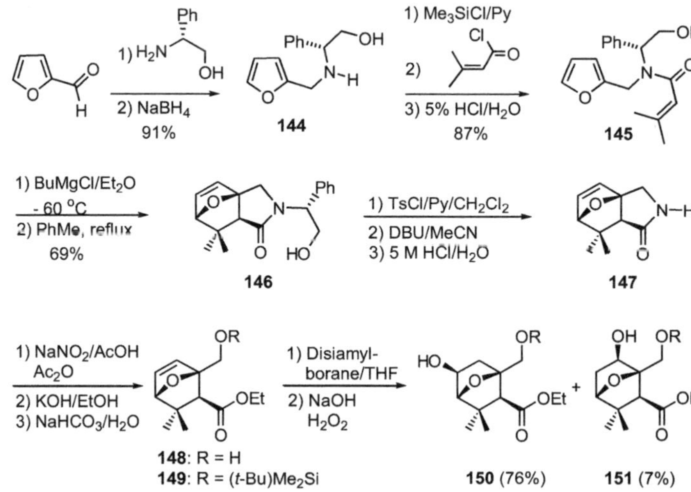

Scheme 24 Asymmetric intramolecular Diels–Alder reaction and regioselective hydroboration of a 7-oxabicyclo[2.2.1]hept-5-ene derivative

diethyl (*E*,*E*)-4-oxohepta-2,5-diene-1,7-dioate, **141**, at 25 °C under 5 kbar, giving a single cycloadduct **142** in high yield. Using one equivalent of monoisopinocampheylborane ((+)-IpcBH_2), desymmetrization of **142** was realized by furnishing alcohol (+)-**143**. Thus, in only two synthetic steps, the two planar compounds **140** and **141** have been converted into a polycyclic system (+)-**143** made of a 7-oxanorbornanol and a 7-oxanorbornene moiety containing 11 stereogenic centers [174] (see also [175]) (Scheme 23).

Another strategy to force furans to undergo Diels–Alder reactions is to limit the cost of the entropy of condensation of intermolecular cycloadditions, making them intramolecular [176–193]. This requires chemical attachment of the dienophile to the furan moiety as illustrated with the following examples.

Condensation of furfural aldehyde with (*R*)-phenylglycinol gives an imine which is reduced with $NaBH_4$ in isopropanol into amine **144** (Scheme 24). Temporary silylation of the primary alcoholic group of **144** as a trimethylsilyl ether and subsequent acetylation of the amine moiety with 3,3-dimethylacryloyl chloride

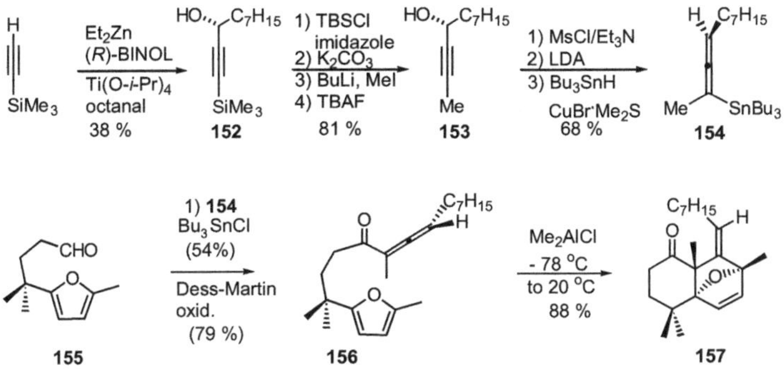

Scheme 25 Chirality transfer in the intramolecular Diels–Alder reaction of an optically active allenic ketone

provides **145** after acidic aqueous work-up. The reaction of **145** with BuMgCl generates the corresponding magnesium alcoholate that is then heated under reflux in toluene furnishing the cycloadduct **146** in 69 % yield [194]. Removal of the chiral auxiliary implies first the formation of an enamine, realized by converting the primary alcohol **146** into the corresponding chloride. The latter eliminates HCl on treatment with DBU. The enamine so-obtained is hydrolyzed under acidic conditions and liberates lactam **147**. The lactam is then converted into the corresponding ethyl 4-hydroxybutyrate **148**, which is etherified as a silyl ether **149**. Regioselective hydroboration of the alkene moiety of 7-oxanorbornene **149** uses a bulky diakylborane such as disiamylborane. It is controlled by the bulk of the silyloxymethyl substituent of the bridgehead center. After oxidative workup, alcohols **150** and **151** are isolated in 76 % and 7 % yield, respectively [195].

In the second example of intramolecular Diels–Alder reaction of furan, the dienophile is an enantiomerically enriched allene derivative (ee = 90 %), the chirality of which is fully transferred to the cycloadduct (Scheme 25). Applying the Pu's method [196–198], trimethylsilylacetylene adds enantioselectively to octanal in the presence of $ZnEt_2$, (*R*)-BINOL, and Ti(*i*-OPr)$_4$, giving propargyl alcohol **152**. After the silylation of the alcohol, silicium/lithium exchange to form the corresponding lithium acetylide, methylation of the latter with MeI, and silyl ether desilylation, **153** was obtained and converted into allenyltin compound **154** via mesylation of the alcohol and S_N2' displacement by tributyltin cuprate. Reaction of **154** with aldehyde **155** and subsequent Dess–Martin oxidation of the allenyl alcohol produces ketone **156**. The intramolecular Diels–Alder reaction requires activation of the dienone dienophile with Me_2AlCl, yielding the tricyclic product **157** [199].

The first total asymmetric synthesis of solanoeclepin A has been presented in 2011 by Tanino et al. [200]. It also features a Me_2AlCl-catalyzed intramolecular Diels–Alder reaction as outlined in Scheme 26. Solanoeclepin A stimulates the hatching of potato cyst nematode, a parasite with destructive effect on several crops. One possible fight against this pest would be to induce early hatching of

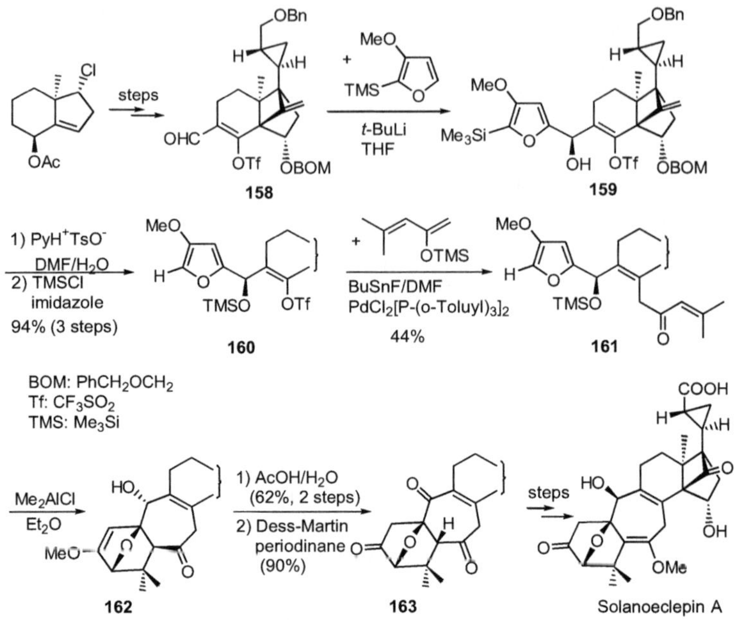

Scheme 26 Total synthesis of solanoeclepin A

the nematode, and this requires disposal of synthetic solanoeclepin A as Nature does not produce this compound in sufficient amount. One of the key steps of the total synthesis is the H/Li exchange of 2-trimethylsilyl-3-methoxyfuran at the C5 position generating an α-furyllithium derivative that adds onto enal **158** to give **159**. After desilylation of the furan moiety and alcohol silylation, **160** is obtained, the vinyl triflate unit present in this compound permits the coupling with the tin enolate derived from 4-methyl-2-trimethylsilyoxypenta-1,3-diene in the presence of a palladium catalyst. The intramolecular Diels–Alder reaction catalyzed by Me_2AlCl generates a single cycloadduct **162**. The alkenyl ether moiety of the latter is hydrolyzed into the corresponding 7-oxanorbornanone; subsequent oxidation of the allylic alcohol provides triketone **163**. Several further steps convert **163** into solanoeclepin A.

5 Reactions of 7-Oxabicylo[2.2.1]heptanes

A large number of synthetic applications of 7-oxanorbornanes have been reported and reviewed [11, 12, 60, 162, 201–209]. Selected examples will be presented here in a non-exhaustive way.

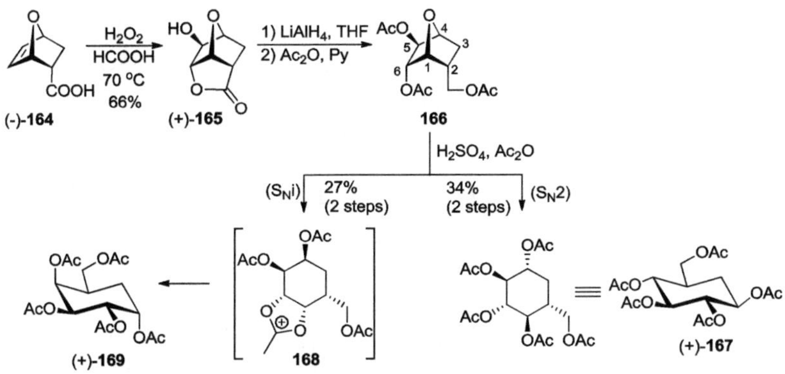

Scheme 27 Ogawa et al.'s synthesis of enantiomerically pure carba-pyranoses

5.1 Acid-Induced Ethereal Bridge Nucleophilic Displacements

At 25 °C aqueous HCl converts the parent 7-oxanorbornane **1** into *trans*-4-chlorocyclohexan-1-ol, and on heating both *trans*- and *cis*-1,4-dichlorocyclohexane are formed [210]. With aqueous HBr, **1** was transformed to *trans*-4-bromocyclohexanol [211]. Oxanorbornane **1** reacts with dinitrogen pentoxide in CH_2Cl_2 giving exclusively *trans*-cyclohexa-1,4-diyl dinitrate [212].

Ogawa et al. [213] have prepared the carba analogues of pyranoses (Scheme 27) by H_2SO_4-induced acetolysis of 7-oxanorbornane **166** derived from the enantiomerically pure 7-oxanorbornene (−)-**164** (ee > 99 %). This compound is obtained by Diels–Alder reaction of furan and acrylic acid. Resolution of the cycloadduct so-obtained uses recrystallization of diastereoisomeric amides derived from (+)-(*R*)-α-methylbenzylamine. Stereoselective epoxidation of (−)-**164** generates an intermediate *exo*-epoxide that undergoes lactonization into (+)-**165** in formic acid at 70 °C. Reduction of the latter with $LiAlH_4$ furnishes a diol that is acetylated into triacetate **166**. In H_2SO_4/Ac_2O **166** undergoes two competitive reactions. In one of them, a direct S_N2 attack of acetic acid onto the least sterically hindered bridgehead center produces β-D-carba-glucopyranose peracetate, (+)-**167**. The other reaction is an S_N1 process with the participation of an *endo* acetoxy group leading to cationic intermediate **168** and finally to the peracetate of α-D-carba-galactopyranose, (+)-**169**.

When **166** is reacted with HBr in AcOH, products of S_N1 acetolysis are not observed; direct S_N2 reactions of bromide anion are highly favored. This generates dibromides that have been converted into pent-*N*,*O*-acetate of (+)-validamine [214]. Ogawa and Takagahi have prepared (+)-pipoxide starting with bromine addition to (−)-**164** and through a multistep synthesis involving HBr reaction of a 7-oxanorbornane intermediate [215].

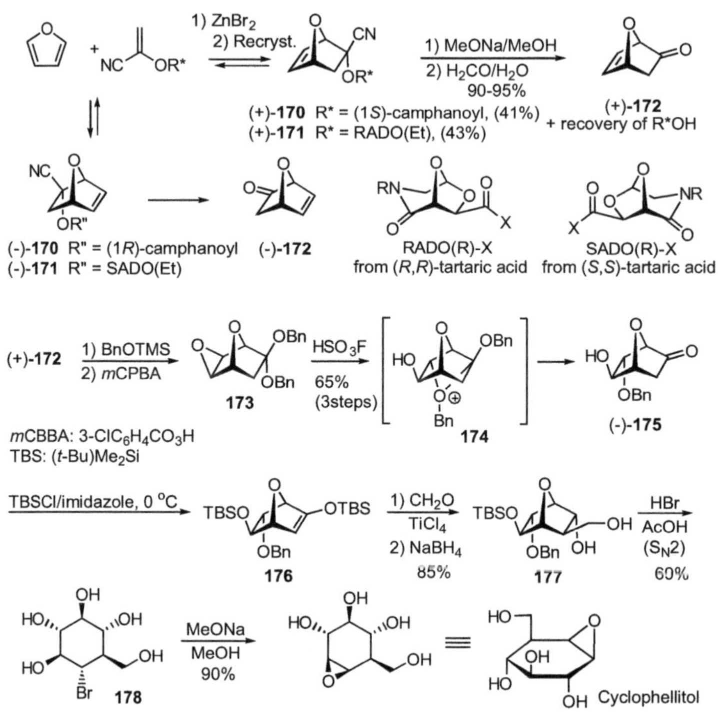

Scheme 28 Total synthesis of cyclophellitol applying the "naked sugar" method, i.e., starting form enantiomerically pure 7-oxabicyclo[2.2.1]hept-5-ene-2-one

A total synthesis of the glucosidase inhibitor cyclophellitol, using the "naked sugar" method [201, 202, 216], is based on the S_N2 ring opening of the oxa bridge of the polysubstituted 7-oxanorbornane [217] (Scheme 28). ZnI_2- or $ZnBr_2$-catalyzed Diels–Alder reaction of furan with (−)-1-cyanovinyl (1*S*)-camphanate [218] or (−)-1-cyanovinyl (1*R*,5*S*,7*R*)-3-ethyl-2-oxo-6,8-dioxa- 3-azabicyclo [3.2.1]octane-7-*exo*-carboxylate [219] (RADO(Et)–O(NC)C=CH_2) derived from (*R*,*R*)-tartaric acid and pyruvonitrile, and recrystallization provides the diastereomerically pure cycloadducts (+)-**170** and (+)-**171**, respectively. The other diastereoisomeric cycloadducts are recycled into (+)-**170** and (+)-**171** via retro-Diels–Alder reactions on heating (reversibility of the furan Diels–Alder reactions). Pure enantiomers (−)-**170** and (−)-**171** are obtained with the same ease using (+)-1-cyanovinyl (1*R*)-camphanate and (+)-cyanovinyl (1*S*,5*R*,7*R*)-3-ethyl-2-oxo-6,8-dioxa-3-azabicyclo[3.2.1]octane-7-exo-carboxylate (SADO(Et)–O–(NC)C=CH_2), respectively. Saponification of (+)-**170** (or (+)-**171**) and treatment with formalin to displace the cyanohydrins furnish enantiomerically pure 7-oxabicyclo[2.2.1] hept-5-en-2-one, (+)-**172**, and allow the recovery of the chiral auxiliaries [(1*S*)-camphanic acid, RADO(Et)-OH]. Similarly, enone (−)-**172** is derived from (−)-**170** and (−)-**171**. The six enantiomerically pure 7-oxanorbornenes (+)- and (−)-**170**, (+)- and (−)-**171**, and (+)- and (−)-**172** are coined "naked sugars" because

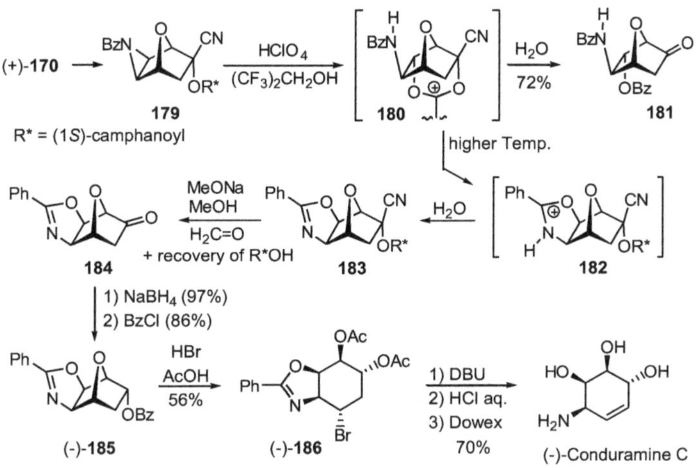

Scheme 29 Asymmetric synthesis of (−)-conduramine C

they are enantiomerically pure like D-glucose, but unlike natural hexoses have three unsubstituted carbon centers that can be substituted in few steps with high regio- and stereoselectivity giving polysubstituted 7-oxanorbornane derivatives. The latter can be converted readily into hexoses [220–224] and analogues, into *C*-glycosides such as *C*-nucleosides [225–227], or into carbahexoses including conduritols [228–230] and aminoconduritols [231–234]. Selected examples are given below. Conversion of enone (+)-**172** into its dibenzyl acetal and subsequent epoxidation provides *exo*-epoxide **173**. Under strongly acidic conditions (HSO_3F), the most strained ethereal moiety, the epoxide, is heterolyzed with the participation of the *endo*-benzyloxy group, leading to the hypothetical intermediate **174** that finally gives **175** with high regio- and stereoselectivity. Noteworthy is the fact that pinacolic rearrangement of the 7-oxanorbonyl system (see below) is not observed under these conditions. After etherification of the *exo*-alcoholic moiety as a silyl ether and conversion of the ketone into enoxysilane **176**, Mukaiyama cross-aldol reaction with formaldehyde generates an *exo*-carbaldehyde that is reduced with $NaBH_4$ into diol **177**. On treatment with HBr in AcOH, **177** is converted into **178**. The process implies acetylation of the diol, exchange of the benzyl and silyl groups for acetates, and displacement of the 7-oxa ethereal bridge by a bromide S_N2 attack at the least sterically hindered bridgehead center. Final treatment with an excess of MeONa in MeOH induces complete methanolysis of the acetates and formation of the epoxide ring of cyclophellitol (for other syntheses of cyclophellitol, see [235]).

Enantiomerically pure (−)-conduramine C has been prepared (in ten steps and 23.3 % overall yield) via S_N2 bromination of 7-oxanorbornane derivative (−)-**185** derived from the "naked sugar" (+)-**170** (Scheme 29). Aziridination of alkene (+)-**170** gives **179**. Subsequent acidic treatment generates a dialkyloxybenzyl cation intermediate **180** that can be quenched with water to give the *trans*-5,6-disubstituted

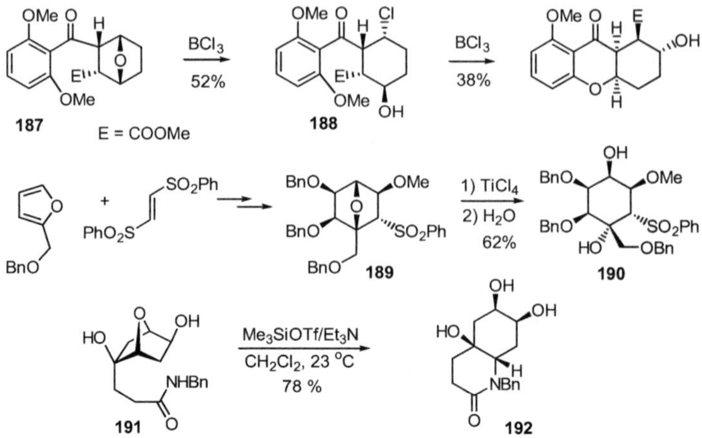

Scheme 30 Heterolytic cleavage of the ethereal bridge of 7-oxanorbornanes

7-oxanorbornanone **181**. In the absence of water and at higher temperature, **180** has the time to be rearranged into the more stable aminoalkyloxybenzyl cation intermediate **182** that reacts with water to give the *cis-exo*-5,6-disubstituted 7-oxanorbornanone **183**. Saponification of camphanate **183** furnishes ketone **184** and (1*S*)-camphanic acid (recovery of the chiral auxiliary). Stereoselective *exo* face reduction of the ketone with $NaBH_4$, followed by benzoylation, provides *endo*-benzoate (−)-**185**. Its reaction with HBr in AcOH produces bromide (−)-**186**. Base-induced elimination of HBr and subsequent hydrolysis of the oxazoline and acetate moieties liberates (−)-conduramine C [236].

Lewis acids can induce heterolytic cleavage of the ethereal bridge of 7-oxanorbornanes as exemplified with the conversion of **187** into **188** [237], and of **189** into **190** [238] that imply intermolecular quenching of the nucleophile, and for the intramolecular displacement of the ethereal bridge of **191** by its *N*-benzylamido moiety to give the bicyclic lactam **192** with high stereoselectivity (S_Ni displacement) [239] (Scheme 30).

5.2 *Base-Induced Ethereal Bridge Opening of 7-Oxabicyclo [2.2.1]heptanes*

When a carbanionic center is engendered β to an ethereal bond, β-elimination follows readily (E_{1cb} type elimination). In the case of 7-oxanorbornane-2-carboxylic esters, of 7-oxanorborn-2-yl alkyl ketones, of 7-oxanorborn-2-yl sulfones, and of 7-oxanorborna-2-ones, their conjugate bases do not undergo quick β-elimination for stereoelectronic reasons (small overlap between the HOMO of the carbanionic center and the LUMO of the adjacent ethereal moiety because of the

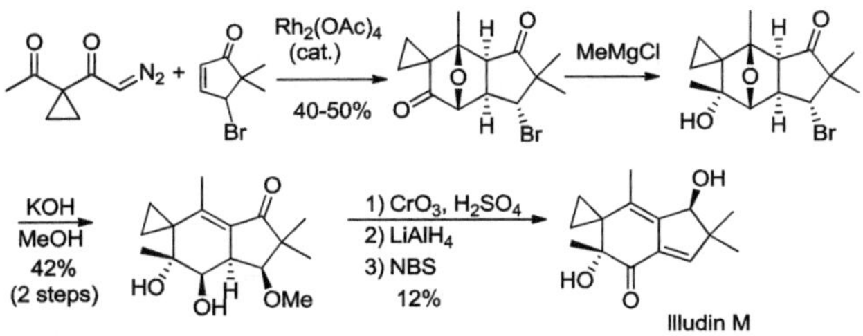

Scheme 31 Synthesis of *rac*-illudin M

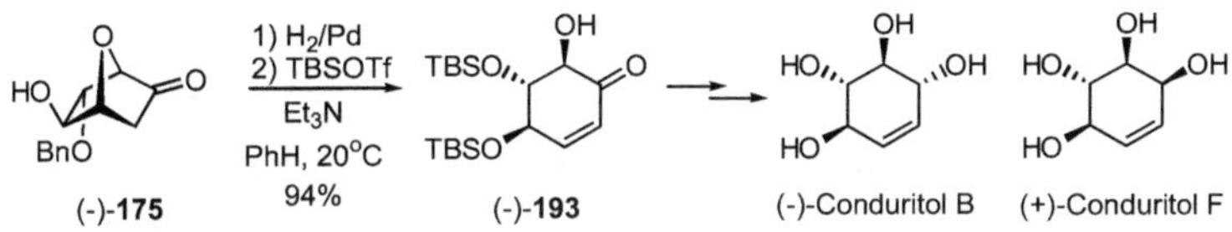

Scheme 32 The conversion of "naked sugars" into conduritols

geometry and rigidity of the bicyclic skeleton). This will be illustrated with the synthesis of an imino-*C*-disaccharide (Scheme 42). However, in the absence of electrophilic reagents and on heating, enolates derived from these 7-oxanorbornane derivatives undergo 7-oxa ring opening with the formation of the corresponding cyclohex-3-en-1-ols [240–247]. This has been exploited in several syntheses of natural products and analogues of biological interest [206, 248–250]. For instance, racemic illudin M (shows in vitro selective toxicity toward tumor cells [251]) has been prepared as outlined in Scheme 31 [249].

We have seen in Scheme 28 that oxanorbornanone (−)-**175** can be converted into the enoxysilane **176** without 7-oxa ring opening when the silylation employs TBSCl and imidazole at 0 °C. The same reaction with TBSOTf (the triflate TBSOTf is a much better oxophilic agent than the chloride TBSCl and thus assist the heterolysis of the ethereal bridge) and Et_3N at 20 °C converts (−)-**175** into cyclohexenone (−)-**193**. This compound is a precursor for the preparation of (−)-conduritol B and of (+)-conduritol F [229, 252, 253] and of cyclitols (Scheme 32).

An example of ethereal bridge opening involving a 7-oxanorborn-2-yl sulfone is given in Scheme 33 [254]. Diels–Alder addition of the difluoroacrylate **194** to furan gives the corresponding cycloadduct **195**. Its alkene moiety at C5–C6 adds to benzenesulfenyl chloride with high *exo* face stereoselectivity and regioselectivity because of the participation of the 2-*endo* carboxylic group that leads to the formation of tricyclic lactone **196**. Reduction of the lactone with $LiAlH_4$ in THF gives a triol that is protected as an acetonide. Then, the 6-*endo*-hydroxy group is converted into the corresponding benzyl ether. Oxidation of the 5-*exo*-phenylthio group with *m*CPBA generates the corresponding sulfone **197**. The treatment of **197** with BuLi in THF at low temperature induces a smooth isomerization into

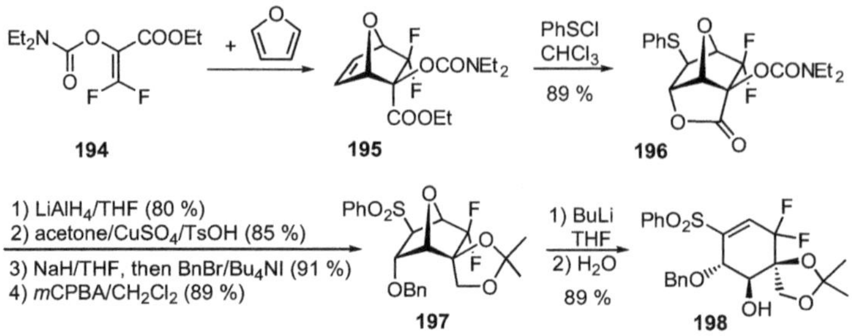

Scheme 33 Selective 7-oxa bridge opening versus β-elimination of a benzyloxy group

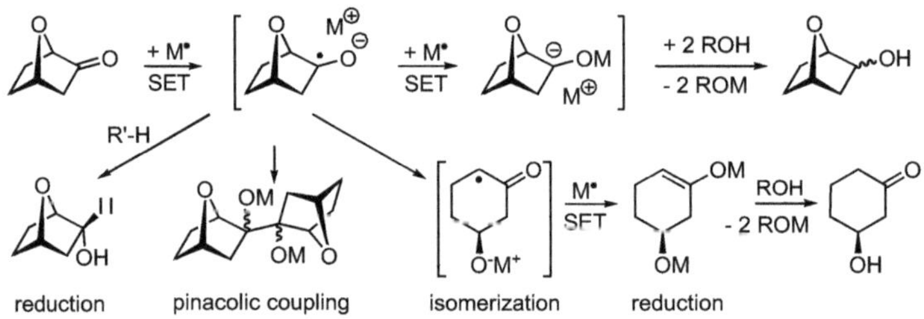

Scheme 34 Principal reactions of ketyl radical anion derived from the single-electron reduction of 7-oxabicyclo[2.2.1]heptan-2-one

cyclohexenol derivative **198**. The reaction is particularly interesting as only the strained 7-oxa bridge undergoes the intramolecular β-elimination, not the 5-*endo*-benzyloxy group, even though the latter elimination is favored entropically.

5.3 *Reductive Ethereal Ring Opening of 7-Oxabicyclo[2.2.1] heptanones*

Single-electron transfer (SET) to 7-oxanorbornanone engenders the corresponding ketyl radical anion that may be reduced further and quenched by the solvent to generate the corresponding *exo*- and *endo*-7-oxanorbornanol. Alternatively, the ketyl radical anion may dimerize into a mixture of *meso*- and *threo*-pinacol or undergo 7-oxa bridge opening producing a radical anion before a second electron is transferred to it to produce finally 3-hydroxycylohexanone (Scheme 34). As in the case of carbanionic species (see above), the 7-oxa ring opening has to overcome a relatively high-energy barrier. Thus, competition between 7-oxa ring opening and ketone reduction, or pinacolic coupling, will depend on the nature of the reducing agent (nature of the counterion M^+), on the substitution of the 7-oxanorbonanone, and on the solvent (radical hydrogen source, protic solvent, etc.).

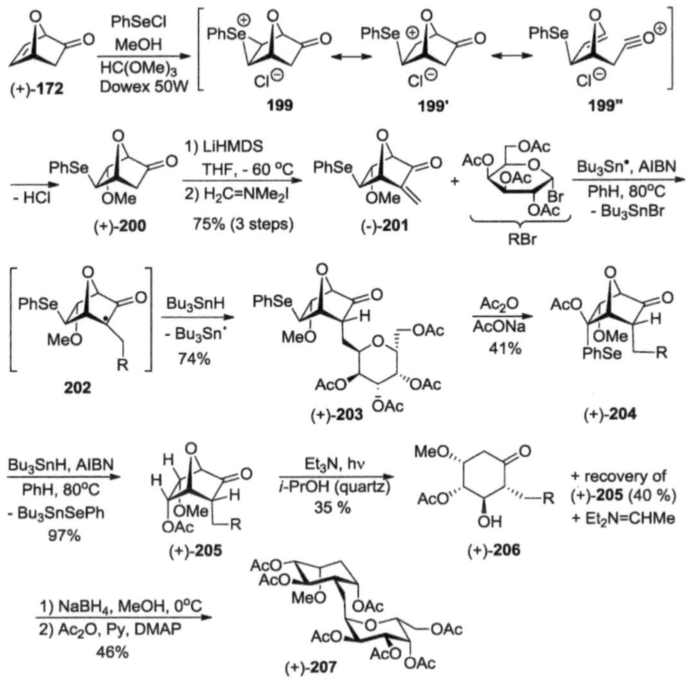

Scheme 35 Synthesis of an α-C-galactopyranoside of a carbapentopyranose

De Clercq et al. [255] and Padwa et al. [155, 156] have used SmI_2 [256] to induce the ethereal bridge opening of 7-oxanorbornanones. When Na/NH_3 is used instead, 7-oxanorbornanols are the main products of reduction. Low-valent titanium salts obtained by mixing $TiCl_4$ with zinc powder in THF lead mostly to products of pinacolic coupling [257]. On their side, Cossy et al. [258] have found that 7-oxanorbornanones are reduced into the corresponding 3-hydroxycyclohexanones by irradiation (low-pressure Hg lamp, quartz vessel) in MeCN in the presence of Et_3N. The triplet excited state of the ketone abstracts an electron from the tertiary amine and forms the corresponding ketyl radical anion that is not tightly bound to the counterion. Indeed, the triethylaminium radical cation is much larger than other counterions such as Na^+, $TiCl_3^+$, or SmI_2^+. This enhances the electron density at the carbonyl group of the ketyl radical anion making it more prone to β-elimination of the ethereal moiety of the 7-oxanorbornanone. The photoinduced electron transfer cannot be done with compounds containing other chromophore than the ketone, but it tolerates polyfunctionality as this will be illustrated below.

The first example of a *C*-glycoside of a carbasugar has been obtained from the "naked sugar (+)-**172** in the way outlined in Scheme 35. Benzeneselenyl chloride adds to enone (+)-**172** with high *exo* face stereoselectivity and high regioselectivity. In the presence of MeOH as nucleophile, adduct (+)-**200** is formed [259]. The reaction involves the generation of selenonium ion **199** which adds the nucleophile stereo-selectively onto its *endo* face at C6, and not at C5, because of the electron-releasing

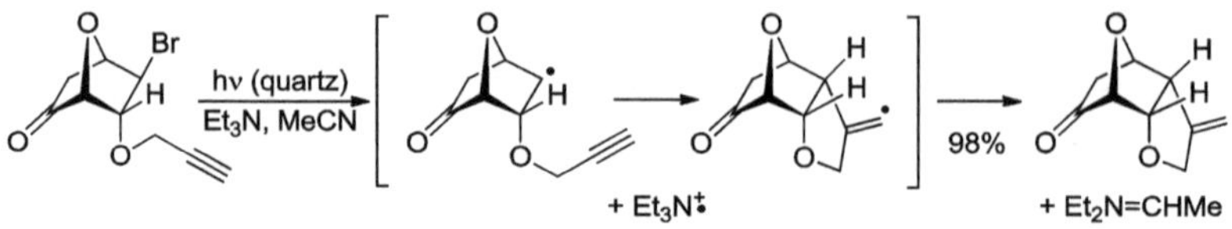

Scheme 36 Cossy et al. photoreductive cyclization

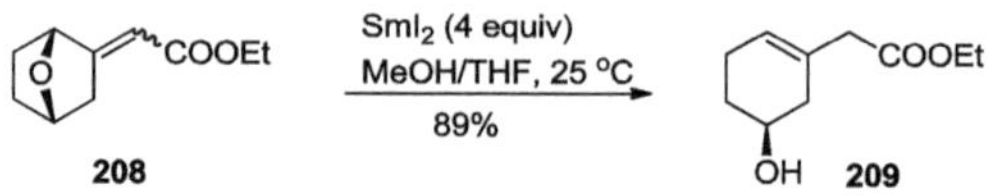

Scheme 37 Yoshida et al. reductive oxa ring opening

ability of the homoconjugated carbonyl group [260–263] which can be expressed with limiting structures **199′** and **199"** (n(C=O)/σ(C–C)/2p(C(6) hyperconjugative interaction). The lithium enolate of ketone (+)-**200** (does not undergo oxa ring opening at −60 °C) reacts with Eschenmoser's salt affording exocyclic enone (−)-**201**. Radical *C*-glycosylation of (−)-**201** with acetobromogalactose and Bu_3SnH/AIBN provides the *endo* α-*C*-galactopyranoside (+)-**203**. The reaction engenders radical intermediate **202** that is quenched exclusively onto its *exo* face by the tin hydride. Oxidation of the selenide moiety with *m*CPBA and subsequent seleno-Pummerer rearrangement gives (+)-**204**. The latter compound is reduced with Bu_3SnH/AIBN into *all-endo* trisubstituted 7-oxanorbornanone (+)-**205**. The Cossy et al. photoreduction of this ketone is stopped after 60 % of conversion, furnishing the 3-hydroxycyclohexanone (+)-**206**. After ketone reduction and acetylation, the α-*C*-pyranoside (+)-**207** is obtained [264].

Interestingly, irradiation of 5-*exo*-bromo-5-*endo*-benzyloxy-7-oxabicyclo[2.2.1] heptan-2-one in the presence of Et_3N neither leads to the corresponding cyclohexenol derivative nor to the reduction of the ketone into a mixture of *exo*- and *endo*-7-oxanorbornan-2-ols. Only the product of hydrogenolysis of the bromide (6-*endo*-benzyloxa-7-oxabicyclo[2.2.1]heptan-2-one) is formed and isolated in 60 % yield. The radical intermediate, formed by single-electron transfer (SET) from Et_3N to the bromide and subsequent C–Br bond cleavage, can be quenched intramolecularly as exemplified with the next reaction [265] (Scheme 36).

The two-electron reduction of 7-oxanorbornane derivative of **208** into cyclohexenol **209** has been realized with SmI_2 in THF using either MeOH or ethylene glycol as the proton source [266] (Scheme 37).

5.4 *Chalconide and Halide/Metal Exchange*

Peduncularine is the principal alkaloid of Tasmanian shrub *Aristotelia peduncularis* that presents anticancer activity [267]. In 2006, Kitamura et al. [268] have presented a synthesis of the racemic alkaloid starting with the Diels–Alder

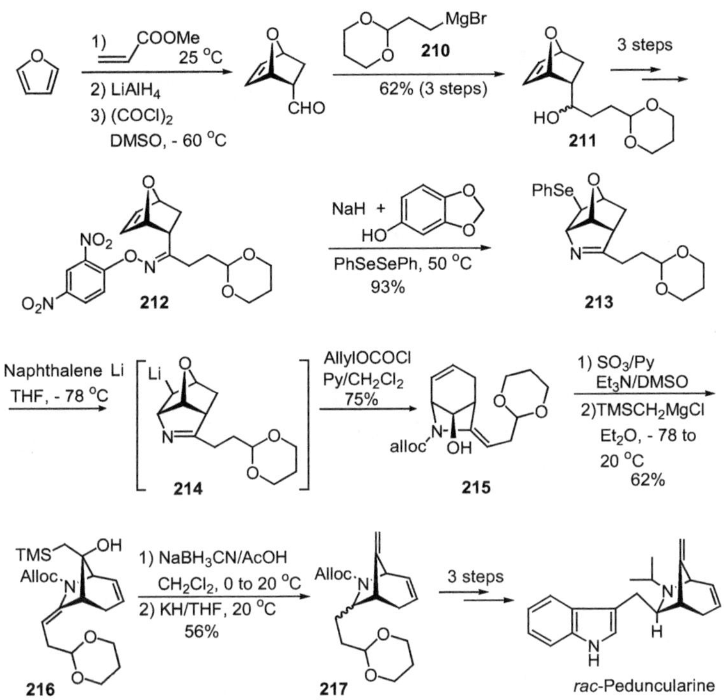

Scheme 38 Synthesis of *rac*-peduncularine by Kitamura et al.

cycloadduct of furan and methyl acrylate (Scheme 38). Reduction of its ester moiety into the corresponding primary alcohol and subsequent Swern oxidation provides 7-oxabicyclo[2.2.1]hept-5-ene-2-*endo*-carbaldehyde. The addition of the Grignard reagent **210** gives alcohol **211**. After oxidation of **211** into the corresponding ketone and oxime formation, compound **212** is obtained. The addition of PhSe radical to the alkene moiety of **212** generates a carbon-centered radical that undergoes cyclization with the oxime and subsequent β-elimination of 2,4-dinitrophenoxy radical producing **213**. Selenium/lithium exchange of **213** engenders a 7-oxanorborn-2-yl lithium species **214** that undergoes ethereal ring opening, giving an intermediate cyclohexenol. Its imine moiety is converted into allylcarbamate (alloc) **215**. Alcohol oxidation and subsequent addition of Me_3SiCH_2MgCl gives **216**. The direct reduction of enamine **216** with $NaBH_3CN$ under acidic conditions and subsequent Peterson olefination results in the formation of a 1:1 mixture of *exo*- and *endo*-**217**. Then, in three steps, the 3-indolyl system is installed in a classic Fischer indole synthesis that produces *rac*-peduncularine.

In 2008, and based on the SmI_2-induced ethereal ring opening of 2-iodo-7-oxa norbornane derivatives (Scheme 39), Carreira et al. presented a strategy for the construction of the azabicyclononane skeleton of banyaside A and B isolated from the cyanobacterium *Nostoc* sp. (IL-235) [269]. Both compounds inhibit the proteolytic activity of trypsin. The synthesis starts with the enantioselective Diels–Alder

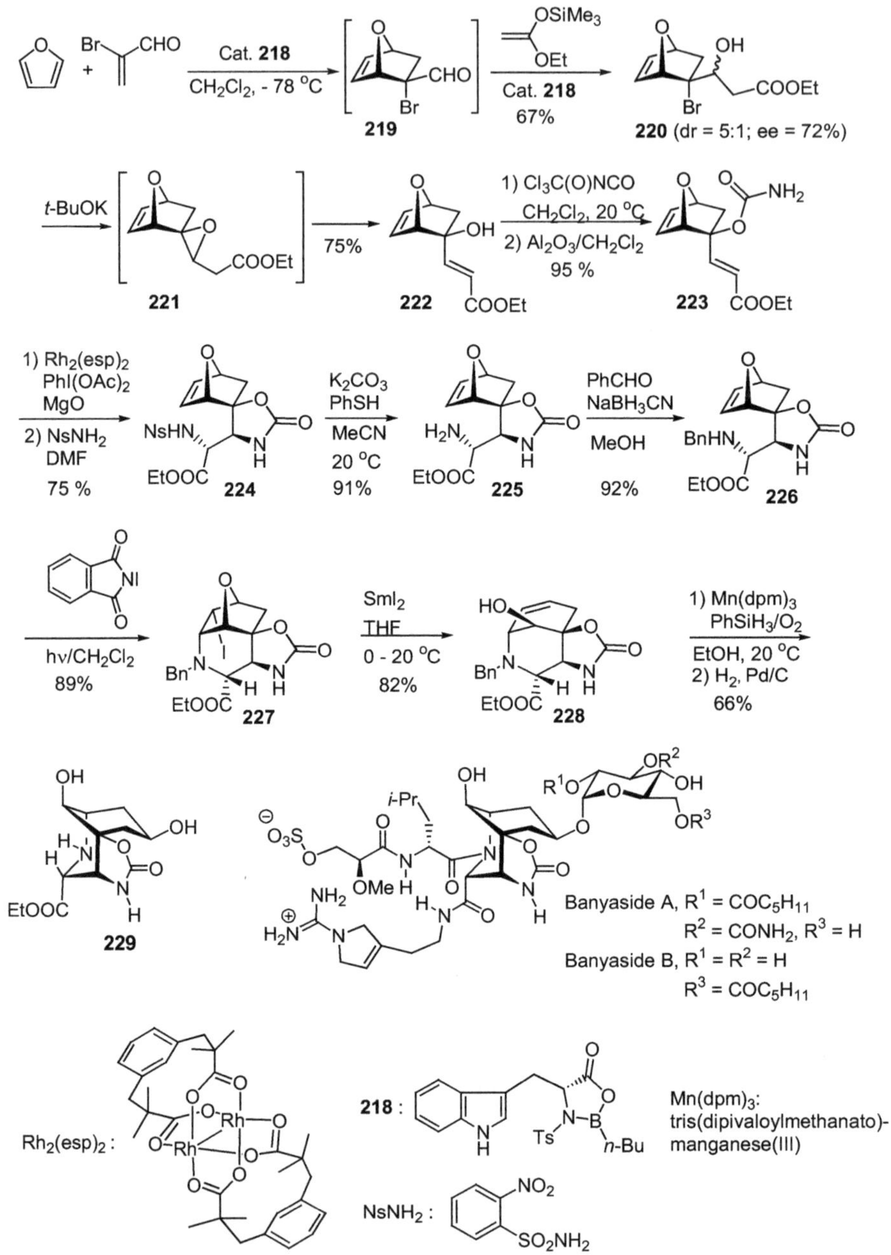

Scheme 39 Carreira et al.'s synthesis of the core element of banyasides

reaction of furan with α-bromocrolein catalyzed by the Corey's oxazaborolidine catalyst **218** [270]. Aldehyde **219** so-obtained is not isolated but reacted directly with 1-ethoxyvinyloxytrimethylsilane in a Mukaiyama condensation also catalyzed by **218**. This gives a 5:1 mixture of diastereoisomeric alcohols **220** (ee = 72 %). Treatment of the latter with *t*-BuOK induces the formation of intermediate epoxide **221** that undergoes a subsequent based-induced isomerization into the *exo*-alcohol **222**. The allylic alcohol **222** is then converted into carbamate **223** upon reaction with trichloroacetyl isocyanate followed by treatment with activated basic alumina. Aziridination of the acrylic moiety of **223** applies the protocol developed by De Bois et al. [271]. This produces an aziridine that is opened readily with the conjugate base of 2-nitrobenzenesulfonamide affording a mixture of diastereo-isomers from which **224** is isolated in 75 % yield after column chromatography. Deprotection of **224** with K_2CO_3 and thiophenol gives the primary amine **225** which is converted as benzyl amine **226**. *endo*-Iodoamination can be carried out by reaction of **226** with either *N*-succinimide or *N*-iodophthalimide under photo-chemical conditions. The *endo*-iodide **227** so-obtained leads to intractable mixtures on treatment with *t*-BuLi or lithium naphthalenide. With Zn in AcOH, the undesired C–N cleavage is favored. However, with SmI_2, **227** is converted into cyclohexenol **228**. The oxidative silylation of the cyclohexene moiety of **228** and subsequent hydrogenolysis of the benzylamine gives aminodiol **229**, the core of banyasides.

5.5 *Cleavage of Carbon–Carbon Bonds of 7-Oxabicyclo [2.2.1]heptan-2-ones*

The Baeyer–Villiger oxidation of 7-oxabicyclo[2.2.1]heptan-2-ones is the most used reaction to cleave a C–C bond of the 7-oxabicyclo[2.2.1]heptanes. Other routes have used retro-Claisen, retro-Diekmann, and Grob fragmentations. They have been reviewed elsewhere [11]. Enoxysilanes derived from 7-oxanorbornan-2-ones can be ozonolyzed into 2,5-anhydrouronic acid derivatives, precursors for *C*-nucleosides [225].

Baeyer–Villiger lactonization of 7-oxanorbornan-2-ones inserts the oxygen atom between the carbonyl and the nearby bridgehead center C1 (migration of the σ(C1–C2) bond). This is due to the electron-releasing ability of the 7-oxa ethereal group (2p(O) HOMO) that makes the C1–C2 bond electron rich and favors its migration. However, when the 7-oxanobornan-2-one is substituted at C3 by a $OSiR_3$, OBn, or OMe group, competitive insertion of the oxygen atom between C2 and C3 is observed (migration of the σ(C3–C2) bond) [271]. The highly regioselective Baeyer–Villiger oxidation of 3-*exo*-methyl-7-oxabicyclo[2.2.1] heptan-2-one, (+)-**230** (derived from the "naked sugar" (+)-**172** by catalytic hydro-genation and ketone α-monomethylation) gives lactone (−)-**231**. Its reaction with acetone trimethylsilyl enol ether is induced by $TiCl_4$ in CH_2Cl_2 and provides a mixture of (+)-**232** (major) and (+)-**233** (minor). The minor ketone (+)-**233** can be

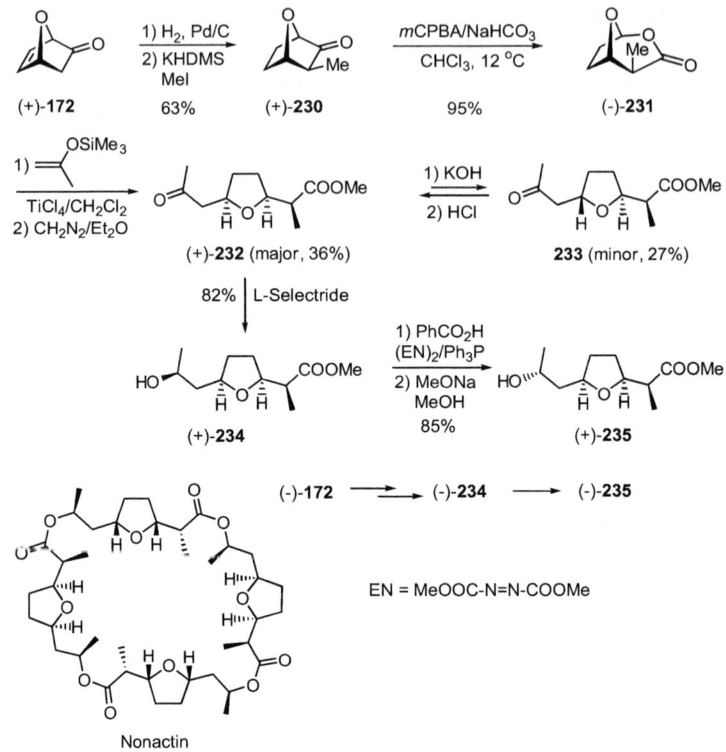

Scheme 40 Warm et al. synthesis of nonactin subunits

equilibrated with a mixture of (+)-**232** and (+)-**233** upon treatment with KOH first and then with HCl. Reduction of (+)-**232** with L-Selectride (LiB(s-Bu)$_3$H) gives (+)-methyl 8-epinonactate, (+)-**234**. Mitsunobu displacement of the latter with benzoic acid followed by Zemplen methanolysis provides (+)-methyl nonactate, (+)-**235**. The enantiomers (−)-**234** and (−)-**235** are prepared with the same ease starting with naked sugar (−)-**172** [272]. According to a procedure developed by Schmidt et al. [273, 274], (+)-**234**, (−)-**234**, (+)-**235**, and (−)-**235** can be combined ("Reverse Coupe du Roi" approach) into natural nonactin (Scheme 40).

A short synthesis of ethisolide and isoavenaciolide has been realized by Cossy et al. in 11 and 12 steps, respectively, from "naked sugar" **172** (Scheme 41). The approach makes use of a regioselective Baeyer–Villiger of a 7-oxanorbornanone intermediate [275]. Enone **172** is converted into its dipropargyl acetal **236** on treatment with propargyl trimethylsilyl ether in the presence of trimethylsilyl triflate. Upon alkene bromination **236**, the *endo*-propargyloxy group migrates regioselectively giving **237** a reaction similar to the acid-promoted conversion of epoxide **173** into (−)-**175** (see Scheme 28). Baeyer–Villiger oxidation of **237** provides lactone **238**. Its acidic methanolysis furnishes a mixture of methyl acetal **239**. Selective reduction of the ester moiety of **239** into a primary alcohol is

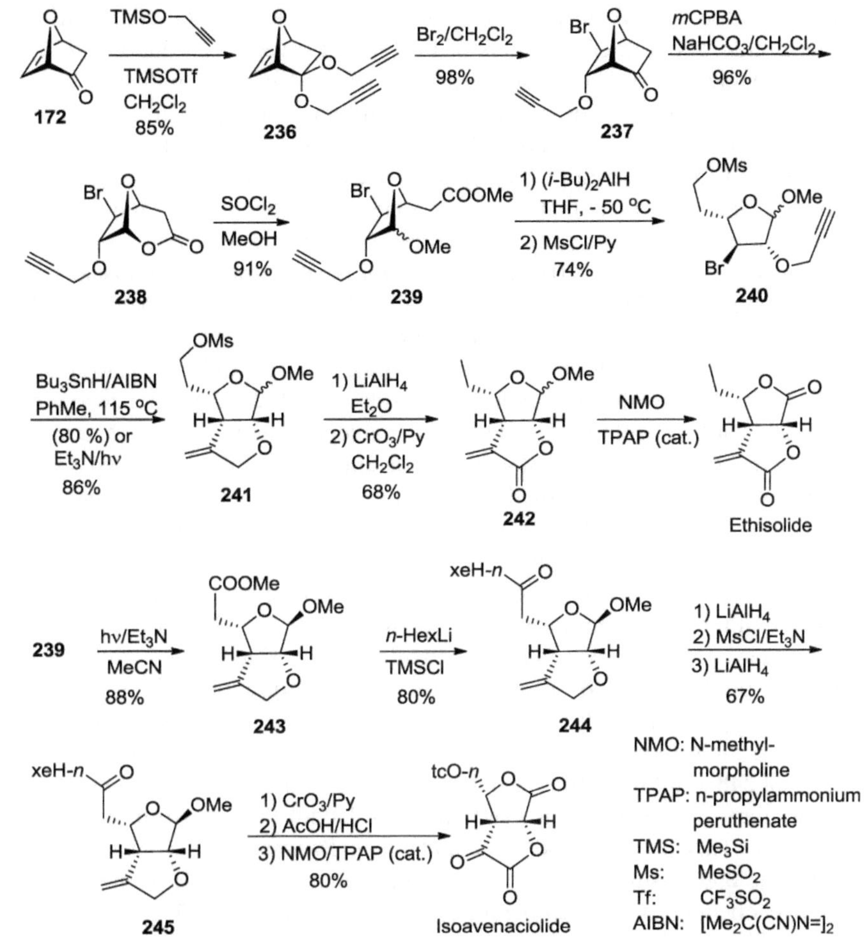

Scheme 41 Cossy et al.'s syntheses of ethisolide and isoavenaciolide

followed by its esterification with methanesulfonyl chloride to give mesylate **240**. Radical debromination with Bu_3SnH gives a radical that undergo cyclization with the formation of allylic ether **241** in 80 % yield. The same reaction can be induced photochemically in the presence of Et_3N in 86 % yield. The mesylate moiety of **241** then is reduced into the corresponding alkane with $LiAlH_4$. Subsequent oxidation of the allylic ether gives the α-methylidene-lactone **242**. The acetal moiety of **242** is oxidized with morpholine *N*-oxide (NMO) in the presence of tetra-*n*-propylammonium perruthenate (TPAP) catalyst, liberating *rac*-ethisolide.

Photoinduced reduction of bromoester **239** provides **243**. The latter compound reacts with one equivalent of *n*-hexyllithium in the presence of trimethylsilyl chloride to give ketone **244**. Its reduction with $LiAlH_4$ and then mesylation of the alcohol mixtures and $LiAlH_4$ reduction of the mixture of mesylates furnishes **245**. Oxidation of allylic ether of **245** into the corresponding lactone and subsequent hydrolysis of the methyl acetal and oxidation of the hemiacetal results in *rac*-isoavenaciolide. As both enantiomers of the "naked sugars" (+)- and (−)-**172** are readily available, the syntheses outlined in Scheme 41 can be applied to prepare ethisolide and isoavenaciolide in both their enantiomerically pure forms.

As the "naked sugars," **170–172** can be converted in few steps into polysubstituted 7-oxabicyclo[2.2.1]heptan-2-ones with high regio- and stereo-selectivity, and the fact that these ketones undergo highly regioselective Baeyer–Villiger oxidations into the corresponding uronolactones makes the "naked sugar" method [202, 207] quite useful for the preparation of unusual sugars including long-chain aldoses and alditols [216, 224, 276, 277], iminoalditols [223] and analogues [278–280], and *C*-disaccharides [281–284]. As an illustration we present in Scheme 42 the total asymmetric synthesis of 1,5-dideoxy-1,5-imino-D-lyxitol C-linked to methyl α-D-glucopyranoside, (−)-**262** [285]. Benzeneselenyl chloride adds at 0 °C onto the *exo* face of the alkene moiety of (+)-**172** in an *anti* fashion. In the absence of external nucleophile, chloride anion is quenched onto the *endo* face of C6 exclusively providing adduct (+)-**246** [286]. The homoconjugated carbonyl group acts as an electron-releasing group in this electrophilic reaction as already discussed above (Scheme 35). When formed at a low temperature, the potassium enolate of (+)-**246** does not undergo 7-oxa ether opening but can be quenched with the Eschenmoser's salt giving enone (−)-**247**. Epoxidation of "naked sugar" (+)-**170** gives an *exo*-epoxide that is ring-opened under acidic conditions producing **248** resulting from the migration of the 2-*endo*-camphanoyloxy group to the 6-*endo* carbon center, forming 7-oxanorbornanone **248** in 71 % yield (considering the recovery of unreacted (+)-**170**). Protection of *exo*-alcohol **248** as a MOM ether and then treatment with MeOH/DBU liberates the chiral auxiliary (camphanic acid) and an *endo*-alcohol that is also protected as a MOM ether. This provides **249** that undergoes Baeyer–Villiger oxidation into uronolactone **250** as single product. The lithium enolate of **250** (does not undergo oxa bridge opening at low temperature) adds to enone (−)-**247** giving a single Michael adduct **252** after acidic work-up. Steric factor controls the *exo* face selectivity for both the Michael reaction of the uronolactone enolate and the proton quenching of intermediate oxanorbornanone enolate **251**. Reduction of ketone **252** is also *exo* face selective. Oxidative removal of the benzeneselenyl group uses *m*CPBA and generates the corresponding chloroalkene. Protection of the *endo*-alcohol as a MOM ether furnishes **253**. BnOLi adds to the lactone moiety of **253**, giving a mixture of furanoses that are silylated. The uronic ester **254** so-obtained is then dihydroxylated on its chloroalkene moiety providing 7-oxanorbornanone **255** after acetylation. Baeyer–Villiger oxidation of **255** leads to **256**. Debenzylation of the uronic ester **256** gives a carboxylic acid that undergoes in situ Curtius rearrangement. This gives an intermediate isocyanate which reacts with benzyl alcohol to provide benzyl carbamate **257**. After desilylation of **257** and hydrogenolysis of the benzyl carbamate, the intermediate aminoaldose **258** is formed. It equilibrates with imine **259** which is reduced into **260** under the conditions of the hydrogenolysis. Final alcohol deprotection into **261** and ester reduction produces aza-*C*-disaccharide (−)-**262**. Enantiomer and stereoisomers of (−)-**262**

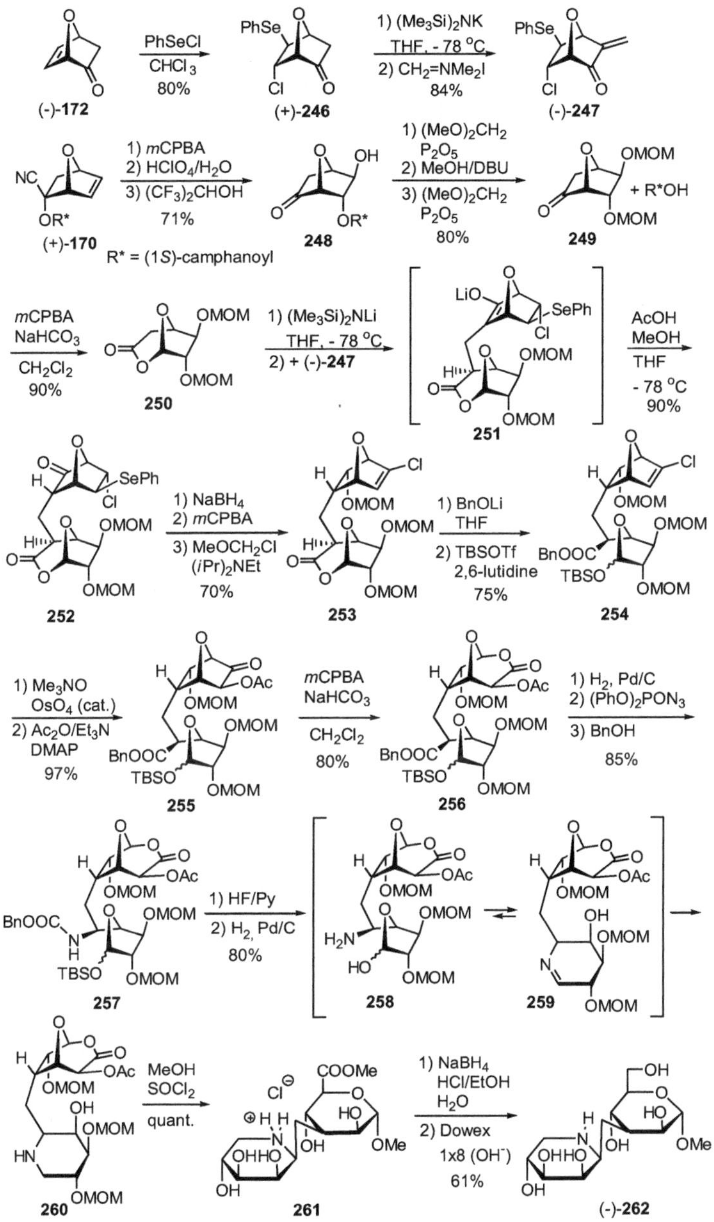

Scheme 42 Synthesis of an aza-*C*-disaccharide

can be made in principle following the same route starting with other naked sugars than (−)-**172** and (+)-**170**. Furthermore, intermediate 3-*endo*-alkylketone **252** can be equilibrated with the more stable 3-*exo*-alkylketone under basic conditions.

Stereoisomers of **249** can be obtained readily, for instance, by *exo*-dihydroxylation of (+)-**170**. As seen in Scheme 29, *trans*- and *cis*-aminohydroxylated derivatives can be prepared also with high stereo- and regioselectivity, thus permitting, in principle, the synthesis of aza-*C*-disaccharides containing amino-iminopentitols.

6 Conclusion

As other cyclic ethers, 7-oxabicyclo[2.2.1]heptanes (7-oxanorbornanes) and alkyl-substituted derivatives have been used to generate all kinds of polymers. A large number of 7-oxanorbornanes are found in Nature, and many of them possess quite interesting biological activities that have stimulated their total synthesis. The 7-oxanorbornane system is a molecular device for the construction of bioactive compounds in which the pharmacophores have to occupy specific positions in space. A large number of enantiomerically pure, or enantiomerically enriched, 7-oxanorbornanes are readily available and can be used to construct all kinds of compounds of biological interest such as rare sugars and analogues and monosaccharide and disaccharide mimetics. The chemistry of 7-oxanorbornanes is quite rich and generates a wide chemodiversity in a highly stereoselective manner.

References

1. Giusti P, Florentino U, Turchi G, Andruzzi F, Magagnini PL (1969) Makromol Chem 128:1
2. Stejny JJ (1973) Macromol Sci Chem 7:1435
3. Saegusa T, Motoi M, Suda H (1976) Macromolecules 9:231
4. Pretula L, Kaluzynski K, Szymanski R, Penczek S (1996) Macromolecules 29:6700
5. Pretula J, Kaluzynski K, Libiszowski J, Szymanski R, Penczek S (1997) J Polym Sci A Polym Chem 35:1733
6. Bai RK, Li S-H, Zou Y-F, Pan C-Y, Uryu T (1994) Macromol Chem Phys 195:119
7. Bai R-K, Li S, Zou Y-F, Pan C-Y, Uryu T (1995) J Polym Sci A Polym Chem 33:1685
8. Buchlauer G, Holbick H (1991) Chemiker-Zeitung 115:141
9. Kita H, Nakaba H, Okoshi A, Watanabe S, Nakadaira K, Oki T, Noguchi J (2000) Jpn Kokai Tokkyo Koho JP 2000128896 A 200000509
10. Maceina M, Slipke JW (2004) J Aquat Plant Manage 42:5
11. Vogel P, Cossy J, Plumet J, Arjona O (1999) Tetrahedron 55:13521
12. Schindler C, Carreira EM (2009) Chem Soc Rev 38:3222
13. Hrebabecky H, Dracinsky M, De Palma AM, Neyts J, Holy A (2009) Collect Czechoslov Chem Commun 74:487
14. Castro CR, Dutler R, Rauk A, Wieser H (1987) J Mol Struct (Theochem) 152:241
15. Adkins H, Billica HR (1948) J Am Chem Soc 70:695
16. Fehnel EA, Goodyear S, Berkowitz J (1951) J Am Chem Soc 73:4978
17. Sikkema DJ, Hoogland P, Bik J, Lam PT (1986) Polymer 27:1441
18. Maegawa T, Akashi A, Yaguchi K, Iwasaki Y, Shigetsura YM, Sajiki H (2009) Chem Eur J 15:6953
19. Olberg RC, Pines H, Ipatieff VN (1943) J Am Chem Soc 65:2260
20. Pines H, Kobynski TP (1970) J Catal 17:394

21. NIST chemistry WebBook, http://www.webbook.nist.gov/chemistry/
22. Chickos JS, Acree WE Jr (2003) J Phys Chem Ref Data 3:519
23. Bedford AF, Beezer AE, Mortimer CT, Springall HD (1963) J Chem Soc: 3823
24. Bozzelli JW, Rajasekaran I, Hur J (2006) J Phys Org Chem 19:93
25. Bozzelli JW, Rajasekaran I (2007) J Phys Chem Ref Data 36:663
26. Robiquet M (1810) Ann Chim 76:302
27. Graber R, Leoni L, Carrel S, Losa GA (1993) Cell Mol Biol 39:45
28. Southcott RV (1989) Med J Aust 151:654
29. Wang G-S (1989) J Ethnopharmacol 26:147
30. Tromovitch TA (1971) J Am Med Assoc 215:640
31. Williams LAD, Moller W, Merisor E, Kraus W, Rosner H (2003) West Indian Med J 52:10
32. Raj RK, Kurup PA (1967) Indian J Chem 5:86
33. Bochis RJ, Fischer MH (1968) Tetrahedron Lett 9:1971
34. Barua AK, Chakrabarti P, Das KG, Nair MSB (1970) Chem Ind: 1376
35. Peter MG, Snatske F, Nagarajan KN, Schmid H (1974) Helv Chim Acta 57:32
36. Deng L, Tang S (2011) Exp Opin Ther Pat 21:1743
37. Liu X-H, Blazsek I, Comisso M, Legras S, Marion S, Quittet P, Anjo A, Wang G-S (1995) Eur J Cancer 31a:953
38. McCluskey A, Sakoff JA (2001) Mini Rev Med Chem 1:43
39. McCluskey A, Sim ATR, Sakoff JA (2002) J Med Chem 45:1151
40. Baba Y, Irukawa N, Tanohira N, Sodeoka M (2003) J Am Chem Soc 125:9740
41. McCluskey A, Ackland SP, Bowyer MC, Baldwin ML, Garner J, Walkom CC, Sakoff JA (2003) Bioorg Chem 31:68
42. Cohen P, Cohen PTW (1989) J Biol Chem 264:21435
43. Shenolikar S, Nairn AC (1991) Adv Sec Mess Phosph Res 23:1
44. Krebs EG (1993) Angew Chem Int Ed Engl 32:1122
45. Lee JE, Cossoy MB, Chau LA, Singh B, Madrenas J (1997) J Immunol 159:61
46. McCluskey A, Taylor C, Quinn RJ, Suganuma M, Fujiki H (1996) Bioorg Med Chem Lett 6:1025
47. Enz A, Zenke G, Pombo-Villar E (1997) Bioorg Med Chem Lett 7:2513
48. Tatlock JH, Linton MA, Hou XJ, Kissinger CR, Pelletier LA, Showalter RE, Tempczyk A, Villafranca JE (1997) Bioorg Med Chem Lett 7:1007
49. Klee CB, Draetta GF, Hubbard MJ (1988) Adv Enzymol Relat Areas Mol Biol 61:149
50. Hubbard MJ, Klee CB (1989) Biochemistry 28:1868
51. McCattrey PG, Perrino BA, Soderling TR, Rao A (1993) J Biol Chem 268:3747
52. Bajsa J, McCluskey A, Gordon CP, Stewart SG, Hill TA, Sahu R, Duke SO, Tekwani BL (2010) Bioorg Med Chem Lett 20:6688
53. Guha PK, Poi R, Bhattacharyya A (1990) Phytochemistry 29:2017
54. Wang Z, Leng H, Sha K, Liu J (1983) Fenzi Kexue Yu Huaxue Yanjiu 3:25
55. Wang Z, Leng H, Sha K, Liu J (1984) Chem Abstr 100:17225h
56. Liu J, Zhang B, Sun J (1983) Yaoxue Xuebao 18:752–759
57. Liu J, Zhang B, Sun J (1984) Chem Abstr 100:120917j
58. Walter WG (1989) J Pharm Sci 78:66
59. Zhou YZ, Jiang T, Ren SM, Yu JS, Xia ZC (2005) J Organomet Chem 690:2186
60. Joshi BS, David J, Gawad DH (1983) Indian J Chem 22:131
61. Bockstahler ER, Weaver LC, Wrigt DL (1968) J Med Chem 11:603
62. Campbell BE, Tarleton M, Gordon CP, Sakoff JA, Gilbert J, McCluskey A, Gasser RB (2011) Bioorg Med Chem Lett 21:3277
63. Kihara N, Morimoto T, Nakanishi T Jpn Kokai Tokkyo Koho (1986) JP61037796 A 19860222
64. Kihara N, Morimoto T, Nakanishi T (1986) Chem Abstr 105:15338c
65. Wallach O (1907) Justus Liebigs Ann Chem 356:197
66. Austerweil MG (1929) Bull Soc Chim Fr: 862
67. Bugarcic ZM, Dunkic JD, Mojsilivic BM (2004) Heteroat Chem 15:468

68. Naves Y-R, Ardizio P (1950) Bull Soc Chim Fr: 673
69. Benn M, Peppard TL (1996) J Agric Food Chem 44:557
70. Avato P, Tava A (1995) Phytochemistry 40:141
71. Gurib-Fakim A, Demarne F (1995) J Essent Oil Res 7:105
72. Busmann D, Berger RG (1994) Z Naturforsch C Biosci 49:545
73. Vaughn SF, Spencer GF (1993) Need Sci 41:114
74. Garg SN, Misra LN, Agarwal SK (1989) Phytochemistry 28:634
75. Taylor WS, Powell JE, Marsilii EJ, Leep KJ (1995) Pestic Sci 44:85
76. Vaughn SF, Spencer GF (1996) Weed Sci 44:7
77. Liu W-G, Goswami A, Steffek RP, Chapman RL, Sariaslani FS, Steffens JJ, Rosazza JPN (1988) J Org Chem 53:5700
78. Rosazza JPN, Goswami A, Liu WG, Sariaslani FS, Steffens JJ, Steffek RP, Beak JN Jr, Chapman RL, Reeg S (1988) Dev Ind Microbiol 29:181
79. Weyerstahl P, Brendel J (1990) Liebigs Ann Chem 1029
80. Mathur RK, Ramaswamy SK, Rao AS, Bhattacharyya SC (1967) Tetrahedron 23:2495
81. Bohlmann F, Zdero C, Kapteyn H (1968) Liebigs Ann Chem 717:186
82. Aziz M, Rouessac F (1988) Tetrahedron 44:101
83. Goldsmith DJ (1962) J Am Chem Soc 84:3913
84. Coates RM, Melvin LS Jr (1970) Tetrahedron 26:5699
85. Barrero AF, Alvarez-Manzaneda EJ, Palomino PL (1994) Tetrahedron 50:13239
86. Sneden AT (1993) Synlett: 313
87. Caglioti L, Naef H, Arigoni D, Jeger O (1958) Helv Chim Acta 41:2278
88. Caglioti L, Naef H, Arigoni D, Jeger O (1959) Helv Chim Acta 42:2557
89. Bohlmann F, Zdero C (1975) Chem Ber 108:1902
90. Marco JA, Sanz JF, Yuste A, Carda M, Jakupovic J (1991) Phytochemistry 30:3661
91. Kaiser R, Lamparsky D (1978) Helv Chim Acta 61:373
92. Mori K, Tamura H (1986) Tetrahedron 42:2643
93. Wahlberg I, Eklund A, Enzell CR, Berg J (1987) Acta Chem Scand B B41:455
94. Takagi Y, Fujimori T, Hata T, Kaneko H, Kato T (1980) Agric Biol Chem 44:705
95. Gmünder MR, Eugster CH (1990) Helv Chim Acta 73:1954
96. Demnitz FWJ, Philippini C, Raphael RA (1995) J Org Chem 60:5114
97. Waraszkiewicz SM, Sun HH, Erickson KL, Finer J, Clardy J (1978) J Org Chem 43:3194
98. Jennings-White CLD, Holmes AB, Raithby PR (1979) J Chem Soc Chem Commun 542
99. Holmes AB, Jennings-White CLD, Kendrick DA (1983) J Chem Soc Chem Commun 415
100. Renaud P, Vionnet J-P (1994) Chimia 48:471
101. Weyerstahl P, Marschall H, Splittgerber U, Wolf D (1996) Liebigs Ann Chem 1195
102. González AG, Martín JD, Norte M, Pérez R, Rivera R, Ruano JZ (1983) Tetrahedron Lett 24:4143
103. Caccamese S, Amico V, Neri P (1990) J Nat Prod 33:1287
104. Cabrera E, García-Granados A, Quecuty MA (1988) Phytochemistry 27:183
105. Jakupovic J, Jaensch M, Bohlmann F, Dillon MO (1988) Phytochemistry 27:3551
106. Fattorusso E, Magno S, Mayol L (1979) Gazz Chim Ital 109:589
107. Rogers C, Keay BA (1993) Can J Chem 71:611
108. Sanz JF, Marco JA (1990) Phytochemistry 29:2913
109. Estrada DM, Ravelo JL, Ruiz-Pérez C, Martín JD (1989) Tetrahedron Lett 30:6219
110. De La Torre MC, Rodríguez B, Bruno M, Savona G, Piozzi F, Perales A, Torres MR, Servettaz O (1990) Phytochemistry 29:2229
111. Gallardo A, Manto E, Martín JD, Pérez C, Rodriguez ML (1988) Rev Latinoam Quím 19:86
112. Sivakoff M, Pore E, Hsueh W, Needleman P (1979) Fed Proc 38:78
113. Gorman RR (1979) Fed Proc 38:83
114. Eggelte TA, de Koning H, Huisman HO (1978) J Chem Soc Perkin 1: 980
115. Kametani T, Suzuki T, Tomino A, Kamada S, Unno K (1982) Chem Pharm Bull 30:796

116. Sprague PW, Heikes JE, Harris DN, Greenberg R (1980) In: Samuelsson B, Ramwell P, Paoletti R (eds) Advances in prostaglandin and thromboxane research, vol 6. Raven, New York, NY, p 493
117. Cross PE, Dickinson RP (1987) Annu Rep Med Chem 22:95
118. Hall SE, Han W-C, Haslanger MF, Harris DN, Ogletree ML (1986) J Med Chem 29:2335
119. Hall SE, Han W-C, Harris DN, Hedberg A, Ogletree ML (1989) J Med Chem 32:974
120. Eggelte TA, De Koning H, Huisman HO (1973) Tetrahedron 29:2491
121. Alder K, Stein G (1937) Angew Chem 50:510
122. Sprague PW, Heikes JE, Gougoutas JZ, Malley MF, Harris DN, Greenberg R (1985) J Med Chem 28:1580
123. Misra RN, Brown BR, Sher PM, Patel MM, Hall SE, Han W-C, Barrish JC, Kocy O, Harris DN, Goldenberg HJ, Michel IM, Schumacher WA, Webb ML, Monshizadegan H, Ogletree ML (1993) J Med Chem 36:1401
124. Barrish JC, Singh J, Spergel SH, Han W-C, Kissick TP, Kronenthal DR, Mueller RH (1993) J Org Chem 58:4494
125. Das J, Vu T, Hall SE, Nakome M, Haslanger MF, Reid JA, Garber D, Chi Truc V, Harris DN, Hedberg A, Ogletree ML (1990) J Med Chem 33:1741
126. Nakane M, Reid JA, Han W-C, Das J, Chi Truc V, Haslanger MF, Garber D, Harris DN, Hedberg A, Ogletree ML, Hall SE (1990) J Med Chem 33:2456
127. Kobayashi S, Eguchi Y, Sato M, Kudo I, Inove K, Ohno M (1992) Chem Pharm Bull 40:2891
128. Hagishita S, Seno K (1989) Chem Pharm Bull 37:327
129. Das J, Vu T, Harris DN, Ogletree ML (1988) J Med Chem 31:930
130. Kusano G, Uchida H, Murakami Y, Sakurai N, Takemoto T (1976) Yakugaku Zasshi 96:321
131. Kusano G, Uchida H, Murakami Y, Sakurai N, Takemoto T (1976) Chem Abstr 85:5907r
132. Li JX, Kadota S, Hattori M, Yoshimachi S, Shiro M, Oogami N, Mizuno H (1993) Chem Pharm Bull 41:832
133. Mo F (1973) Acta Crystallogr B 29:1796
134. Mo F (1977) Acta Crystallogr B 33:641–649
135. Fakuyama Y, Kaneshi A, Tani N, Kodama M (1993) Phytochemistry 33:483
136. Masamune T, Anetai M, Takasugi M, Katsui N (1982) Nature 297:495
137. Fukuzawa A, Furusaki A, Ikura M, Masamune T, Hall LD, Rajanayagam V (1985) J Chem Soc Chem Commun 748
138. Tsutsumi M, Sakurai K (1966) Jpn J Appl Entomol Zool 10:129
139. Murai A, Tanimoto N, Sakamoto N, Masamune T (1988) J Am Chem Soc 110:1985
140. Murai A (1989) Pure Appl Chem 61:393
141. Bredereck H, Simchen G, Rebsdat S, Kantlehner W, Horn P, Wahl R, Hoffman H, Grieshaber P (1968) Chem Ber 101:41
142. Mori K, Watanabe H (1989) Pure Appl Chem 61:543
143. Watanabe H, Mori K (1991) J Chem Soc Perkin 1: 2919
144. Corey EJ, Houpis IN (1990) J Am Chem Soc 112:8997
145. Corey EJ, Hong B-C (1994) J Am Chem Soc 116:3149
146. Shiina Y, Tomata Y, Miyashita M, Tanino K (2010) Chem Lett 39:835
147. Brown HC, Chandrasekharan J, Ramachandran PV (1988) J Am Chem Soc 110:1539
148. Parkes KEB, Pattenden G (1986) Tetrahedron Lett 27:2535
149. Matsuno T, Tani Y, Maoka T, Matsuo K, Komori T (1986) Phytochemistry 25:2837
150. Bjornland T, Borch G, Liaaen-Jensen S (1986) Phytochemistry 25:201
151. Fiksdahl A, Bjornland T, Liaaen-Jensen S (1984) Phytochemistry 23:649
152. Moss QP, Ooi CK (1992) J Chem Soc Chem Commun 342
153. Coe PL, Maunder P, Frampton CS, Martin JA, Thomas WA, Whitcombe IWA (2002) J Fluor Chem 115:13
154. Marco JA, Sanz-Cervera JF, Garciá-Lliso V, Domingo LR, Carda M, Rodríguez S, López-Ortiz F, Lex J (1995) Liebigs Ann Chem 1837
155. Padwa A, Sandanayaka VP, Curtis EA (1994) J Am Chem Soc 116:2667

156. Padwa A, Curtis EA, Sandanayaka VP (1997) J Org Chem 62:1317
157. Weingarten MD, Prein M, Price AT, Snyder JP, Padwa A (1997) J Org Chem 62:2001
158. Padwa A, Price AT (1998) J Org Chem 63:556
159. Warrener RN, Schultz AC, Butler DN, Wang S, Mahadevan IB, Russell RA (1997) J Chem Soc Chem Commun 1023
160. Warrener RN, Johnston MR, Gunter MJ (1998) Synlett: 593
161. Fringuelli F, Taticchi A (eds) (1990) Dienes in the Diels-Alder reaction. Wiley, New York, NY
162. Kappe OC, Murphree S, Padwa A (1997) Tetrahedron 53:14179
163. Crowley PJ, Fawcett J, Griffith GA, Moralee AC, Percy JM, Salafia V (2005) Org Biomol Chem 3:3297
164. Diels O, Alder K (1929) Chem Ber 62:554–558
165. Anet FAL (1962) Tetrahedron Lett 3:1219
166. Lee MW, Herndon W (1986) Chem Rev 86:795
167. Berson JA, Swidler R (1953) J Am Chem Soc 75:1721
168. Nudenberg W, Butz LW (1944) J Am Chem Soc 66:307
169. Beuster PH, Aben RWM, Seerden JPG, Smits JMM, Sceeren HW (1998) Eur J Org Chem: 2483
170. Dauben WG, Kessel CR, Takemura KH (1980) J Am Chem Soc 102:6893
171. Dauben WG, Gerdes JM, Smith DB (1985) J Org Chem 50:2576
172. Dauben WG, Lam JYL, Guo ZR (1996) J Org Chem 61:4816
173. Ancerewicz J, Vogel P (1996) Helv Chim Acta 79:1393
174. Marchionni C, Vogel P, Roversi P (1996) Tetrahedron Lett 37:4149
175. Marchionni C, Meilert KT, Vogel P, Schenk K (2000) Synlett: 1111
176. Feringa BL, Gelling OJ, Meesters L (1990) Tetrahedron Lett 31:7201
177. Harwood LM, Ishikawa T, Phillips H, Watkin D (1991) J Chem Soc Chem Commun 527
178. Fallis AG (1999) Can J Chem 77:159
179. Keay BA, Hunt IR (1999) Adv Cycloaddition 6:173
180. Heiner T, Michalski S, Gerke K, Kuchta G, Buback M, de Meijere A (1995) Synlett: 355
181. Heiner T, Kozhushkov SL, Noltemeyer M, Haumann T, Boese R, de Meijere A (1996) Tetrahedron 52:12185
182. Zubkov FI, Nikitina EV, Varlamov AV (2005) Russ Chem Rev 74:639
183. Buser S, Vasella A (2005) Helv Chim Acta 88:3151
184. Padwa A, Wang Q (2006) J Org Chem 71:3210
185. Padwa A, Crawford KR, Straub CS, Pieniazek SN, Houk KN (2006) J Org Chem 71:5432
186. Domingo LR, Aurell MJ, Arno M, Saez JA (2007) J Org Chem 72:4220
187. Zubkov FI, Ershova JD, Orlova AA, Zaytsev VP, Nikitina EV, Peregudov AS, Gurbanov AV, Atash V, Borisov RS, Khrustalev VN, Maharramov AM, Varlamov AV (2009) Tetrahedron 65:3789
188. O'Keefe BM, Mans DM, Kaelin DE Jr, Martin SF (2010) J Am Chem Soc 132:15528
189. Chen Y, Wang L, Liu Y, Li Y (2011) Chem Eur J 17:12582
190. Hill B, Jordan R, Qu Y, Assoud A, Rodrigo R (2011) Can J Chem 89:1457
191. Samanta K, Kar GK, Sarkar AK (2012) Tetrahedron Lett 53:1376
192. Read ML, Krapp A, Miranda PO, Gundersen L-L (2012) Tetrahedron 68:1869
193. Wang C-C, Li W-DZ (2012) J Org Chem 77:4217
194. Mukaiyama T, Iwasawa N (1981) Chem Lett: 29
195. Benningshof JCJ, Blauuw RH, van Ginkel AE, van Maarseveen JH, Rutjes FPJT, Hiemstra H (2002) J Chem Soc Perkin 1: 1693
196. Gao G, Moore D, Xie R-G, Pu L (2002) Org Lett 4:4143
197. Moore D, Pu L (2002) Org Lett 4:1855
198. Xu M-H, Pu L (2002) Org Lett 4:4555
199. Jung ME, Min S-J (2005) J Am Chem Soc 127:10834

200. Tanino K, Takahashi M, Tomata Y, Tokura H, Uehara T, Narabu T, Miyashita M (2011) Nat Chem 3:484
201. Vogel P, Auberson Y, Bimwala M, De Guchteneere E, Vieira E, Wagner J (1989) In: Horton D, Hawkins LD, McGarvey GJ (eds) Trends in synthetic carbohydrate chemistry, ACS symposium series 286. American Chemical Society, Washington, DC, p 197, Chapt. 13
202. Vogel P, Fattori D, Gasparini F, Le Drian C (1990) Synlett: 173
203. Hudlicky T, Entwistle DA, Pitzer KK, Thorpe AJ (1996) Chem Rev 96:1195
204. Woo S, Keay BA (1996) Synthesis: 669
205. Chiu P, Lautens M (1997) Topics in current chemistry, vol 190. Springer, Berlin, 1
206. Jiang S, Singh G (1988) Tetrahedron 54:4697
207. Vogel P (1998) Curr Org Chem 2:255
208. Vogel P (2008) Top Curr Chem 282:187
209. Masesane IB (2010) Trends Org Chem 14:13
210. Bennett EL, Niemann C (1952) J Am Chem Soc 74:5076
211. Olberg RC, Pines H, Ipatieff VN (1944) J Am Chem Soc 66:1096
212. Dormer JC, Hylands KA, Moodie RB (1998) J Chem Soc Perkin II: 243
213. Ogawa S, Iwasawa Y, Nose T, Suami T, Ohba S, Ito M, Saito Y (1985) J Chem Soc Perkin I: 903
214. Ogawa S, Kasahara T, Suami T (1979) Bull Chem Soc Jpn 52:118
215. Ogawa S, Takagahi T (1985) J Org Chem 50:2356
216. Vogel P (2000) Curr Org Chem 4:455
217. Moritz V, Vogel P (1992) Tetrahedron Lett 33:5243
218. Vieira E, Vogel P (1983) Helv Chim Acta 66:1865
219. Reymond J-L, Vogel P (1990) Tetrahedron Asymmetry 1:729
220. Warm A, Vogel P (1986) J Org Chem 51:5348
221. Auberson Y, Vogel P (1989) Helv Chim Acta 72:278
222. Wagner J, Vogel P (1991) Carbohydr Res 222:151
223. Wagner J, Vogel P (1991) Tetrahedron 47:9641
224. Jeganathan S, Vogel P (1991) J Org Chem 56:1133
225. Bimwala RM, Vogel P (1989) Helv Chim Acta 72:1825
226. Jeanneret V, Gasparini F, Péchy P, Vogel P (1992) Tetrahedron 48:10637
227. Péchy P, Gasparini F, Vogel P (1992) Synlett: 676
228. LeDrian C, Vogel P (1988) Helv Chim Acta 71:1399
229. LeDrian C, Vieira E, Vogel P (1989) Helv Chim Acta 72:338
230. LeDrian C, Vionnet J-P, Vogel P (1990) Helv Chim Acta 73:161
231. Lysek R, Schutz C, Vogel P (2005) Helv Chim Acta 88:2788
232. Lysek R, Schutz C, Vogel P (2005) Bioorg Med Chem Lett 15:3071
233. Lysek R, Vogel P (2006) Tetrahedron 62:2733
234. Lysek R, Favre S, Vogel P (2007) Tetrahedron 63:6558
235. Marco-Contelles J (2001) Eur J Org Chem: 1607
236. Allemann S, Vogel P (1994) Helv Chim Acta 77:1
237. Borthwick AD, Curry DJ, Poynton A, Whalley WB, Hooper JW (1980) J Chem Soc Perkin I: 2435
238. Arjona O, Iradier F, Mañas RM, Plumet J, Grabuleda X, Jaime C (1998) Tetrahedron 54:9095
239. Schindler CS, Diethelm S, Carreira EM (2009) Angew Chem Int Ed 48:6296
240. Kozokowski AP, Kuniak MP (1978) J Org Chem 43:2083
241. Brion F (1982) Tetrahedron Lett 23:5299
242. Chung K, Lee HG, Choi I, Choi J (2001) J Org Chem 66:5937
243. Gustafsson J, Sterner O (1994) J Org Chem 59:3994
244. Campbell MM, Kaye AD, Sainsbury M, Yavarzadeh R (1984) Tetrahedron 40:2461
245. Arjona O, de Dios A, de la Pradilla RF, Plumet J (1991) Tetrahedron Lett 32:7209
246. Arjona O, Lorenzo G, Medel R, Plumet J (2001) Tetrahedron Lett 42:7041
247. McMorris TC, Staake MD, Kelner MJ (2004) J Org Chem 69:619

248. Grootaert WM, De Clerq PJ (1986) Tetrahedron Lett 27:1731
249. Kinder FR Jr, Bair KW (1994) J Org Chem 59:6965
250. Van Royen LA, Mijngheer R, De Clerq PJ (1983) Tetrahedron Lett 24:3145
251. Kelner MJ, McMorris TC, Taetle R (1990) J Natl Cancer Inst 82:1562
252. Jeanneret V, Vogel P, Renaut P, Millet J, Theveniaux J, Barberousse V (1998) Bioorg Med Chem Lett 8:1687
253. Renaut P, Millet J, Sepulchre C, Theveniaux J, Barberousse V, Jeanneret V, Vogel P (1998) Helv Chim Acta 81:2043
254. Crowley PJ, Fawcett J, Kariuki BM, Moralee AC, Percy JM, Salafia V (2002) Org Lett 4:4125
255. De Schrijver J, De Clercq PJ (1993) Tetrahedron Lett 34:4369
256. Molander GA (1992) Chem Rev 92:29
257. McMurry JE (1989) Chem Rev 89:1513
258. Cossy J, Aclinou P, Bellosta V, Furet N, Baranne-Lafont J, Sparfel D, Souchaud C (1991) Tetrahedron Lett 32:1315
259. Arvai G, Fattori D, Vogel P (1992) Tetrahedron 48:10621
260. Eschenmoser A, Frey A (1952) Helv Chim Acta 35:1660
261. Carrupt P-A, Vogel P (1984) Tetrahedron Lett 25:2879
262. Carrupt P-A, Vogel P (1990) J Org Chem 55:5696
263. Sordo JA, Varela-Alvarez A, Giani S, Vogel P (2008) Appl Catal A Gen 336:72
264. Cossy J, Ranaivosata J-L, Bellosta V, Ancerewicz J, Ferrito R, Vogel P (1995) J Org Chem 60:8351
265. Cossy J, Bellosta V, Ranaivosata J-L, Gille B (2001) Tetrahedron 57:5173
266. Yoshida A, Takayama H (2001) Tetrahedron Lett 42:3603
267. Ros H-P, Kyburtz R, Preston NW, Galagher IR, Bock C, Hesse M (1979) Helv Chim Acta 62:481
268. Kitamura M, Ihara Y, Uera K, Narasaka K (2006) Bull Chem Soc Jpn 79:1552
269. Schindler CS, Stephenson CRJ, Carreira EM (2008) Angew Chem Int Ed 47:8852
270. Corey EJ, Loh T-P (1993) Tetrahedron Lett 34:3979
271. Espino CG, Williams Fiori K, Kim M, De Bois J (2004) J Am Chem Soc 126:15378
272. Warm A, Vogel P (1987) Helv Chim Acta 70:690
273. Zak H, Schmidt U (1975) Angew Chem Int Ed Engl 87:432
274. Schmidt H, Gombos J, Haslinger E, Zak H (1976) Chem Ber 109:2628
275. Cossy J, Ranaiviosata JL, Bellosta V (1996) Tetrahedron 52:629
276. Jeganathan S, Vogel P (1989) J Chem Soc Chem Commun 993
277. Neff DP, Chen YW, Vogel P (1991) Helv Chim Acta 74:508
278. Reymond J-L, Pinkerton AA, Vogel P (1991) J Org Chem 56:2128
279. Chen YW, Vogel P (1994) J Org Chem 59:2487
280. Schaller C, Vogel P (1999) Synlett: 1219
281. Bimwala RM, Vogel P (1992) J Org Chem 57:2076
282. Ferritto R, Vogel P (1994) Tetrahedron Asymmetry 5:2077
283. Jeanneret V, Meerpoel L, Vogel P (1997) Tetrahedron Lett 38:543
284. Baudat A, Vogel P (1997) J Org Chem 62:6252
285. Marquis C, Picasso S, Vogel P (1999) Synthesis: 1441
286. Black KA, Vogel P (1986) J Org Chem 51:5341

Synthesis of 5,6- and 6,6-Spirocyclic Compounds

Margaret A. Brimble and Louise A. Stubbing

Contents

Abstract The selective and efficient synthesis of spiroacetals has attracted attention from the synthetic community, both because of the synthetic challenge of

M.A. Brimble (✉) • L.A. Stubbing
School of Chemical Sciences, The University of Auckland, 23 Symonds St, Auckland Central 1010, New Zealand
e-mail: m.brimble@auckland.ac.nz

J. Cossy (ed.), *Synthesis of Saturated Oxygenated Heterocycles I*, Topics in Heterocyclic Chemistry 35, DOI 10.1007/978-3-642-41473-2_5,

complex spiroacetal natural product scaffolds, as well as the drive to develop and improve existing methods. A number of recently reported methods for the synthesis of spiroacetals are discussed, including their application in the synthesis of natural products containing the spiroacetal scaffold.

Keywords [3+2]-cycloaddition • [4+2]-cycloaddition • Anionic homo-Fries rearrangement • Benzannulated • Chiral phosphoric acids • Cyclopropane ring opening • Dehydrative spirocyclization • DIHMA • Electrophilic cyclization • Enol ether • Epoxide ring opening • Ferrier rearrangement • Furan oxidation • Hydroalkoxylation • Intramolecular hydrogen abstraction • *o*-quinone methide • Oxa-Michael • Oxonium ion • Oxymercuration • RCM • Rearrangement • Reductive cyclization • Spiroacetal • Spirocyclization • Transition metal catalysis • Wacker cyclization

1 Introduction

The selective and efficient synthesis of spiroacetals has attracted much attention in the synthetic community [1–9] both because of the synthetic challenge of complex spiroacetal natural products, as well as the drive to develop and improve existing methods. This review aims to cover recent reports (2008 onwards) of methods for the synthesis of spiroacetals and their application in natural product synthesis.

1.1 Stereochemistry of Spiroacetals and the Anomeric Effect

The stereochemistry of spiroacetals (both synthetic and naturally occurring) can largely be divided into two categories—the so-called anomeric or axial/axial conformers, and non-anomeric or axial/equatorial conformers. In 6,6-spiroacetals, four possible conformations exist (Fig. 1a–2d) with varying relative stabilities, which are interchangeable via ring flipping [2].

Most naturally occurring spiroacetals have a doubly anomeric configuration, representing the lowest energy conformer, where both acetal oxygen atoms are axial with respect to each other. Such an arrangement has been calculated to contribute an estimated 2.4 kcal/mol decrease in energy per axial substituent in 6,6-spiroacetals [10]. The anomeric effect, as this is known, is thought to be due to overlap of the axial nonbonding orbital of the acetal oxygen atom with the C–X antibonding σ^* orbital of the electronegative substituent (in the case of spiroacetals, the C–O σ^* orbital, Fig. 1) [11–14]. Where the electronegative substituent is in the equatorial position (i.e., **1b**), this overlap is not possible. The conformers are also occasionally referred to as the "thermodynamic" and "contra-thermodynamic" isomers; however, as there are cases where the doubly anomeric conformer is not the most thermodynamically favored this terminology can become confusing.

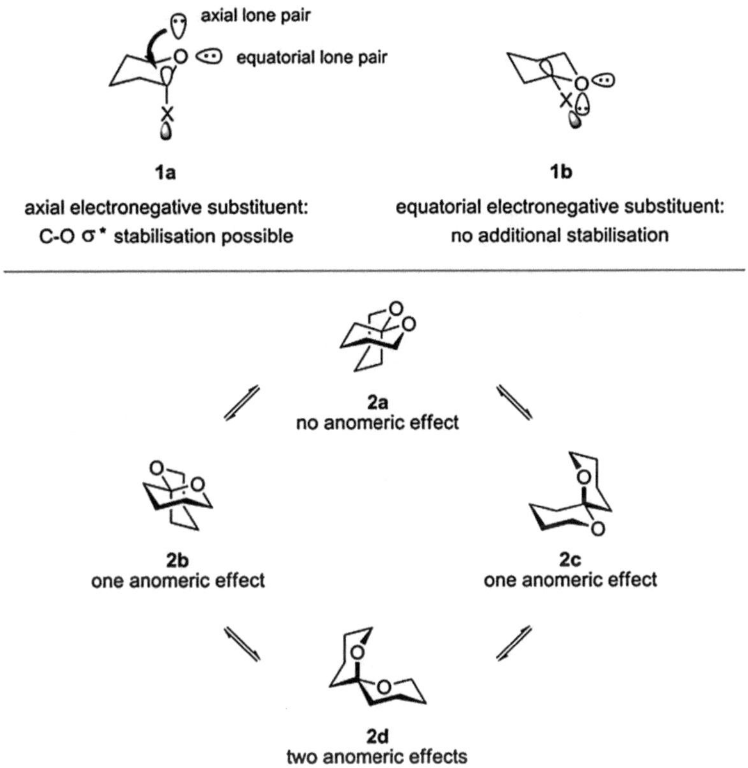

Fig. 1 Conformations of spiroacetals and the anomeric effect

In 5,5-spiroacetals the required antiperiplanar arrangement of the oxygen atoms is not obtainable; thus conformational preferences in these systems are far more complex and difficult to predict [13].

1.2 Methods for the Construction of 5,6- and 6,6-Spiroacetals

Methods for the synthesis of spiroacetals include dehydrative spirocyclization of a dihydroxyketone (or surrogate thereof); additions to enol ethers; nucleophilic addition via an oxonium ion; hydroalkoxylation of internal alkynes; furan oxidation; electrophilic cyclization; ring opening of epoxides; oxa-Michael additions; cycloaddition processes; ring-closing metathesis; intramolecular hydrogen abstraction reactions; reductive cyclizations; a Wacker-type cyclization; ring opening of cyclopropanes; and a variety of rearrangements (Scheme 1). Many of these methods may be catalyzed by transition metal complexes, a review of which has been published recently [15] (Scheme 1).

Methods for the selective formation of nonthermodynamic spiroacetals are less common, particularly those that can effectively overcome the inherent preference

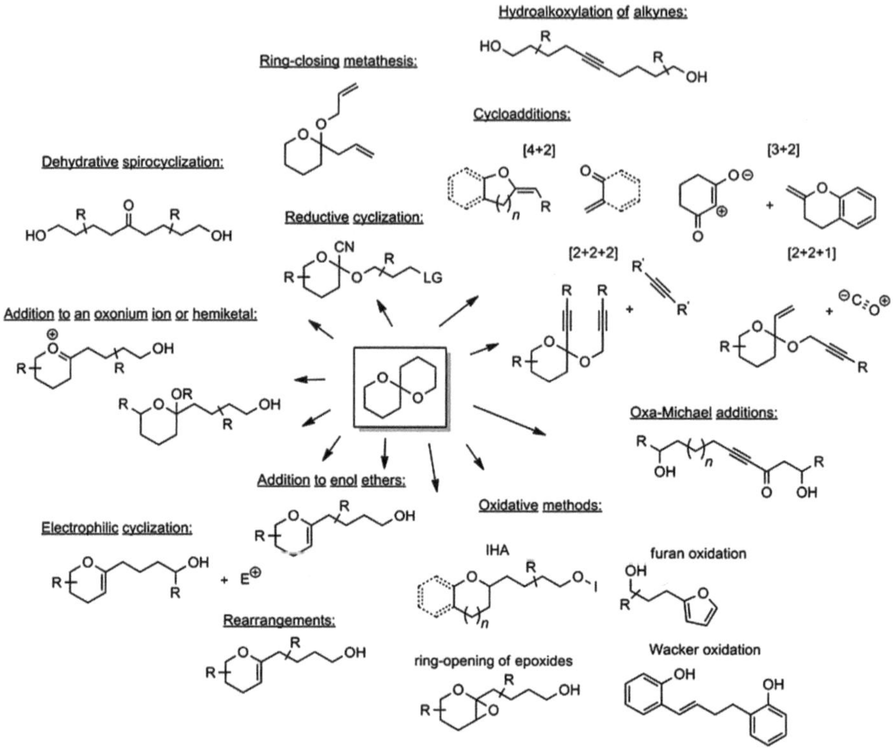

Scheme 1 Overview of methods for the synthesis of spiroacetals

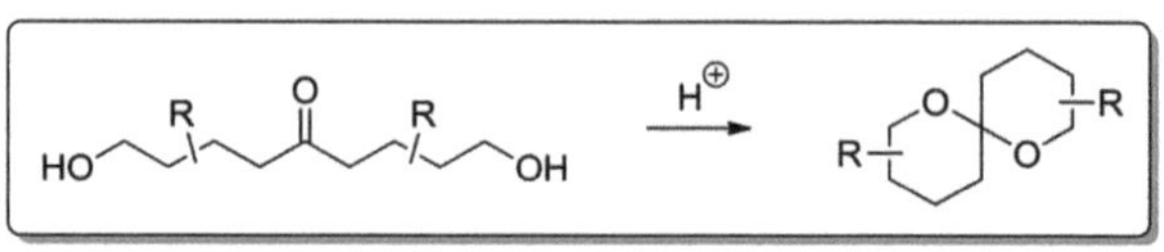

Scheme 2 Dehydrative spirocyclization of a dihydroxyketone

for the most thermodynamically stable isomer. These methods have been reviewed recently [2, 16].

2 Recent Advances and Application in the Synthesis of 5,6- and 6,6-Spiroacetals

2.1 Dehydrative Spirocyclization of a Dihydroxyketone

Dehydrative spirocyclization of a dihydroxyketone (Scheme 2) leads to the thermodynamically most stable spiroacetal as the major isomer. As discussed earlier, in most cases this also corresponds to the doubly anomerically stabilized spiroacetal.

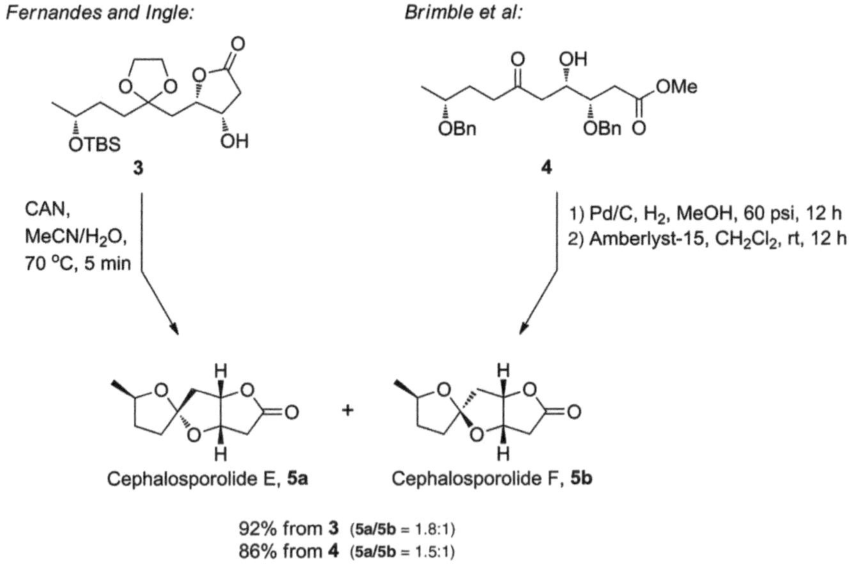

Scheme 3 Fernandes and Ingle's [46] and Brimble et al.'s [47] syntheses of cephalosporolides E and F

In cases where substitution of the spiroacetal means that there is little difference between the relative energies of the conformations, a mixture is usually obtained. Given that most spiroacetal-containing natural products are doubly anomerically stabilized, this method is ideal for, and has found extensive use in, their synthesis.

The requisite dihydroxyketones are commonly assembled via iterative aldol coupling reactions [1], but other methods including Nef reactions [17, 18], acetylide additions [19, 20], 1,3-dipolar nitrile oxide cycloadditions [21], iterative alkylation of dithianes [22–28], hydrazones [29], oximes [30], nitriles [31], or dihalomethylene species [32–34], cross-metathesis/hydroboration/oxidation [35], iterative substitution of a xanthate [36], dihydroxylation/desymmetrization of alkenes [37], Horner–Wadsworth–Emmons olefinations [38, 39], allylmetallations [40], and alkyne–alkyne cross-coupling [41] have also been reported.

2.1.1 Aliphatic Spiroacetals

Recent examples of the application of this strategy for the synthesis of aliphatic natural product spiroacetals have been demonstrated by the syntheses of spirastrellolide A [42], spirofungins A and B [43–45], cephalosporolides E and F [46, 47], spirangien A [48, 49], and pteridic acids A and B [50].

Cephalosporolides E and F have been synthesized independently by Fernandes and Ingle [46] and Brimble et al. [47]. Both groups used a dehydrative spirocyclization strategy to assemble the spiroacetals in high yield as a mixture of the two epimers (Scheme 3).

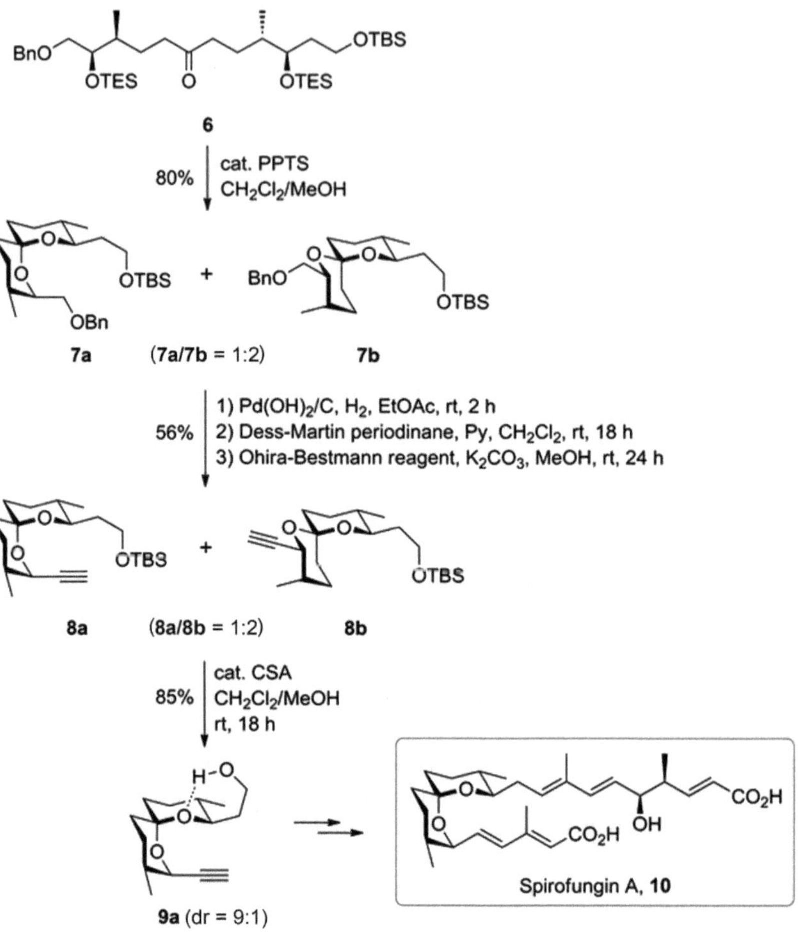

Scheme 4 Rizzacasa et al.'s synthesis of spirofungin A [45]

Rizzacasa et al. [45] used this method in their synthesis of the *Streptomyces violaceusniger* metabolite spirofungin A, **10** (Scheme 4). However, dehydrative spirocyclization of ketone **6** afforded the undesired singly anomeric isomer **7b** as the major product, as it is less sterically hindered than the desired isomer **7a**, and therefore the thermodynamically favored isomer. Elaboration to the alkyne spiroacetals **8a** and **8b** did not significantly change the inherent thermodynamic preference for the singly anomeric isomer; however upon removal of the TBS-ether, intramolecular hydrogen bonding (that was not possible in the singly anomeric isomer) enabled equilibration of the mixture to access predominantly the desired doubly anomeric isomer **9a**.

Smith and coworkers have used a Ca(II)-mediated equilibration strategy to access the singly anomeric CD spiroacetal of spongistatin-1, **15** [26] (Scheme 5). Oxidative removal of the dithiane **11** and subsequent spirocyclization afforded spiroacetals **12a** and **12b** in excellent yield, favoring the undesired doubly anomeric

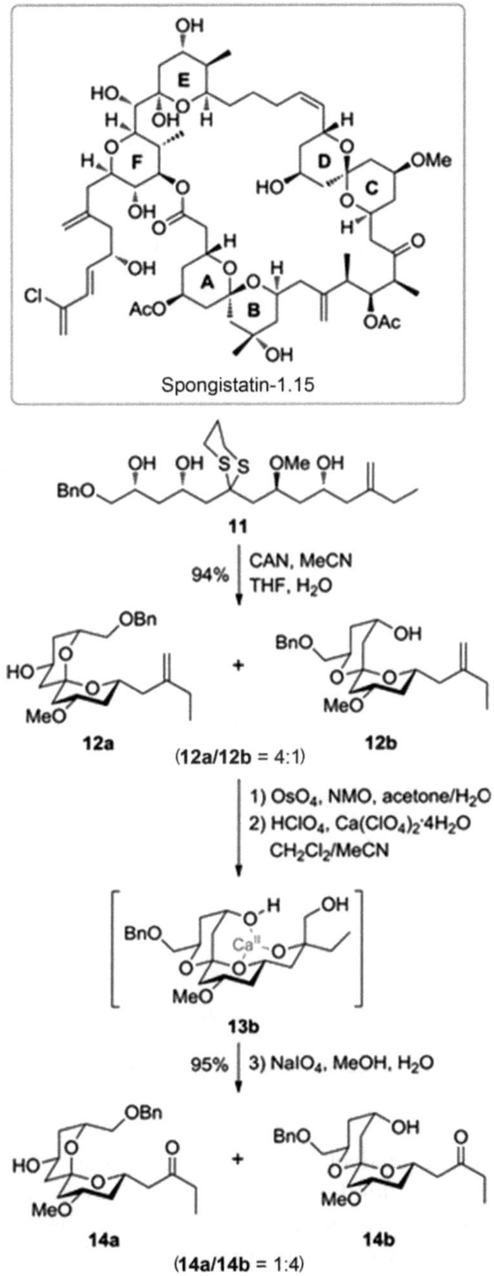

Scheme 5 Smith et al.'s synthesis of the CD spiroacetal of spongistatin-1 [26]

isomer. However, transformation to the triol species **13b** (doubly anomeric isomer not shown) allowed selective equilibration to the mono-anomeric spiroacetal **14b** due to favorable chelation of the calcium ion.

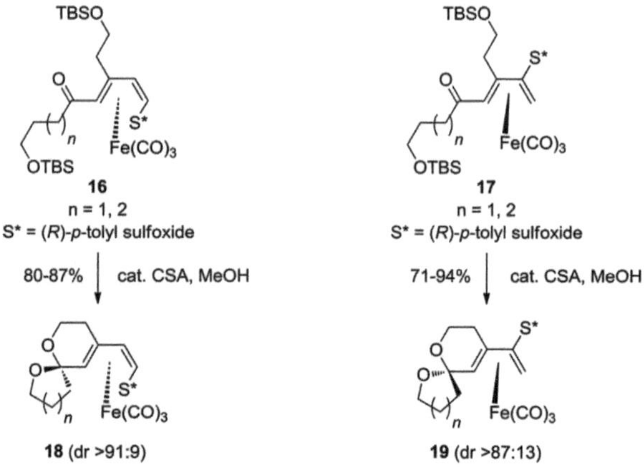

Scheme 6 Paley et al.'s use of an iron(0)-tricarbonyl diene as a chiral-directing element [51]

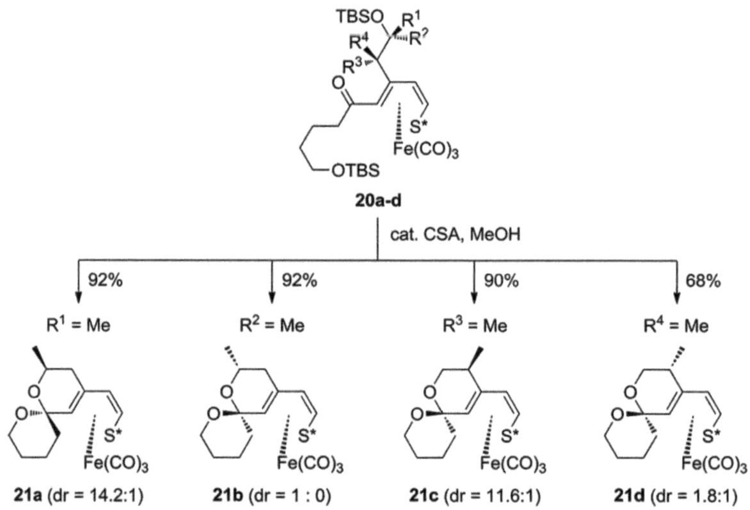

Scheme 7 Paley et al.'s use of an iron(0) tricarbonyl diene for spirocyclization in the presence of an existing chiral center [51]

A recent attempt to selectively control the dehydrative spirocyclization of a dihydroxyketone using a chiral-directing element has been reported by Paley et al. [51]. Incorporation of an iron(0)-tricarbonyl diene complex into the dihydroxyketone skeleton was used to direct spirocyclization (Scheme 6). Thus, the unsubstituted spiroacetals **18** and **19** were obtained selectively in high yield from the dehydrative spirocyclization of sulfinyl iron(0) diene complexes **16** and **17**, respectively.

The results were more complex for substrates containing an additional chiral element (i.e., **20a–d**) (Scheme 7). These differences were attributed to

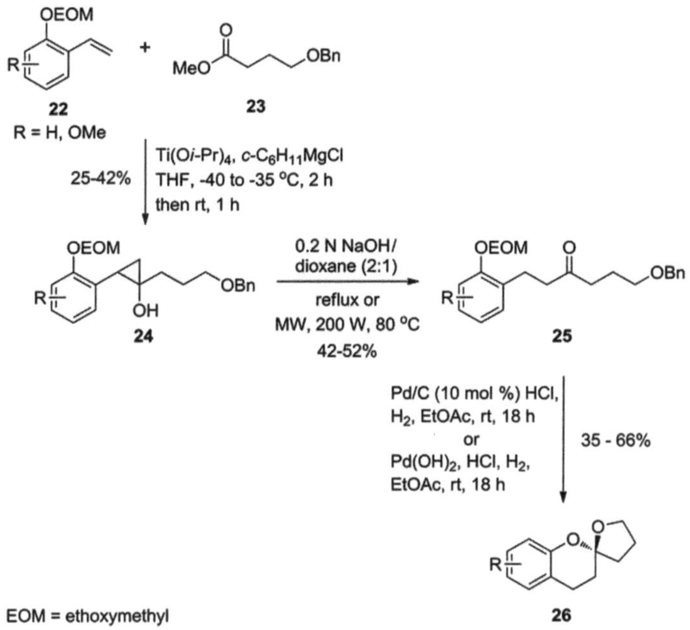

Scheme 8 Kulinkovich coupling/ring-opening strategy toward mono(benzannulated) spiroacetals [52]

minimization of an unfavorable dipole–dipole interaction between the oxygen atom of the "B" ring of the spiroacetal and the $Fe(CO)_3$ moiety.

2.1.2 Mono(benzannulated) Spiroacetals

Brimble and Haym have used dehydrative spirocyclization to achieve the synthesis of a range of simple mono(benzannulated) spiroacetals. The requisite dihydroxy-ketones were prepared via a Kulinkovich coupling/ring-opening strategy [52] (Scheme 8). Subsequent removal of the protecting groups under acidic conditions afforded the mono(benzannulated) spiroacetals **26** in moderate yield.

This approach is relatively low yielding. A more efficient approach [53] involved Horner–Wadsworth–Emmons (HWE) union of the phosphonates **27** with aromatic aldehyde **28** to give the cyclization precursors **29** in generally high yield (Scheme 9). Chemoselective reduction followed by treatment with either DDQ or 1 M HCl then afforded the spiroacetals **30–32**.

Alternatively, using a benzyl ether-protecting group, an efficient tandem hydrogenation/deprotection/cyclization sequence was demonstrated [53] (Scheme 10). Direct hydrogenation of the HWE products unmasks the dihydroxyketones from which the mono(benzannulated) spiroacetals **36** are derived.

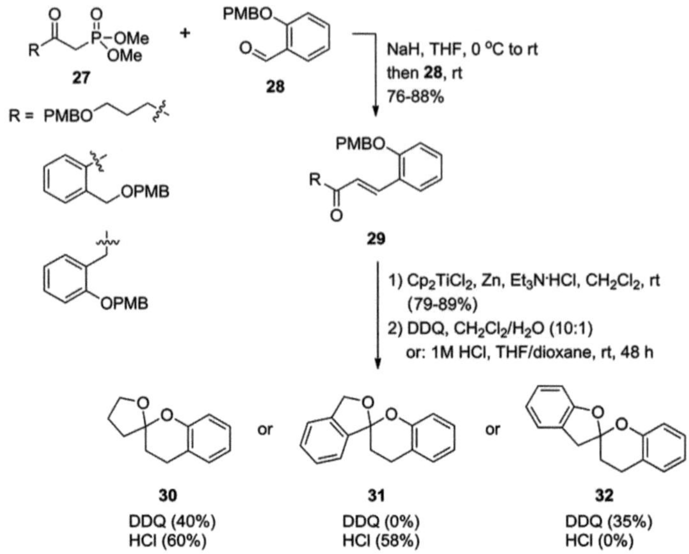

Scheme 9 HWE/chemoselective reduction strategy toward mono(benzannulated) spiroacetals [53]

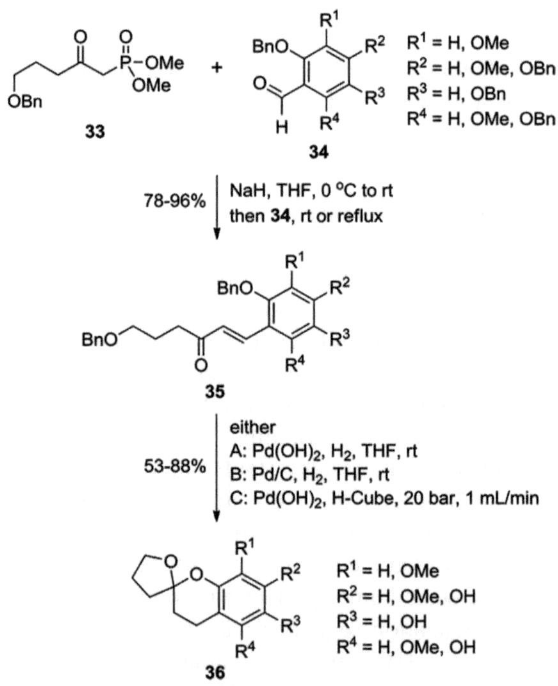

Scheme 10 HWE/deprotection strategy toward mono(benzannulated) spiroacetals [53]

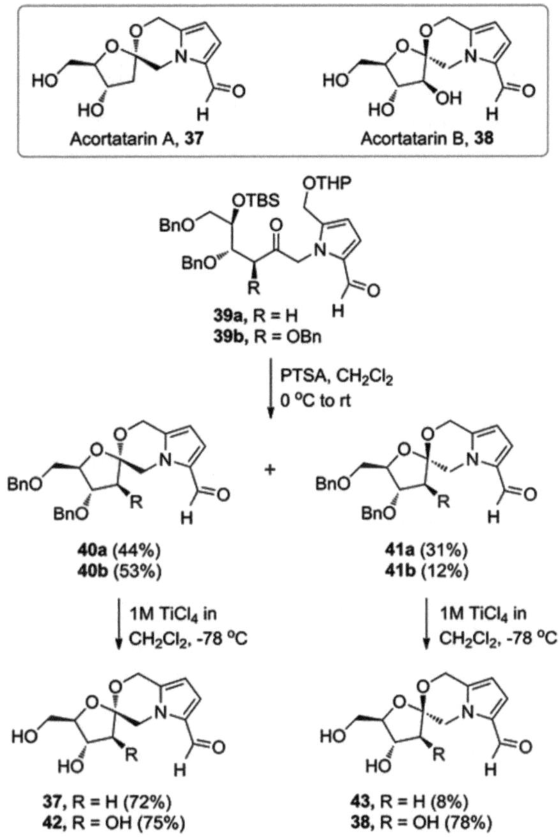

Scheme 11 Sudhakar et al.'s synthesis of acortatarins A and B [54]

A recent example of the use of dehydrative spirocyclization for the synthesis of a heteroannulated spiroacetal natural product is Sudhakar et al.'s synthesis of the acortatarins **37** and **38** [54] (Scheme 11).

Treatment of ketone **39** with PTSA resulted in deprotection of both the TBS and THP ethers. Subsequent spirocyclization of the unmasked dihydroxyketone afforded spiroacetals **40** and **41** as a mixture of anomers. Global deprotection with $TiCl_4$ then gave the natural products **37** and **38** in good yield, along with their spiro-epimers.

2.1.3 Bis(benzannulated) Spiroacetals

Dehydrative spirocyclization of a dihydroxyketone has also been successfully applied to the synthesis of bis(benzannulated) spiroacetals. A recent example is the total synthesis of paecilospirone **46** by Brimble et al. [55] (Scheme 12).

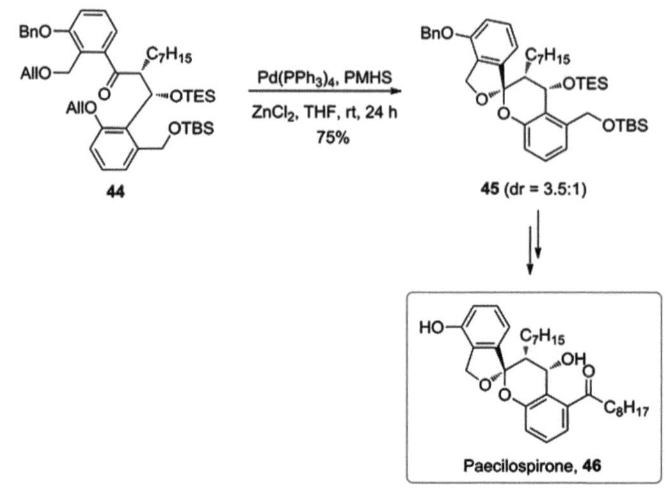

Scheme 12 Brimble et al.'s synthesis of paecilospirone [55]

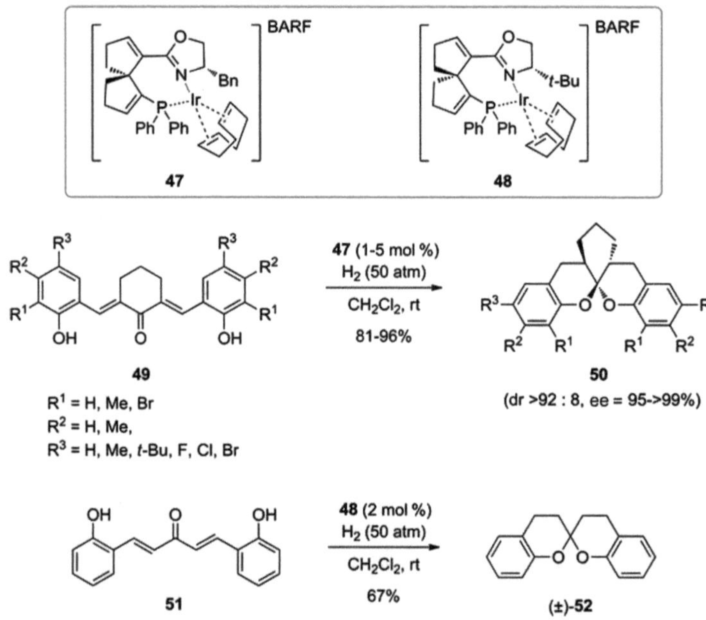

Scheme 13 Ir/SpinPHOX as catalysts for the synthesis of bis(benzannulated) spiroacetals [56]

In this case the use of allyl ether-protecting groups allowed their removal under pH neutral conditions. Thus, treatment of **44** with catalytic Pd(0) in the presence of a polymethylhydrosiloxane–zinc chloride complex (PMHS–$ZnCl_2$) afforded the spiroacetal core of paecilospirone in good yield, as a 3.5:1 mixture of anomers.

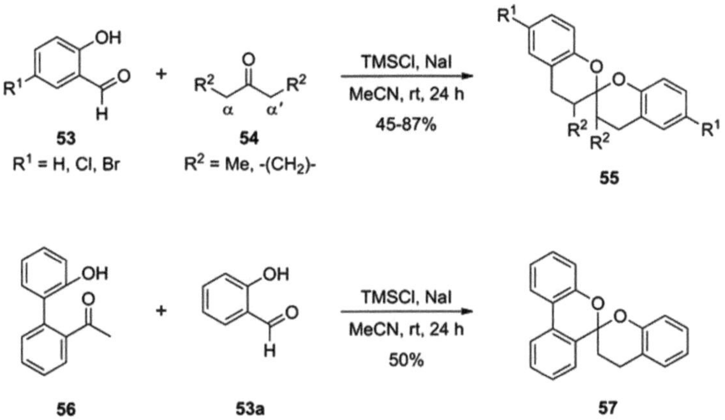

Scheme 14 TMSI-promoted spirocyclization of dihydroxyketones

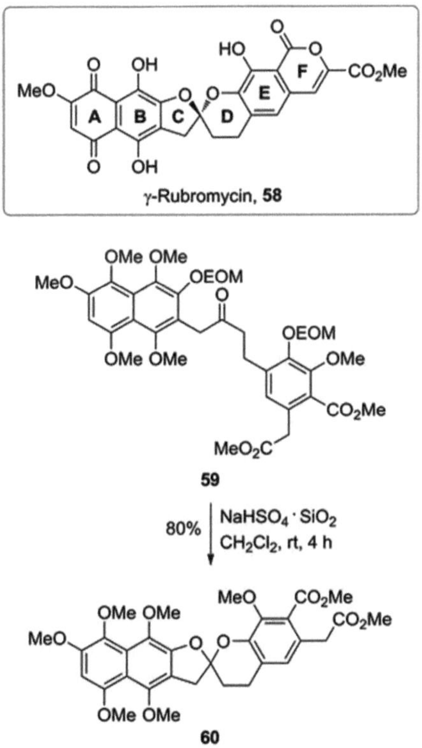

Scheme 15 Dehydrative spirocyclization in Brimble et al.'s synthesis of γ-rubromycin [61]

Wang et al. [56] have recently described the use of Ir/SpinPHOX catalysts **47** and **48** for the synthesis of bis(benzannulated) spiroacetals under asymmetric hydrogenation conditions (Scheme 13). Treatment of a variety of substituted

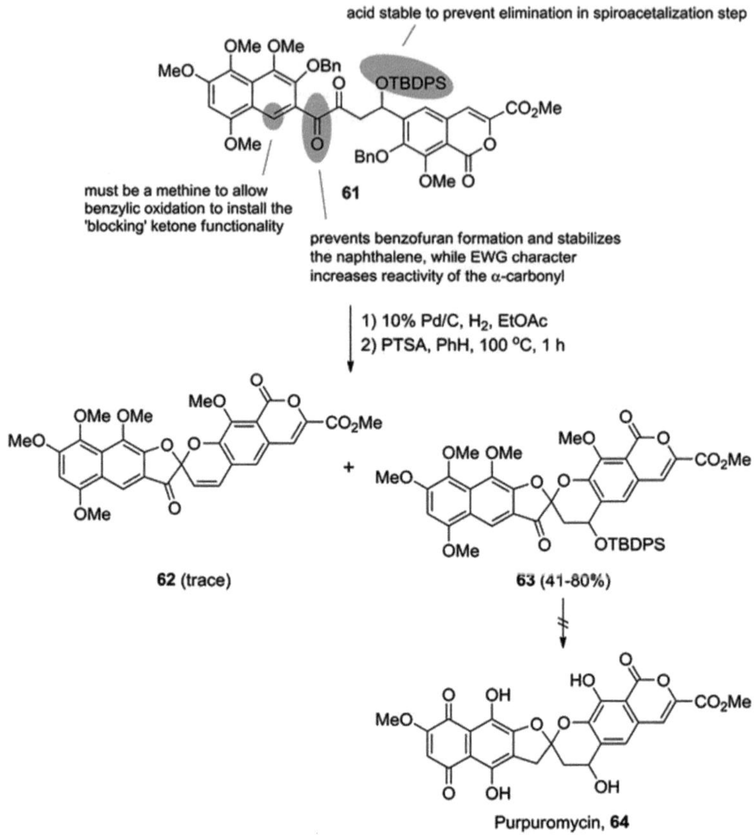

Scheme 16 Kozlowski et al.'s dehydrative spirocyclization studies toward purpuromycin [62]

phenols **49** with Ir(I) catalyst **47** under hydrogen at a high pressure afforded the bis (benzannulated) spiroacetals **50** in high yield and diastereoselectivity, in addition to excellent enantioselectivity. This high level of selectivity may be attributed to the restraining effect of the alkyl tether. On the other hand, hydrogenation of the acyclic ketone **51** with **48** afforded the corresponding spiroacetal **52** as a racemic mixture.

Shi et al. have recently reported a TMSI-promoted approach toward heteroannular acetals which has also been applied to spiroacetals [57]. Double arylmethylation of a ketone at the α-position with an aromatic aldehyde affords a dihydroxyketone that then undergoes dehydrative cyclization to give the desired acetal. Spiroacetals were accessed using symmetric ketones, where both the α and α′ sites are available for alkylation (Scheme 14). Thus, coupling of salicylic aldehydes **53** and ketones **54** with excess TMSI and NaI afforded the spiroacetals **55**. Alternatively, coupling of a salicylic aldehyde **53a** with an unsymmetrical ketone **56** afforded the dihydroxyketone **57**.

The synthesis of members of rubromycin family of bis(benzannulated) spiroacetals under acid-catalyzed conditions have been hampered by the complex

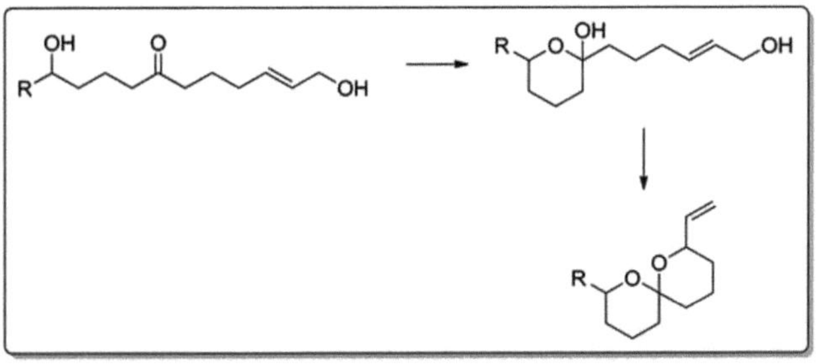

Scheme 17 Metal-catalyzed addition/elimination of allylic alcohols toward spiroacetals

stereoelectronic properties of naphthalene and/or isocoumarin precursors [58–60]. In their formal synthesis of γ-rubromycin **58**, Brimble et al. [61] successfully balanced these factors to access the bis(benzannulated) core via dehydrative spirocyclization of a dihydroxyketone **59** (Scheme 15).

Key to this strategy was removal of the isocoumarin F ring in the spirocyclization precursor, thereby increasing the nucleophilicity of the E ring phenol such that cyclization could occur under mildly acidic conditions. Thus, treatment of precursor **59** with silica-supported sodium hydrogen sulfate afforded the bis(benzannulated) spiroacetal **60** in high yield.

Additionally, in a recent report of their studies toward the related purpuromycin **64**, Kozlowski and coworkers [62] found that the use of diketone **61** enabled spirocyclization to take place under acidic conditions (Scheme 16).

Hydrogenolysis of the benzyl ethers, followed by treatment with PTSA afforded the bis(benzannulated) spiroacetal **63** in moderate to good yield, with only trace amounts of the benzopyran byproduct **62**. Unfortunately, the authors were unable to selectively reduce the "blocking" ketone functionality nor successfully oxidize the naphthalene to the dihydroquinone and thus convert the spirocyclic core structure to purpuromycin.

2.2 Metal-Catalyzed Addition/Elimination of Allylic Alcohols

Activation of an allylic alcohol by metal catalysts can be used to facilitate addition of a variety of nucleophiles. This method has recently been adapted for the synthesis of spiroacetals via hemiacetals substituted with an allylic alcohol side chain (Scheme 17).

Hirai et al. [63] reported the use of this strategy for the synthesis of 5,5-benzannulated spiroacetal **66** (Scheme 18). The major (doubly anomeric) isomer **66a** arises from ring opening/closing of the hemiacetal **65**, followed by oxypalladation to give **69** (doubly stabilized by the anomeric effect). Subsequent *syn* elimination then provides **66a** as the major isomer.

This approach has also recently been utilized by Borrero and Aponick [64] in their synthesis of acortatarin A (Scheme 19). Unfortunately, the Pd-catalyzed

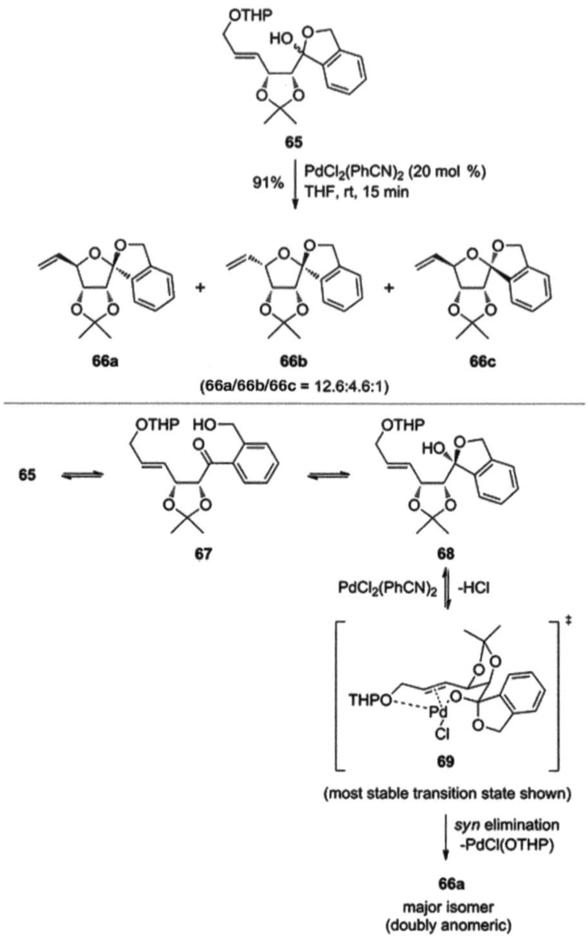

Scheme 18 Hirai et al.'s Pd-catalyzed synthesis of 5,5-spiroacetals [63]

spirocyclization of precursor pyrrole **70** was not selective, affording spiroacetals **71** and *epi*-**71** as a separable 1:1 mixture. Spiroacetal **71**, containing the desired stereochemistry, was then elaborated to acortatarin A **37** in 5 steps in moderate yield.

Alternatively, Cossy et al. [65] have recently disclosed a Fe(III)-catalyzed spirocyclization of a ζ-hydroxy allylic acetate, **73** (Scheme 20). Formation of hemiacetal **74**, followed by Fe(III)-catalyzed addition of the alcohol and elimination gives the doubly anomeric spiroacetal **75** in good yield and selectivity. This method has subsequently been used in synthetic studies toward the cytotoxic marine metabolites, the bistramides [66]. Thus, treatment of the hemiacetal **76** under the same conditions afforded the bistramide-like spiroacetal **77** in moderate yield.

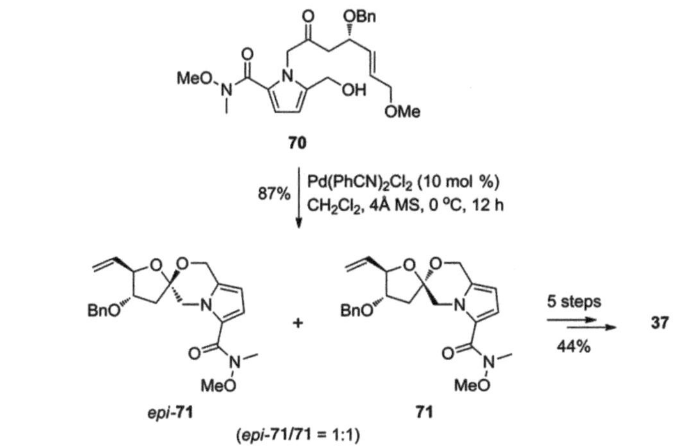

Scheme 19 Borrero and Aponick's Pd-catalyzed spirocyclization approach toward acortatarin A [64]

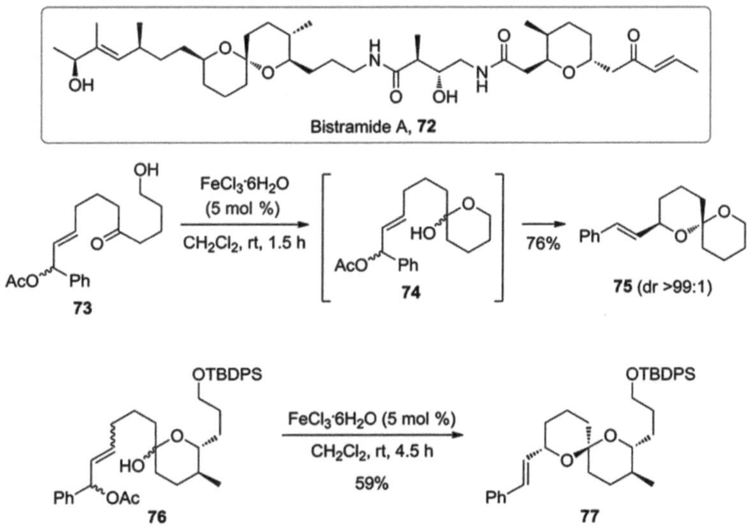

Scheme 20 Cossy et al.'s Fe(III)-catalyzed spirocyclization approach [65, 66]

2.3 Acid-Catalyzed Spirocyclization of a Hemiacetal

Dehydrative spirocyclization may also be performed on hemiacetals, whereby treatment with a Brønsted or Lewis acid gives rise to an oxonium ion intermediate. Subsequent attack of the pendant alcohol thus affords the spiroacetal (Scheme 21).

In their studies toward the avermectin 2b spiroacetal **81**, Henryon and Férézou [67] have utilized an acid-catalyzed spirocyclization of hemiacetal **80** (Scheme 22).

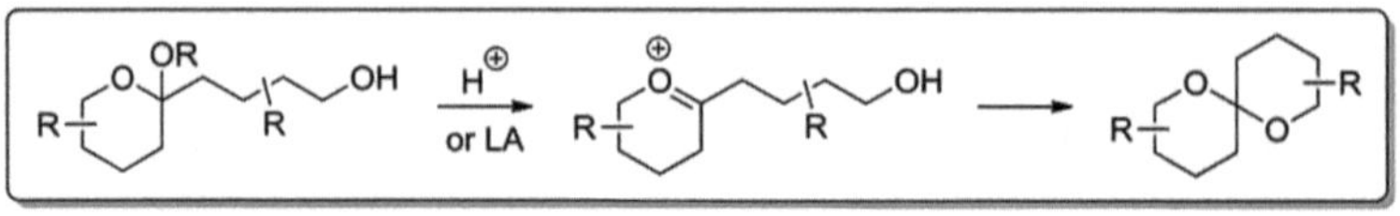

Scheme 21 Acid-catalyzed spirocyclization of a hemiacetal

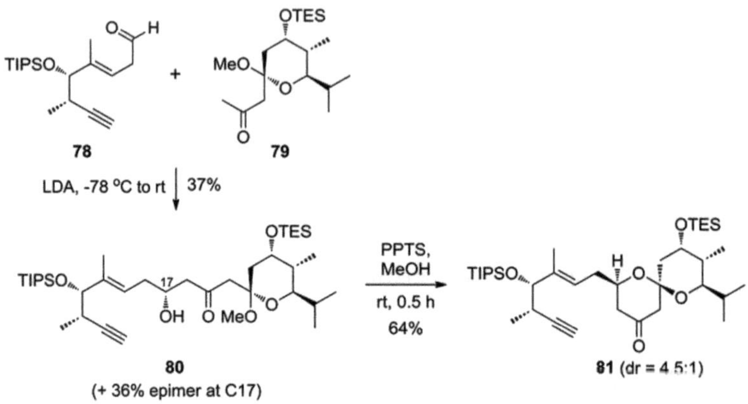

Scheme 22 Henryon and Férézou's approach toward the avermectin 2b spiroacetal [67]

The requisite hemiacetal was prepared via coupling of the enolate of **79** with aldehyde **78**. Subsequent treatment with PPTS in methanol afforded the desired spiroacetal **81** in moderate yield and selectivity.

Jung et al. used a similar strategy in their syntheses of the cytotoxic spiroacetals, auripyrones A **84a** [68] and B **84b** [69] (Scheme 23). Oxidation of alcohol **82**, followed by deprotection of the PMB ether resulted in cyclization to the stable hemiacetals **83a** and **83b** in good yield. Further oxidation and spirocyclization of the hemiacetal under mild acidic conditions afforded the auripyrones **84a** and **84b**.

Yadav et al. [70] have also used this strategy in their total synthesis of pteridic acid A **85** (Scheme 24). Addition of the lithium acetylide derived from **87** to lactone **86**, followed by treatment with catalytic CSA and methanol afforded the hemiacetal. Partial hydrogenation over Lindlar's catalyst then gave the *Z*-alkene hemiacetal **88**. The spirocyclization was affected under mild acidic conditions, affording the spiroacetal framework of pteridic acid A in excellent yield as a single (doubly anomeric) isomer.

This method has also been utilized by Brimble and coworkers [71] in their formal synthesis of berkelic acid **96** (Scheme 25). Addition of trimethylsilyl enol ether **92** to oxonium ion **93** (generated by treatment of hemiacetal **91** with boron trifluoride diethyl etherate) and subsequent cyclization provided the hemiacetal **94**. Finally, hydrogenolysis of the benzyl ethers in the presence of catalytic acid induced spirocyclization, affording the tetracyclic core of berkelic acid, **95**, in moderate yield over 3 steps.

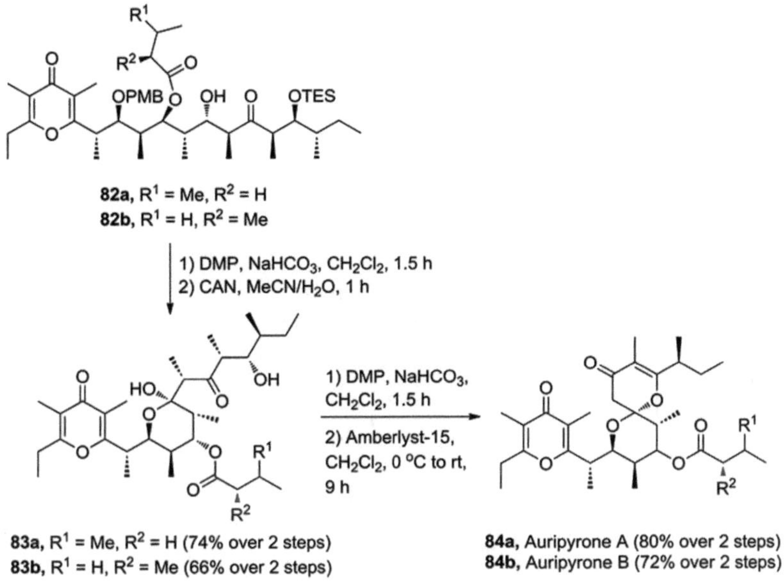

Scheme 23 Jung et al.'s syntheses of auripyrones A and B [68, 69]

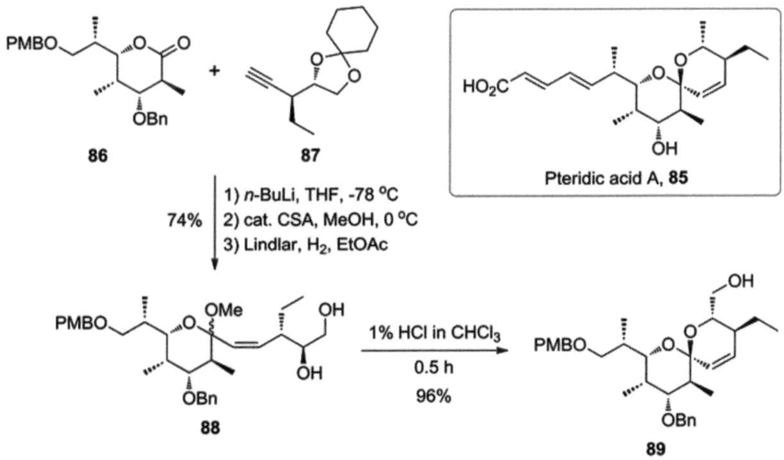

Scheme 24 Yadav et al.'s synthesis of pteridic acid A [70]

2.4 *Spirocyclization of an Endocyclic Enol Ether*

Spirocyclization of an appropriately substituted dihydropyran under acidic conditions (Scheme 26) is another popular method for the synthesis of spiroacetals. The dihydropyrans themselves are obtained via a variety of methods, the most recent of which include Suzuki–Miyaura strategies [72–74]. Other recent developments of

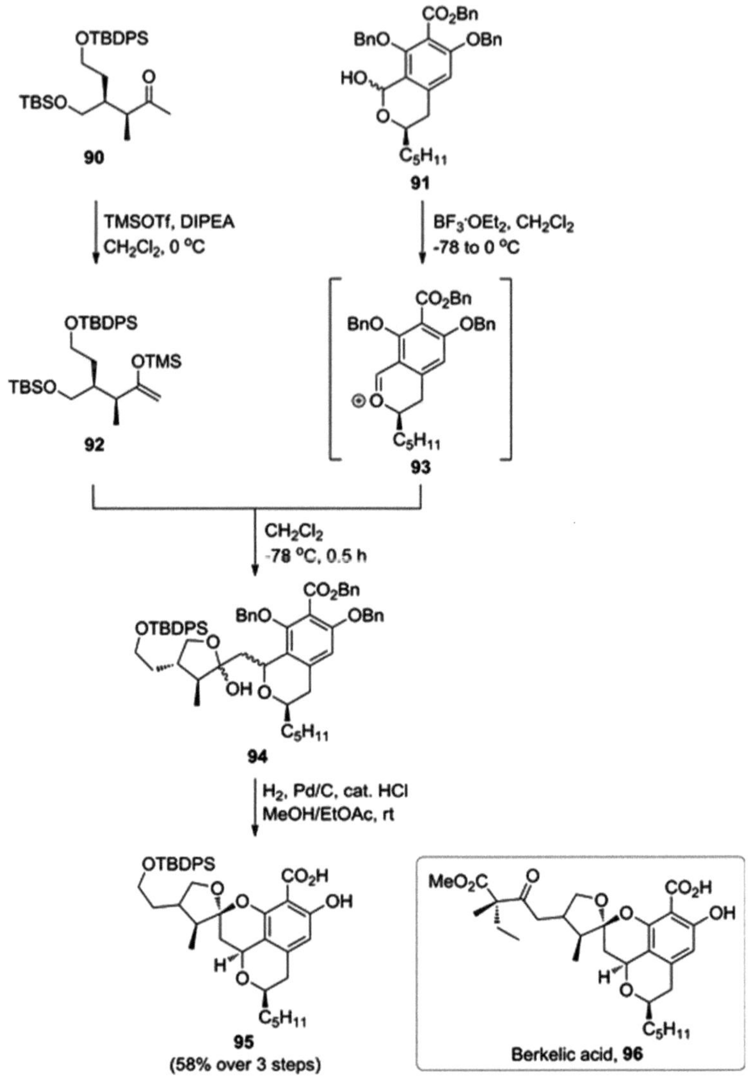

Scheme 25 Brimble et al.'s formal synthesis of berkelic acid [71]

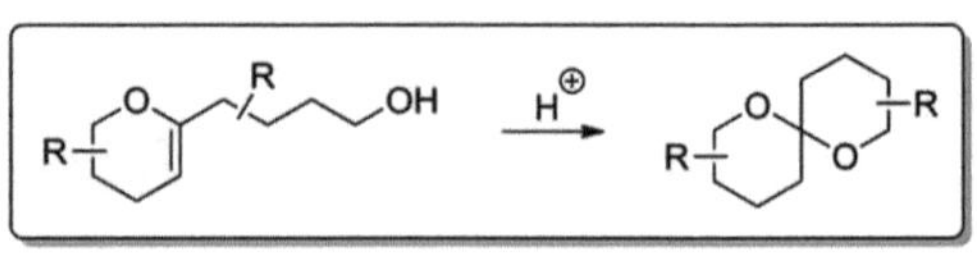

Scheme 26 Spirocyclization of an endocyclic enol ether

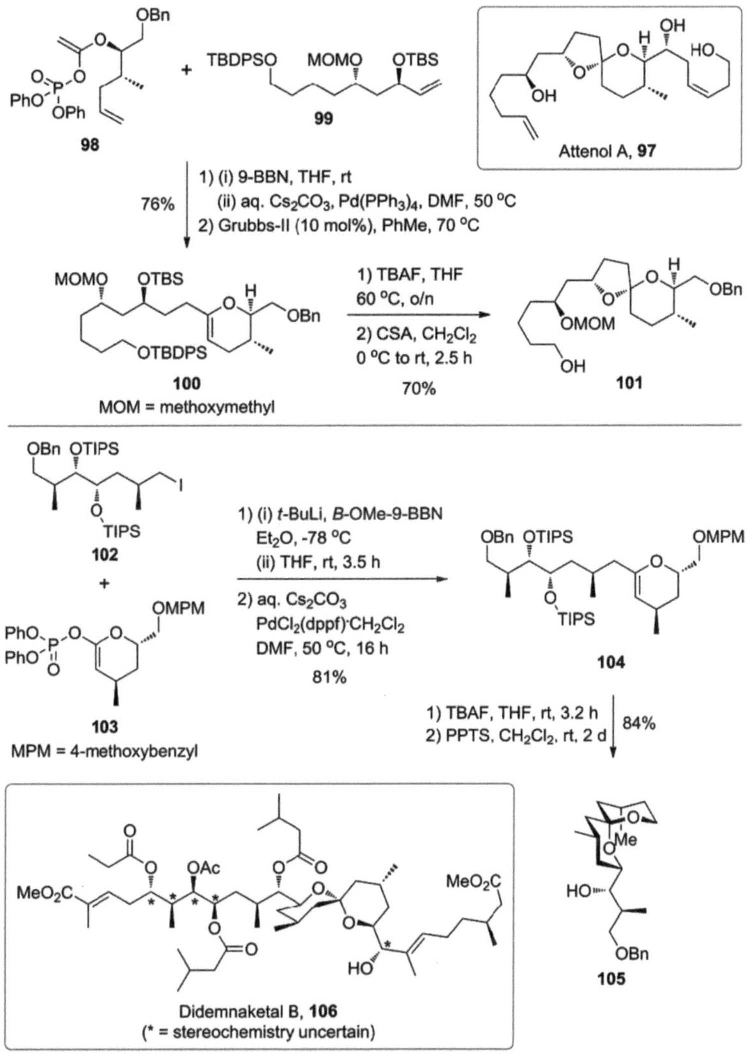

Scheme 27 Fuwa and Sasaki's syntheses of attenol A and didemnaketal B spiroacetal [72, 73]

this method include the use of chiral phosphoric acids for the selective synthesis of mono-anomeric spiroacetals [75, 76] and an intramolecular oxymercuration/reduction sequence [77, 78].

2.4.1 Acid-catalyzed Cyclizations

Fuwa and Sasaki [72, 73] have used this approach in their syntheses of attenol A **97** and the spiroacetal portion of didemnaketal B **106** (Scheme 27). En route to

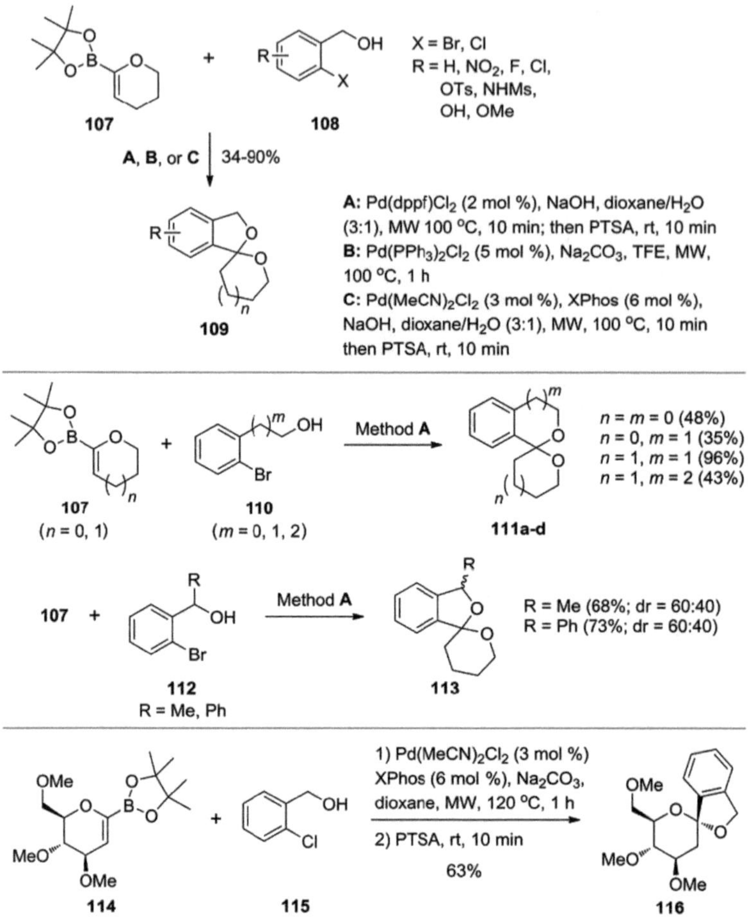

Scheme 28 Cossy et al.'s Suzuki–Miyaura coupling approach toward benzannulated spiroacetals [74]

attenol A, the authors constructed the requisite endocyclic enol ether **100** via a two-step Suzuki–Miyaura coupling/RCM sequence. Deprotection of the silyl ethers with TBAF and subsequent acidic treatment afforded the 5,6-spiroacetal **101** in good yield.

Synthesis of the C9–C28 spiroacetal fragment of didemnaketal B was achieved using a similar strategy. Suzuki–Miyaura coupling of an alkyl borate (derived from iodide **102** and *B*-methoxy-BBN) to the phosphate **103** afforded endocyclic enol ether **104**. Spirocyclization took place after cleavage of the silyl ethers with subsequent treatment with mild acid (PPTS), affording the doubly anomerically stabilized spiroacetal **105** in high yield.

A related Suzuki–Miyaura coupling/acidic spirocyclization sequence has recently been extended to the synthesis of benzannulated spiroacetals by Cossy et al. [74] (Scheme 28). The boronate **107** in this case was derived from the desired endocyclic enol ether framework and was coupled to an aryl halide (bromide is preferred). Electron-rich substrates gave lower yields of product due to side reactions in the spirocyclization step. A range of ring sizes could be accessed, though the yields of 5,5-, 6,5-, and 7,6-spiroacetals were lower than the more common 5,6- and 6,6-spiroacetals.

Papulacandin-like spiroacetal **116** could also be accessed using this method, starting from boronate **114**. Coupling with aryl chloride **115**, followed by acidic treatment gave the anomerically stabilized spiroacetal **116** in moderate yield.

2.4.2 Chiral Acid-Catalyzed Spirocyclization

The use of chiral phosphoric acids and their derivatives (e.g., **117–119**) has recently emerged [7] as a novel method for the stereoselective spirocyclization of simple dihydropyrans. These Brønsted acids act as bifunctional catalysts, providing both acidic and basic sites for complexation to the electrophile and nucleophile, respectively (Fig. 2). Two studies using such catalysts for the stereoselective synthesis of spiroacetals have been reported.

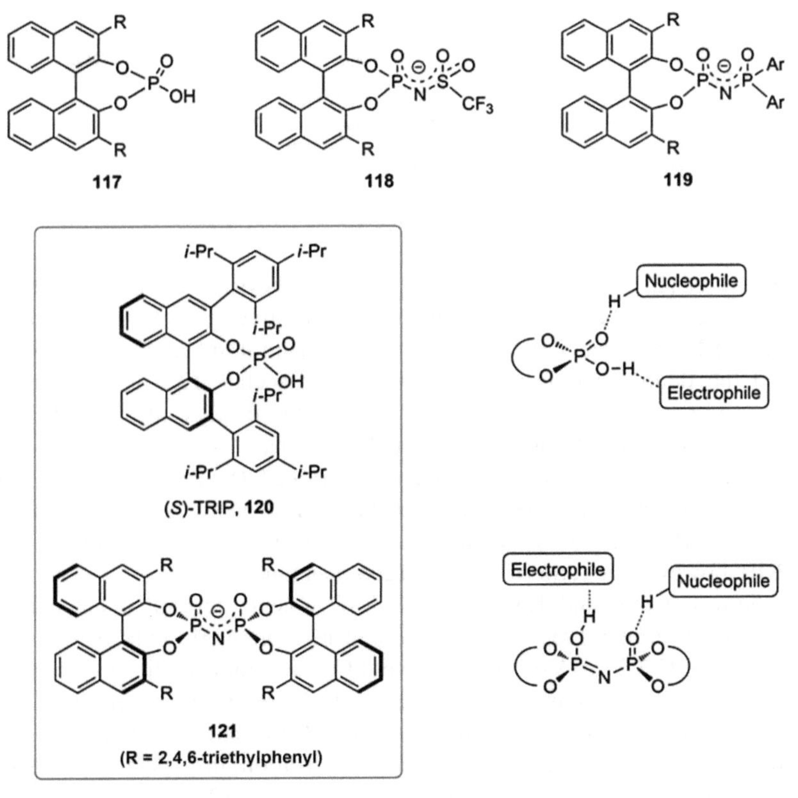

Fig. 2 Chiral phosphoric acids

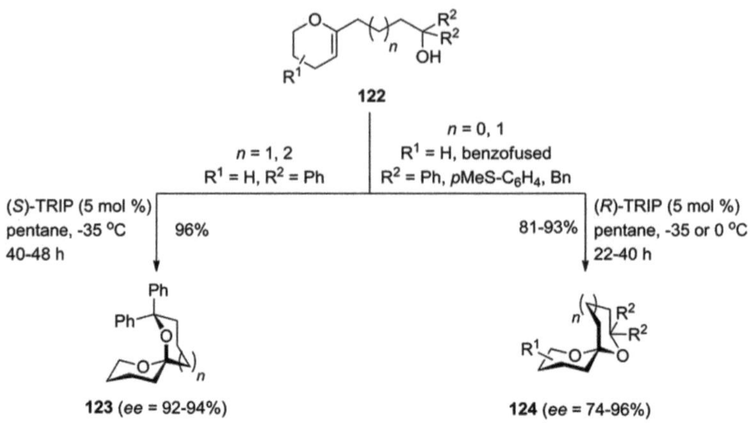

Scheme 29 Nagorny et al.'s use of (*S*)- and(*R*)-TRIP for spirocyclization of an enol ether [75]

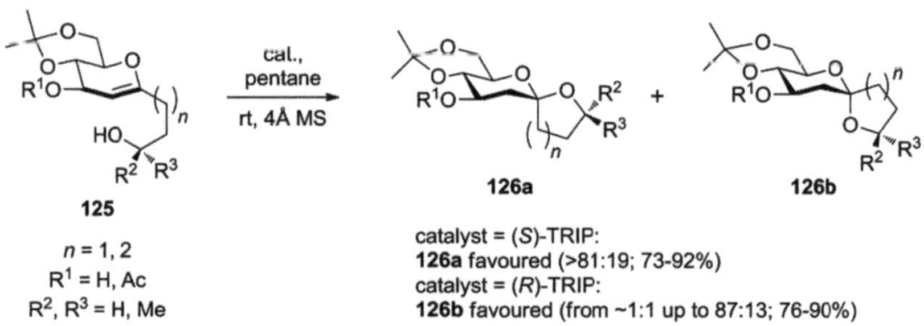

Scheme 30 Use of (*S*)- and(*R*)-TRIP in the spirocyclization of chiral substrates [75]

Nagorny et al. [75] have used the chiral phosphoric acids (*S*)-TRIP **120** and its enantiomer to catalyze the asymmetric dehydrative spirocyclization of dihydropyrans **122** (Scheme 29). Treatment of the achiral substrates with (*S*)-TRIP afforded the doubly anomeric 6,6- and 6,7-spiroacetals **123** in excellent yield and selectivity. Conversely, use of (*R*)-TRIP as the catalyst afforded the mono-anomeric spiroacetals **124** in high yield and good enantioselectivity.

The use of chiral substrates was also examined (Scheme 30). The contra-thermodynamic mono-anomeric spiroacetals **126a** were selectively obtained when (*S*)-TRIP was used as the catalyst. The selectivity could be reversed using (*R*)-TRIP, however the ratio of **126b** to **126a** was much lower.

In contrast, Čorić and List approach [76] was to develop a C2-symmetric imidophosphate dimer **121**, whereby the considerable steric bulk of the BINOL backbone with C-3,3′ aryl substituents would confer both rigidity to the complex and restrict the dimensions of the active site to increase the selectivity. Interestingly, during preliminary optimization studies, Čorić and List found that while (*S*)-TRIP **120** effectively catalyzed the spirocyclization, the selectivity was not as high as that observed using their imidophosphate dimer **121** (Scheme 31).

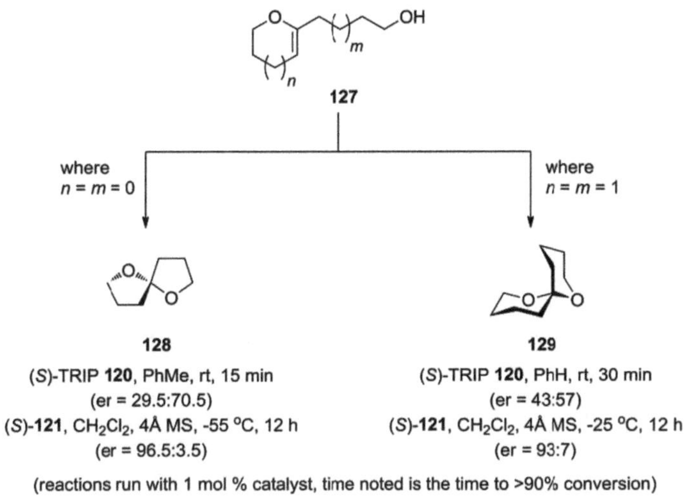

Scheme 31 Comparison of imidophosphate dimer **121** and (*S*)-TRIP for the spirocyclization of enol ethers

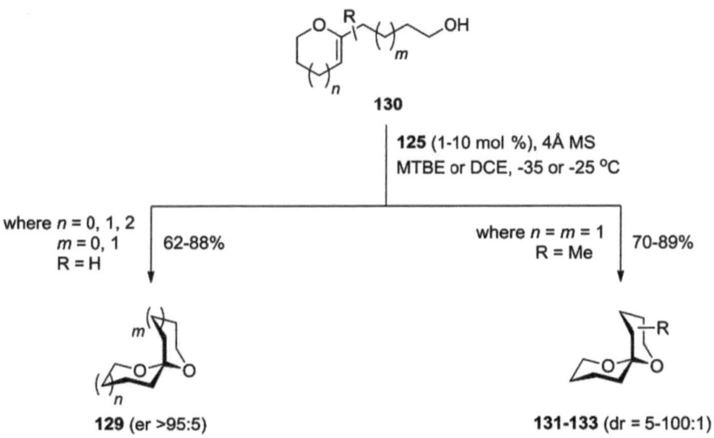

Scheme 32 Use of imidophosphate dimer **121** for the spirocyclization of enol ethers [76]

Using their imidophosphate dimer **121**, unsubstituted dihydropyran and dihydrofuran alcohols **130** gave the corresponding spiroacetals **129** in good yield with excellent stereocontrol, favoring the contra-thermodynamic mono-anomeric spiroacetal (Scheme 32). The thermodynamic spiroacetal could also be obtained using *ent*-**121** in comparable yield and enantioselectivity.

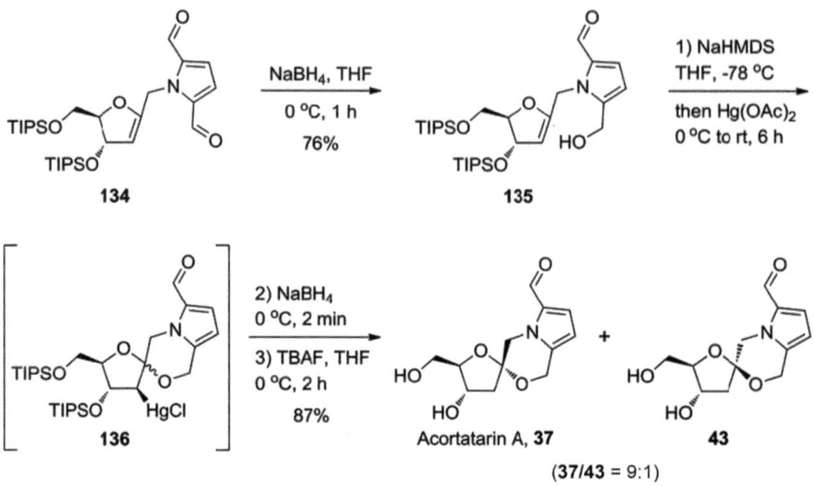

Scheme 33 Tan et al.'s synthesis of acortatarin A [78]

2.4.3 Intramolecular Oxymercuration

Alternatively, an intramolecular oxymercuration/reduction sequence can be used to effect spirocyclization with endocyclic enol ethers [77]. This method was recently used by Tan and coworkers [78] in their synthesis of acortatarin A (Scheme 33).

Reduction of the advanced intermediate dialdehyde **134** provided the requisite alcohol for the oxymercuration. Pretreatment with base (giving the alkoxide) was necessary to increase selectivity for the desired spiroacetal. Reduction of the 2-mercurio spiroacetals **136**, followed by deprotection provided acortatarin A (**37**) in good yield together with its epimer **43** in a 9:1 ratio.

2.5 Spirocyclization of an Exocyclic Enol Ether

Alternatively, an exocyclic enol ether may be used for the synthesis of spiroacetals, the spirocyclization once again taking place via generation of an oxonium ion intermediate (Scheme 34).

An alternative method for the synthesis of spiroacetals involves cyclization of a pendant alcohol to an exocyclic enol ether. A recent example of an acid-catalyzed cyclization of an exocyclic enol ether has been reported by Goekjian et al. [79] in their synthesis of the cytotoxic marine metabolite bistramide A, **72** (Scheme 35). Lactone **137** and benzothiazole **138** were coupled in a modified Julia–Koscienski olefination [80] to give the requisite exocyclic enol ether **139**; subsequent treatment with catalytic PTSA in dichloromethane afforded the spiroacetal **140** in good yield over two steps.

Similarly, Rodriguez et al. [81] have used an iterative sulfone lithiation approach to synthesize a range of aliphatic 5,5-spiroacetals (Scheme 36). Acylation and

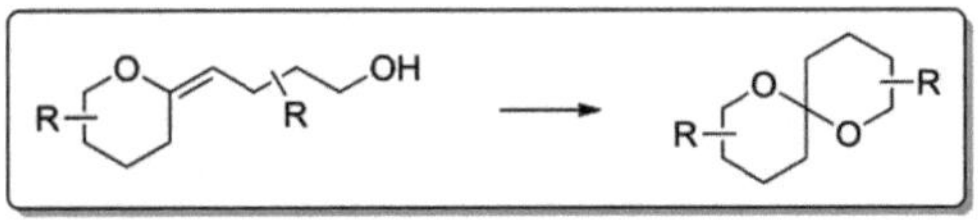

Scheme 34 Spirocyclization of an exocyclic enol ether

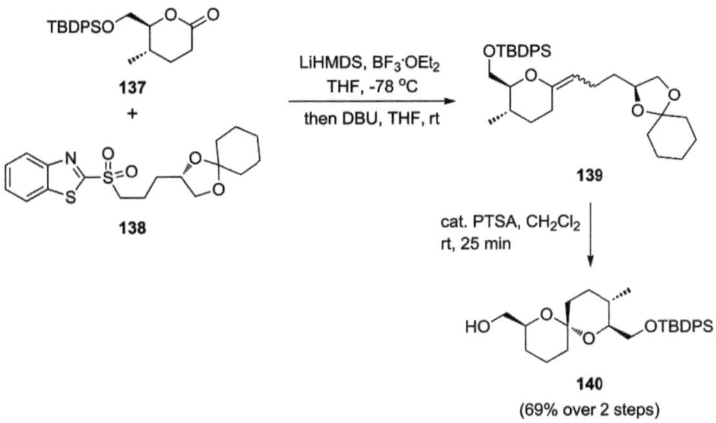

Scheme 35 Goekjian et al.'s approach toward the spiroacetal of bistramide A [79]

substitution of the sulfoximines **141** generates the exocyclic enol ethers **143**. α-Lithiation and subsequent addition of an epoxide incorporates the alkyl chain, and finally, an intramolecular oxa-Michael addition affords the 5,5-spiroacetals **147**.

The sulfoximine-substituted spiroacetals **147** were obtained in good yield using this approach. Reductive removal of the sulfoximine over Al/Hg amalgam afforded the simple 5,5-spiroacetals **149** in moderate to high yield, as approximately 1:1 mixtures of diastereomers.

Alternatively, Mootoo and coworkers [82, 83] have developed a two-step iodoetherification/Ag-catalyzed cyclization method for the synthesis of spiroacetals from homoallylic alcohols **150** (Scheme 37). Treatment of the diol with iodonium dicollidine perchlorate (IDCP), followed by exposure to silver triflate in the presence of collidine affords the spiroacetals **154** in moderate to good yield as diastereomeric mixtures. This method was most recently applied in their synthesis of a simplified monensin analogue, polyether spiroacetal **156** [84]. Thus, two-step treatment of diol **155** under the established conditions provided spiroacetal **156** as a 2:1 mixture of epimers.

2.6 *Transition Metal-Catalyzed Hydroalkoxylation of Internal Alkynes*

The transition metal-catalyzed intramolecular hydroalkoxylation approach for the synthesis of spiroacetals has increased in popularity in the last decade. A variety of

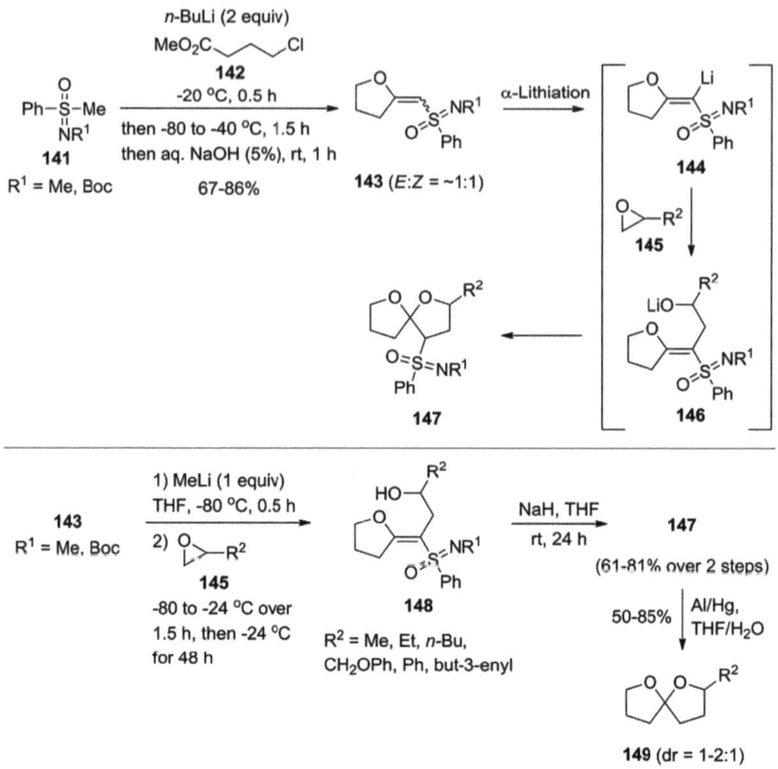

Scheme 36 Rodriguez et al.'s sulfone lithiation approach toward spiroacetals [81]

transition metal complexes may be used to affect the spirocyclization, and both aliphatic and benzannulated spiroacetals may be accessed via this method (Scheme 38).

The hydroalkoxylation of an internal alkyne to afford a spiroacetal was first described by Utimoto in 1983 [85]. Using $PdCl_2$ or $PdCl_2(PhCN)_2$ as catalyst, the spiroacetals **158** were obtained in high yield (Scheme 39). The method was not adopted as a general tool for the synthesis of spiroacetals for some time, with only a few reports exploiting the approach [86, 87]. Recent applications of palladium-catalyzed intramolecular alkyne hydroalkoxylation include syntheses of spirolaxine methyl ether [88] and enantiomers of the natural cephalosporolides [89].

There was an explosion of work described on gold(I) and gold(III) catalysis in the early 2000s, and its application in the hydroalkoxylation of alkynes was no exception [90, 91]. Beginning with studies toward Au(I)- or Au(III)-catalyzed synthesis of bridged acetals [92], these catalysts were soon applied to the synthesis of spiroacetals [93, 94]. This method has been widely embraced by the synthetic community and used in the syntheses of a variety of complex spiroacetal natural products, including the cephalosporolides [95], okadaic acid [96], and ushikulide A [97].

Iridium complexes have also been shown to catalyze the tandem cycloisomerization/hydroalkoxylation of bis-homopropargylic alcohols to give furanyl and

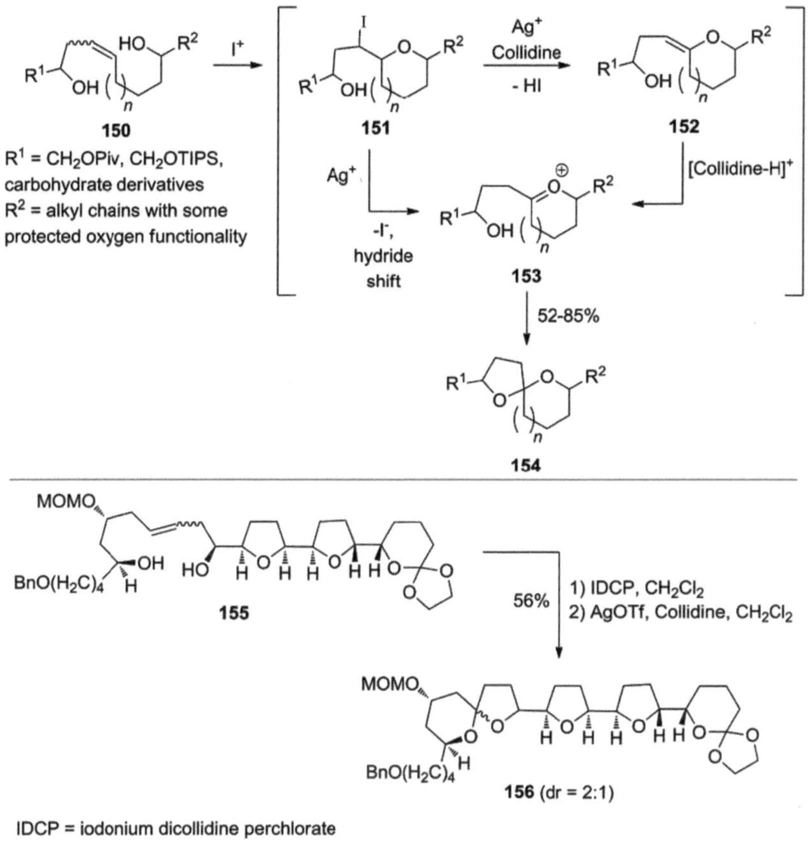

Scheme 37 Mootoo et al.'s synthesis of spiroacetals from allylic alcohols [82–84]

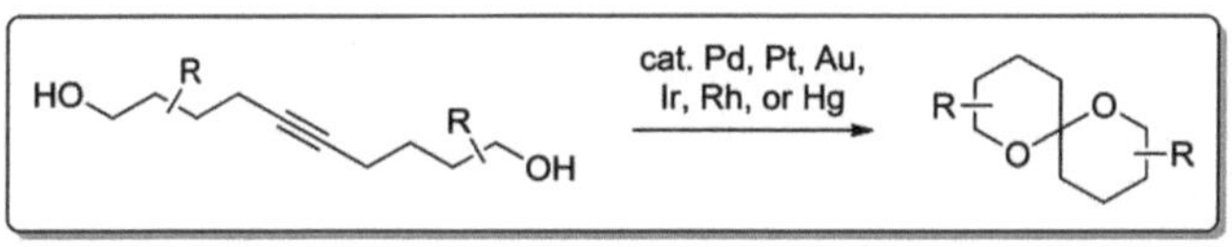

Scheme 38 Transition-metal catalyzed hydroalkoxylation of internal alkynes in the synthesis of spiroacetals

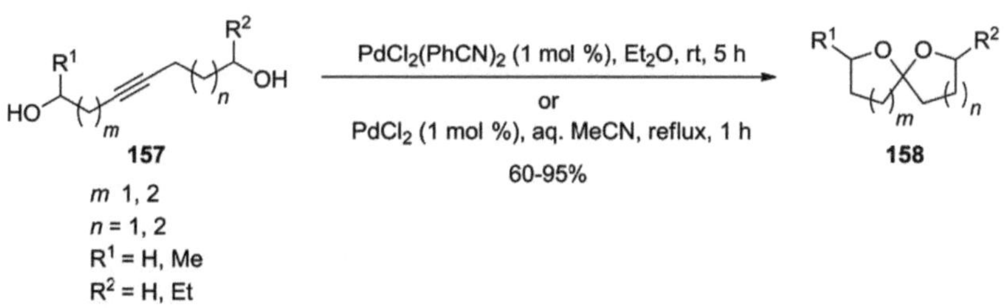

Scheme 39 Utimoto's Pd-catalyzed hydroalkoxylation of internal alkynes [85]

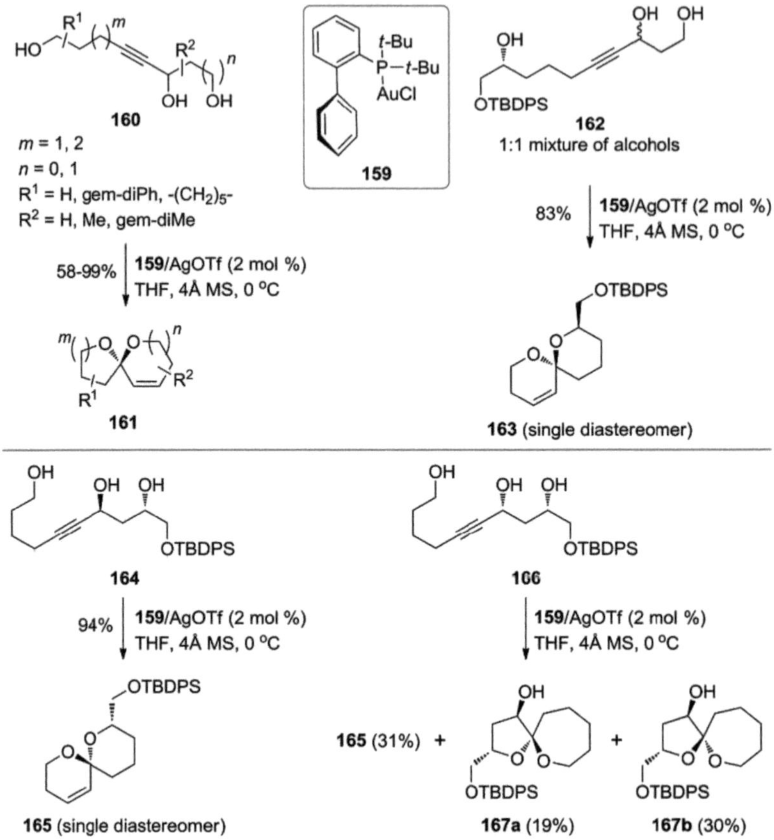

Scheme 40 Aponick et al.'s gold(I)-catalyzed hydroalkoxylation approach toward unsaturated spiroacetals [106]

pyranyl acetals [98]. Messerle and coworkers [99–103] have extended this work to the synthesis of spiroacetals, through the development of several Ir- and Rh-based catalysts for the hydroalkoxylation of internal alkynes (Fig. 4).

Additionally, other metals catalysts have also been exploited for this transformation, including a platinum complex used for the synthesis of spirastrellolide B [104] and a mercury(II) catalyst for the synthesis of hippuristanol [105].

2.6.1 Aliphatic Spiroacetals: Gold, Platinum, Palladium, and Mercury Catalysts

Recently, Aponick et al. [106] reported a study wherein alkyne triols **160, 162, 164,** and **166** underwent spirocyclization with a Au(I) catalyst to give the unsaturated spiroacetals **161, 163,** and **165** (Scheme 40). The effect of the stereochemistry of the propargylic alcohol was also investigated; it was found that 1,3-*anti* relative

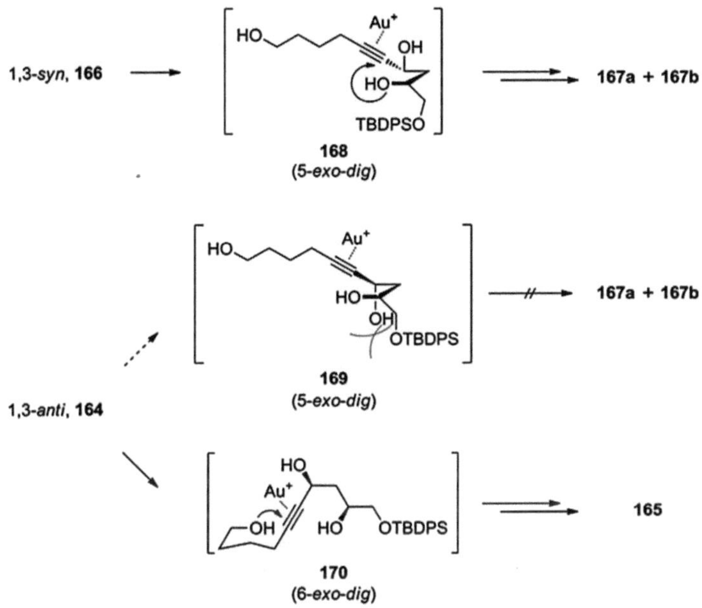

Fig. 3 Preferred cyclization modes of 1,3-*anti* and 1,3-*syn* propargylic triols

stereochemistry resulted in exclusive formation of the 6,6-spiroacetal **165**, while the 1,3-*syn* isomer gave a mixture of 6,6- and 5,7-spiroacetals **167**.

The observed difference in selectivity is presumably due to preferential 5-*exo-dig* cyclization of **166** (Fig. 3). In the 1,3-*anti* triol **164**, however, steric hindrance in the five-membered transition state **169** results in a preference for 6-*exo-dig* cyclization, leading to the formation of desired 6,6-spiroacetal **165**.

Forsyth et al. [96] have recently used Au(I)-catalyzed spirocyclizations for the synthesis of both spiroacetal fragments of okadaic acid **176** (Scheme 41). Synthesis of the C15–C27 spiroacetal was accomplished by treating alkyne **171** with catalytic AuCl in dichloromethane. Hydrolysis of the PMP-acetal then afforded **172** in high yield over two steps.

Attention then turned to synthesis of the C28–C38 spiroacetal from alkyne triol **173**. Alkyne triol **173** was obtained as a 1:1.5 mixture of 1,3-*anti*/1,3-*syn* epimers. As seen in Aponick et al.'s earlier work [106], the regioselectivity of the Au(I)-catalyzed spirocyclization was profoundly influenced by the relative stereochemistry of the 1,3-diol. Thus, the 1,3-*syn* triol gave a mixture of the 6,6- and 5,7-spiroacetals **174–175**, while the 1,3-*anti* triol gave the desired 6,6-spiroacetal **174** selectively.

Similarly, Trost and O'Boyle [97] found during their total synthesis of (–)-ushikulide A **177** that gold(I)-catalyzed spirocyclization of 1,3-*anti*-triol **178a** afforded the unsaturated 6,6-spiroacetal **179** in good yield (Scheme 42). The unsaturated product was, however, undesired (cf. **179** to ushikulide A, **177**) and protection of the alcohol as a benzoyl ether was required to prevent elimination.

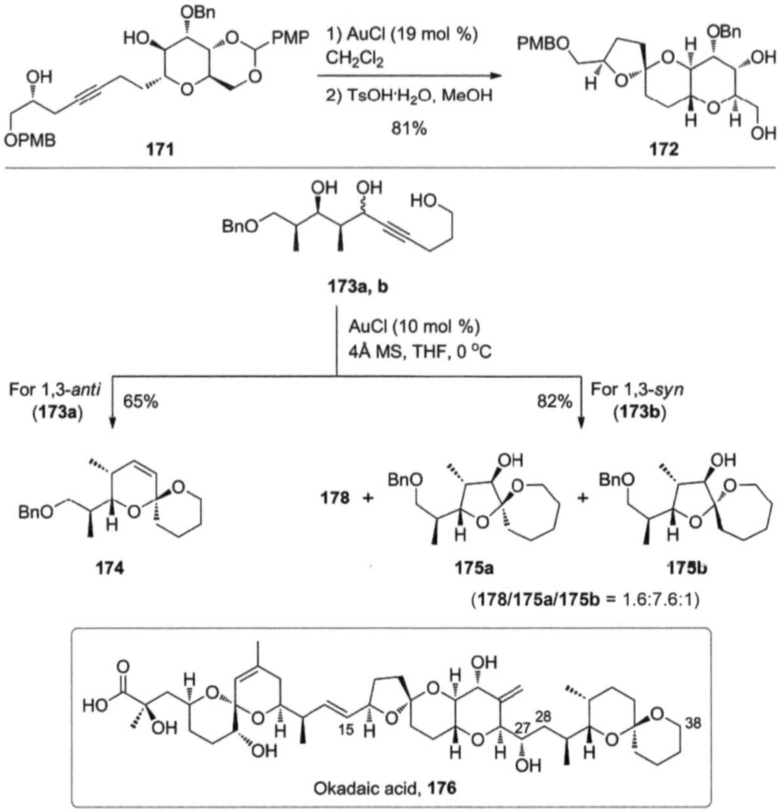

Scheme 41 Forsyth et al.'s gold(I)-catalyzed synthesis of okadaic acid spiroacetals [96]

In their work toward the cephalosporolides **5a–b** and **181**, Dudley and Tlais [95, 107] used a gold(I)-catalyzed intramolecular hydroalkoxylation strategy to form the 5,5-spiroacetal (Scheme 43). Treatment of alkyne **182** with a high catalyst loading of AuCl afforded a mixture of spiroacetals **183a** and **183b** in good yield. Similarly, in their successful synthesis of cephalosporolide E **5a**, spirocyclization with AuCl, followed by removal of the TBS-ether afforded the 5,5-spiroacetals **185** as an approximately 1:1 mixture. Epimerization with zinc chloride and magnesium oxide, followed by oxidation to the lactone completed the synthesis of **5a** in moderate yield.

Ramana et al. [89] opted for a Pd(II) catalyst for the intramolecular hydroalkoxylation of alkyne **186** in their synthesis of the enantiomeric cephalosporolides **189a** and **189b** (Scheme 44). The spirocyclization afforded a 1:1 mixture of the spiroacetals **187**, which were then elaborated to the *ent*-natural products via hydrolysis of the isopropylidene acetal, followed by oxidation and Barton–McCombie deoxygenation.

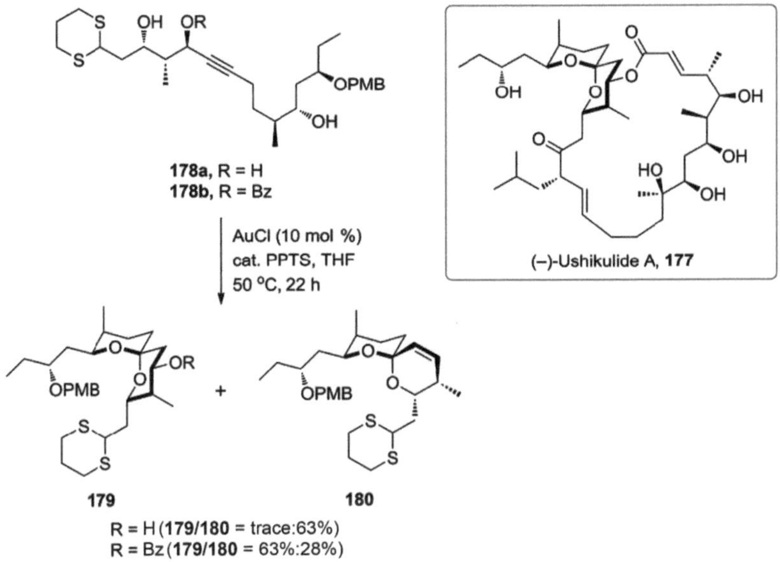

Scheme 42 Trost and O'Boyle's gold(I)-catalyzed synthesis of ushikulide A [97]

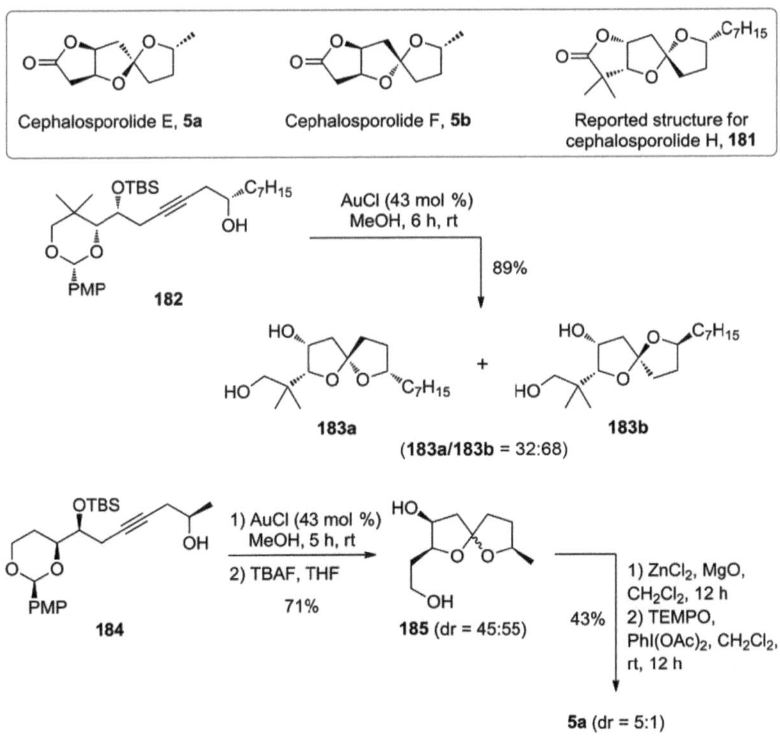

Scheme 43 Dudley and Tlais' gold(I)-catalyzed synthesis of cephalosporolide E [95, 107]

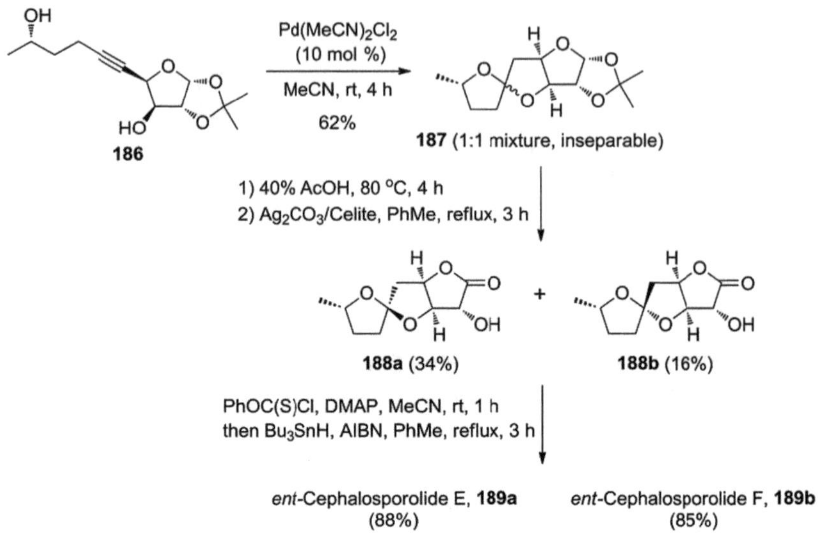

Scheme 44 Ramana et al.'s Pd(II)-catalyzed synthesis of *ent*-cephalosporolides E and F [89]

Mercury(II) triflate has emerged as an effective catalyst for a variety of cyclizations [108]. Its use in the hydroalkoxylation of alkynediols has been investigated by Deslongchamps and coworkers [109] (Scheme 45).

Treatment of the alkynes **190** with $Hg(OTf)_2$ in acetonitrile afforded the spiroacetals **191** in excellent yield. In cases where both 5-*exo-dig* and 6-*exo-dig* cyclization was possible ($m = 2$, $n = 2$), 6-*exo-dig* was favored, giving only 6,6-spiroacetals as the product. However, in substrates where both 6-*exo-dig* and 5-*endo-dig* cyclizations were possible ($m = 1$, $n = 1, 2$), 5-*endo-dig* was favored. The unsaturated spiroacetals **193** could be accessed from triol alkynes **192** in similarly excellent yield via a cyclization/elimination sequence upon treatment with $Hg(OTf)_2$. Deslongchamps et al. [105] had also used this method in their synthesis of hippuristanol, where treatment of **194** with $Hg(OTf)_2$ afforded **195** in excellent yield.

2.6.2 Aliphatic bis(spiroacetals)

Lee et al. [110] have reported the synthesis of bis(spiroacetals) using an intramolecular alkyne hydroalkoxylation strategy (Scheme 46). Treatment of diynes **196** with catalytic $Ph_3PAuCl/AgOTf$ under microwave irradiation afforded the bis (spiroacetals) **197–198** in moderate to good yield, generally favoring the trans-bis (spiroacetal) **197**.

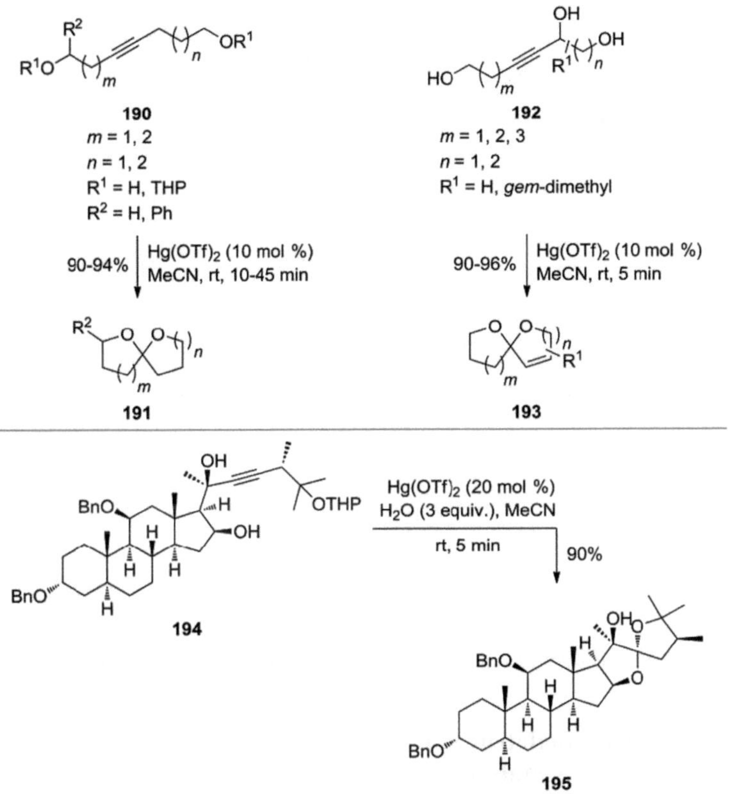

Scheme 45 Deslongchamps et al.'s Hg(II)-catalyzed hydroalkoxylation of internal alkynes [105, 109]

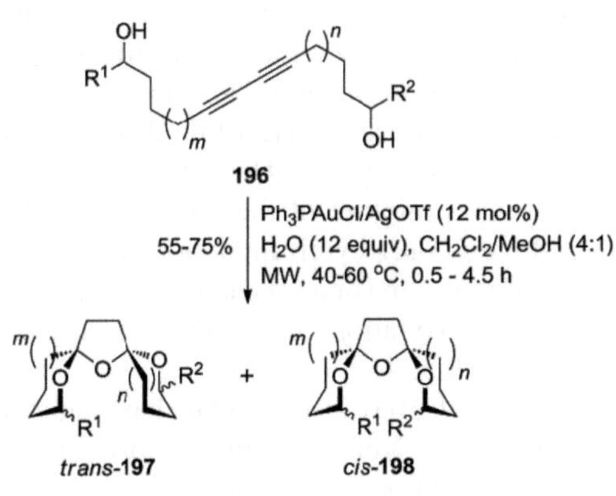

Scheme 46 Lee et al.'s gold(I)-catalyzed synthesis of bis(spiroacetals) [110]

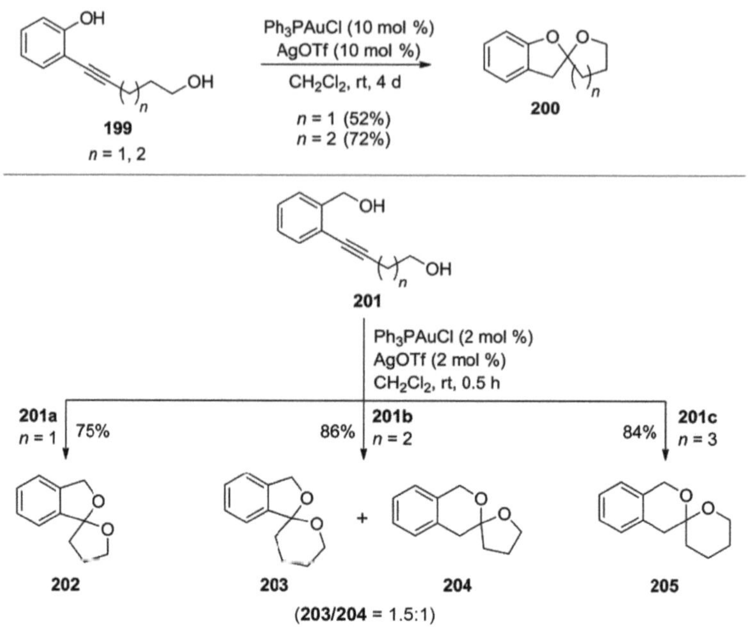

Scheme 47 Xue et al.'s gold(I)-catalyzed synthesis of mono(benzannulated) spiroacetals [111]

2.6.3 Mono(benzannulated) Spiroacetals: Gold, Rhodium, Iridium, and Mercury Catalysts

The synthesis of mono(benzannulated) spiroacetals has been studied by several research groups. Xue et al. [111] reported the use of a Au(I)/Ag(I) catalytic system for the spirocyclization of aromatic alkynes **199** and **201** (Scheme 47). The phenolic substrates **199** underwent spirocyclization to afford the 5,5- and 5,6-mono (benzannulated) spiroacetals **200** in moderate to good yield. Alkyne **201a**, where $n = 1$, gave the 5,5-spiroacetal exclusively, while **201c**, where $n = 3$, gave the 6,6-spiroacetal only, both in good yield. The benzylic alcohol substrate **201b**, where $n = 2$, however, gave a mixture of the regioisomers **203** and **204**, due to competing 5-*exo-dig* and 6-*endo-dig* cyclization pathways.

In addition, several Rh(I)- and Ir(I)-based catalysts have been found to catalyze the intramolecular hydroalkoxylation of aromatic alkynes to give mono (benzannulated) spiroacetals (Fig. 4).

Messerle and coworkers [99–103] in particular, have studied a range of these catalysts (Scheme 48). As seen in Xue et al. study, substrates where 5-*exo-dig* and 6-*endo-dig* cyclization can compete (i.e., where $n = 2$) give a mixture of spiroacetals. In addition, where $n = 1$, the benzopyran **220** is often obtained as a minor product, with varying selectivity.

Crabtree et al. [112] have also reported the use of Ir(I) catalyst **213** for this transformation, affording the 5,5-spiroacetal **202** in good yield (Scheme 49).

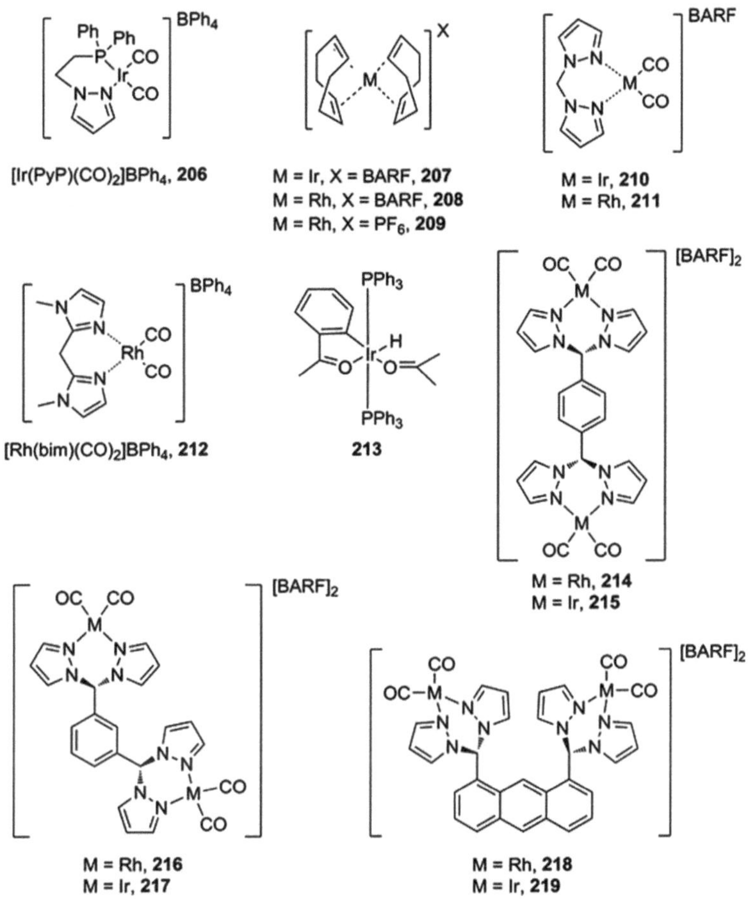

Fig. 4 Rh(I)- and Ir(I)-based catalysts that have found use in hydroalkoxylation reactions toward spiroacetals

However, treatment of the substrate **201b** with catalyst **213** afforded a 11:1 mixture of the 6,5- and 5,6-benzannulated spiroacetals **204** and **203**.

Deslongchamps et al. [109] have also applied their mercury(II) protocol to mono (benzannulated) spiroacetals (Scheme 50). Thus, treatment of alkyne diol **201** with catalytic $Hg(OTf)_2$ afforded the 6,6-spiroacetal **205** in excellent yield. The THP-protected alcohol could also be used directly in the reaction with only a small drop in yield.

2.6.4 Bis(benzannulated) Spiroacetals: Gold, Rhodium, and Iridium Catalysts

For the bis(benzannulated) system **221**, Ir(I)-catalyst **207** and Rh(I)-catalyst **209** were the most effective for the spirocyclization, with the reaction reaching 98 %

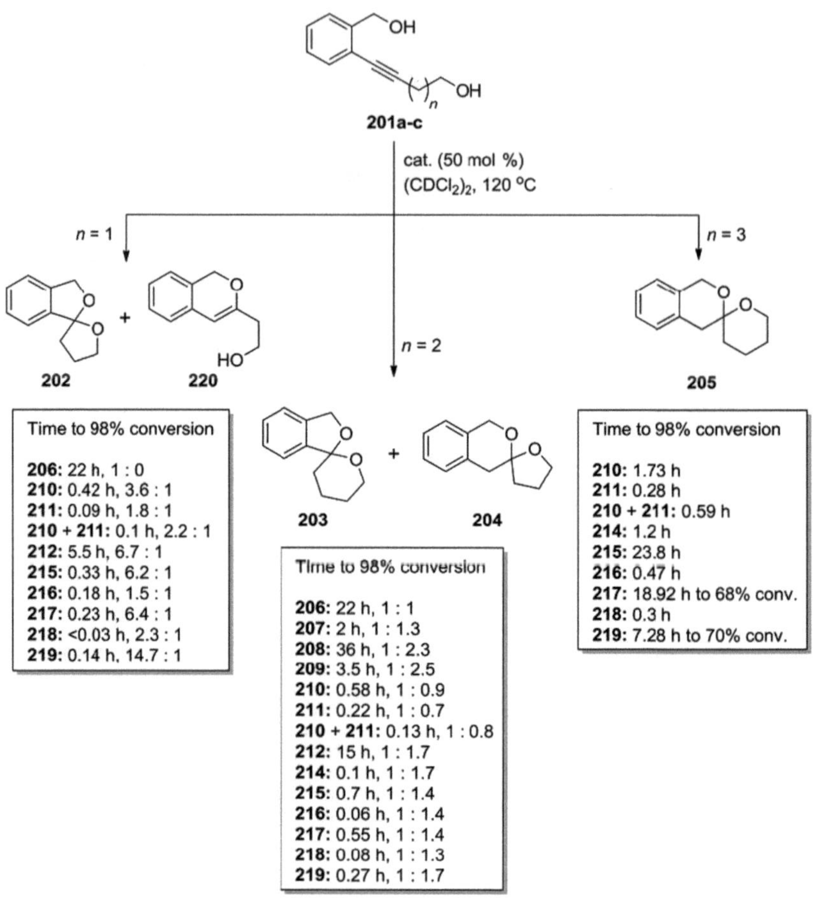

Scheme 48 Messerle et al.'s studies of catalysts **206–219** in the hydroalkoxylation of internal alkynes

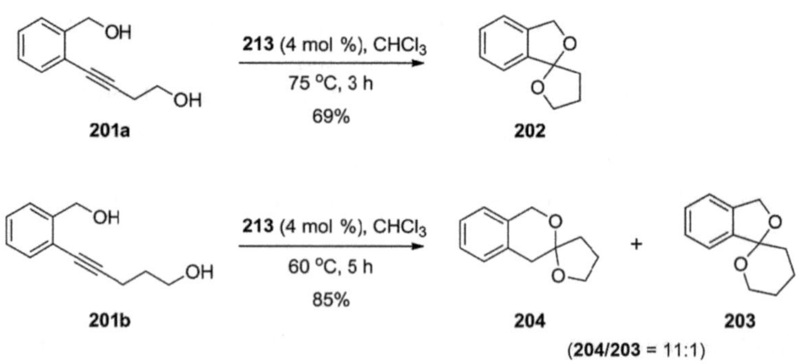

Scheme 49 Crabtree's Ir(I)-catalyzed hydroalkoxylation of an internal alkyne [112]

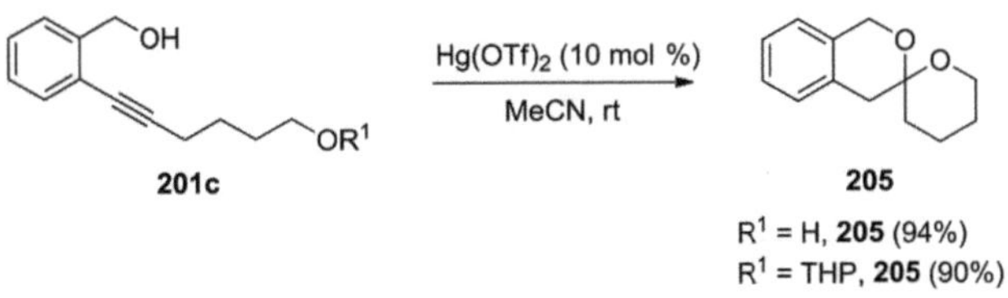

Scheme 50 Deslongchamps et al.'s Hg(II)-catalyzed synthesis of mono(benzannulated)-spiroacetals [109]

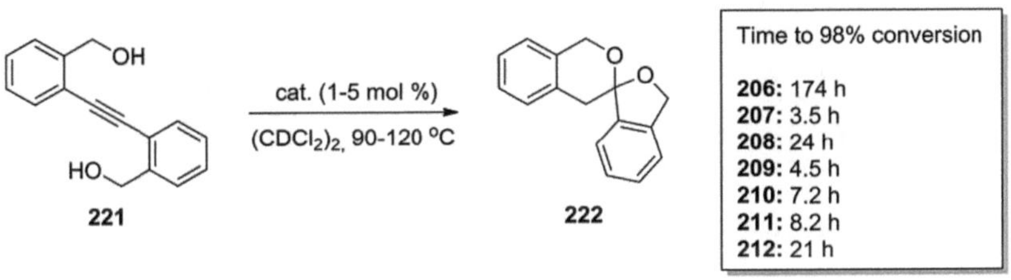

Scheme 51 Messerle et al.'s Rh(I)- and Ir(I)-catalyzed synthesis of bis(benzannulated)-spiroacetals

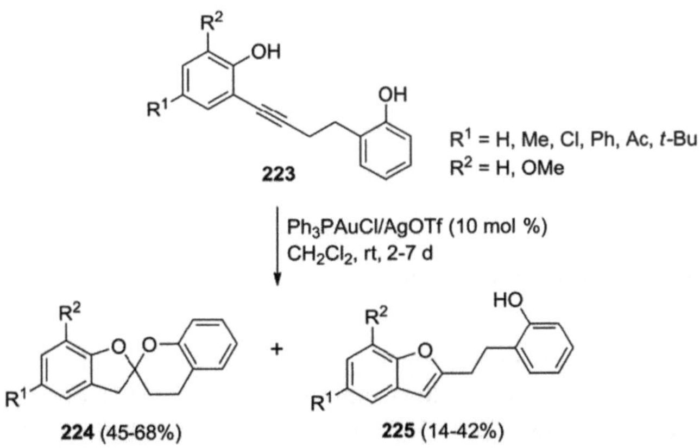

Scheme 52 Xue et al.'s gold(I)-catalyzed synthesis of bis(benzannulated)-spiroacetals [111]

conversion in 3–5 h (Scheme 51). Regioselectivity is, of course, not a problem as both 6-*endo-dig* and 5-*exo-dig* cyclizations will give the same product, **222**.

Xue et al.'s studies on the gold(I)-catalyzed spirocyclization of bis(aromatic) alkynes **223** [111] found that addition of AgOTf was required to suppress formation of the undesired benzofuran side product **225** (Scheme 52). In the absence of the silver(I) salt, benzofuran **225** only was obtained in excellent yield (95 %). Electron-donating groups at R^1 afforded the spiroacetal **224** in good yield (62-68 %), while electron-withdrawing groups at either R^1 or R^2 resulted in lower yields of the spiroacetal. In one case ($R^1 = H$, $R^2 = OMe$), benzofuran **225** was obtained in comparable yield to the spiroacetal product (42 % and 45 %, respectively).

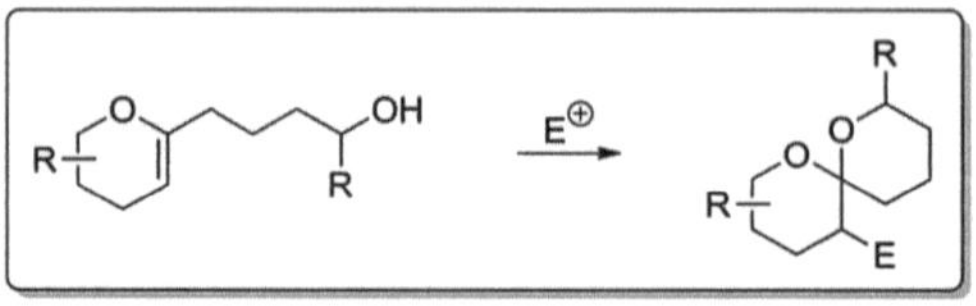

Scheme 53 Electrophilic cyclization methods for the synthesis of spiroacetals

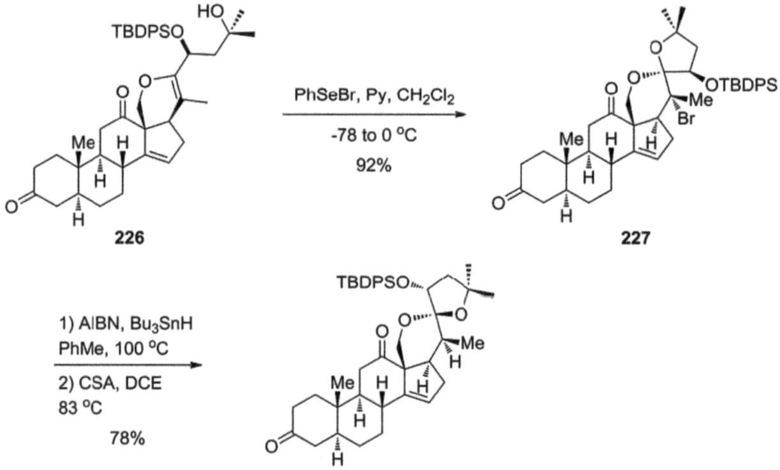

Scheme 54 Shair et al.'s synthesis of the western half of cephalostatin 1 [123]

2.7 *Electrophilic Cyclization*

Classically, treatment of an endocyclic enol ether with an electrophilic halide reagent such as *N*-bromo- or *N*-iodosuccinimide [72, 113–117], or organoselenium reagents (PhSeX) [118–122], affords electrophile-substituted spiroacetals (Scheme 53). The halide is then removed using reductive methods or used for further elaboration of the spiroacetal toward the target molecule.

Shair et al. [123] have used an electrophilic spirocyclization in their synthesis of cephalostatin 1 (Scheme 54). The spiroacetal of the western half, **228**, was obtained by treatment of the endocyclic enol ether **226** with PhSeBr, giving the mono-anomeric spiroacetal **227** in excellent yield. Transformation to the desired doubly anomeric spiroacetal **228** was achieved by reductive removal of the bromide, followed by epimerization with CSA.

Electrophilic spirocyclization may also be performed on hemiacetals. This method has been used most recently by Roush et al. [124] in the synthesis of a model spiroacetal fragment toward integramycin (Scheme 55). Thus, a 1:1.3 mixture of ketone **231** and hemiacetal **232** was treated with NIS in dichloromethane to

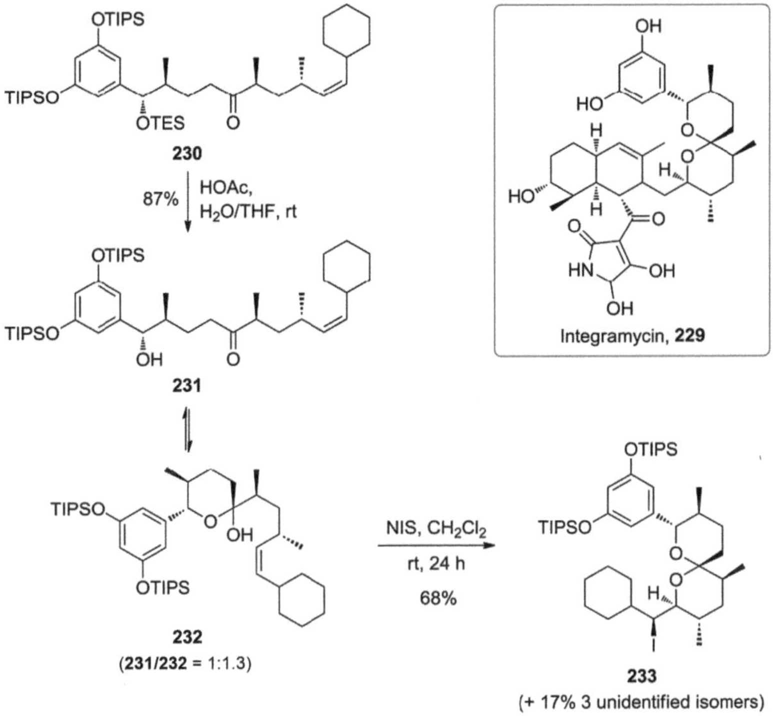

Scheme 55 Roush et al.'s synthesis of integramycin model spiroacetal **233** [124]

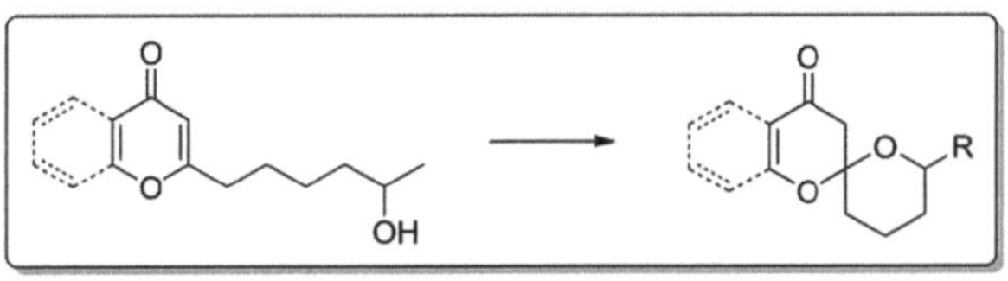

Scheme 56 Oxa-Michael cyclizations in the synthesis of spiroacetals

give the desired doubly anomeric spiroacetal **233** as the major product, along with four unidentified isomers in a 10:7:3:3 ratio.

2.8 *Oxa-Michael Cyclization*

The intramolecular addition of a pendant alcohol to a pyrone or chromone has long been recognized [125] as an effective method for the spirocyclization of α,β-unsaturated ketone spiroacetals (Scheme 56).

Recently, this approach has been exploited in the synthesis of norhalichondrin B **235** [126] (Scheme 57). Additionally, an oxa-Michael strategy has been used by

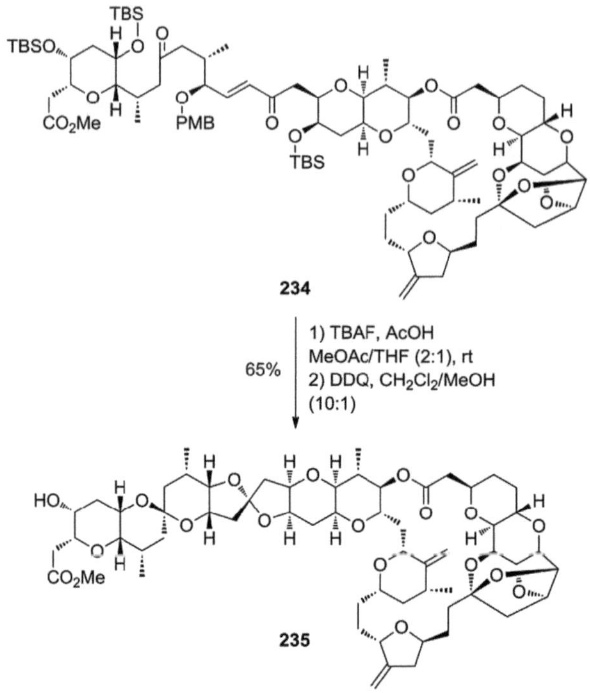

Scheme 57 Oxa-Michael approach towards norhalichondrin B [126]

Brimble et al. [127] for the synthesis of bis(benzannulated)-spiroacetals **238** (Scheme 58). Deprotection of the EOM group in chromones **236** prompted intramolecular oxa-Michael addition of the resulting phenol in most cases. In the two examples where only the free phenol **237** was recovered, the spirocyclization could be induced by heating the neat phenol with dry potassium carbonate under microwave irradiation. The resulting bis(benzannulated) 6,6-spiroacetals were obtained in low to moderate yields.

Cascade sequences involving Michael additions have also been reported recently for the synthesis of spiroacetals. An oxa-Michael initiated spirocyclization cascade sequence has been used in the synthesis of berkelic acid by both Fürstner et al. [128] and Brimble et al. [129] (Scheme 59).

Brimble and coworkers used a three-step HWE/oxa-Michael/acid-catalyzed cyclization sequence to access the simplified spirocyclic core **243**. Unfortunately, application of this method to a more complex substrate in order to obtain the substituted tetrahydrofuran was unsuccessful [130]. However, Fürstner and coworkers were able to effect a similar cascade from enone **244** in one step, affording tetracyclic spiroacetal **245** in excellent yield as a single isomer.

Alternatively, the oxa-Michael addition can be used as the final step in the spirocyclization cascade. Wulff et al. [131] have reported a tandem desymmetrization/oxa-Michael addition in their studies toward the spiroacetal

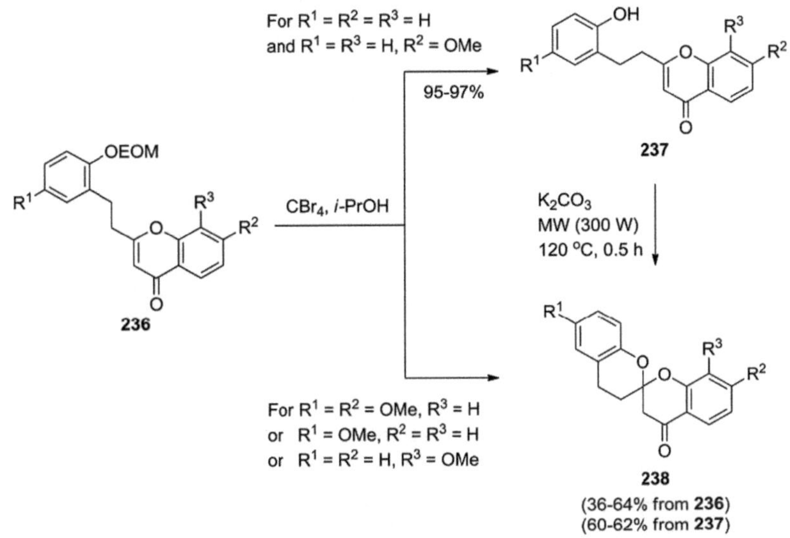

Scheme 58 Brimble et al.'s oxa-Michael approach toward bis(benzannulated) spiroacetals [127]

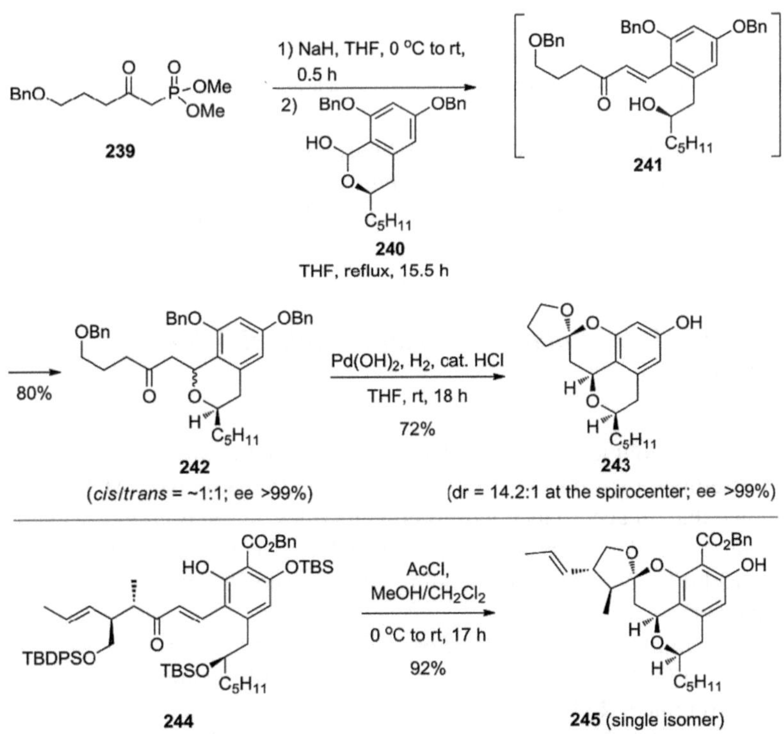

Scheme 59 Oxa-Michael initiated cascade approaches toward spiroacetals

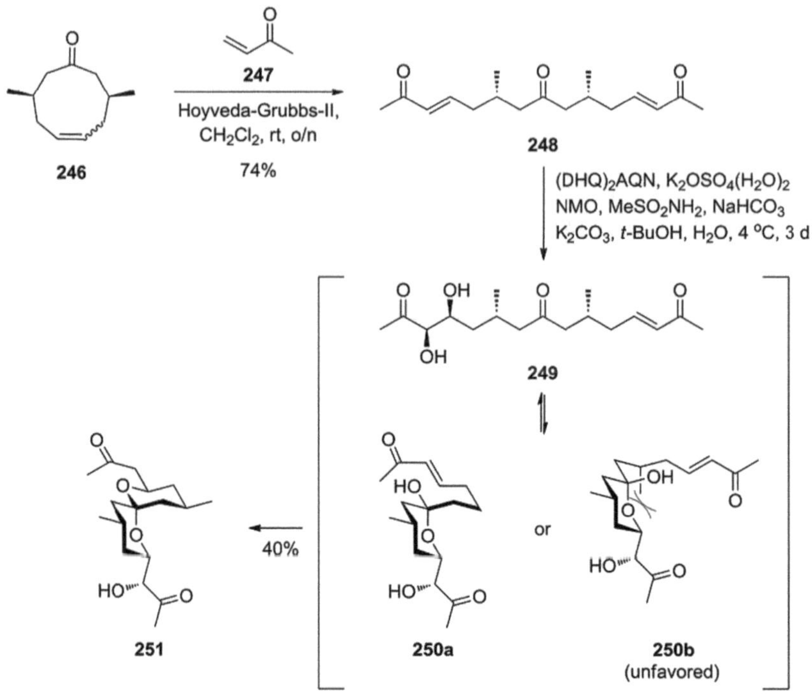

Scheme 60 Wulff et al.'s oxa-Michael terminated cascade approach toward didemnaketal A [131]

portion of didemnaketal A (Scheme 60). Sharpless asymmetric dihydroxylation of the *meso* bis-enone **248** and in situ hemiacetalization followed by conjugate addition gave the singly anomeric spiroacetal **251** as the only product.

2.9 *Double Intramolecular Hetero-Michael Addition (DIHMA)*

Alternatively, a double intramolecular hetero-Michael addition (DIHMA) can be used to assemble a spiroacetal from an α,β-ynone (Scheme 61). ^{1}H NMR studies have shown that, in accordance with Baldwin's rules, initial 6-*endo-dig* cyclization rather than 6-*exo-dig* is favored. Subsequent 5-*exo-trig* cyclization furnishes the 5,6-spiroacetal [132]. The corresponding 6,6-spiroacetals are obtained in an analogous manner.

Recent examples of its application include Koskinen et al.'s work toward calyculin C [132–134] and Forsyth et al.'s report of a simplified okadaic acid spiroacetal fragment **255** [135] (Scheme 62). Acidic hydrolysis of the TES-ethers and consequent DIHMA of the ynones afforded the spiroacetals **253** and **255**, respectively.

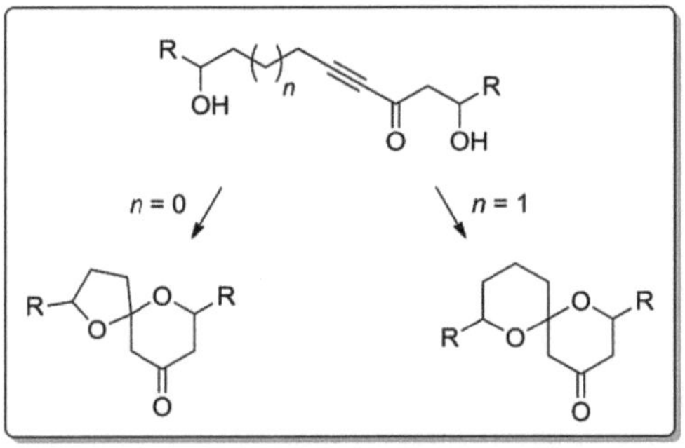

Scheme 61 Double intramolecular hetero-Michael addition in the synthesis of spiroacetals

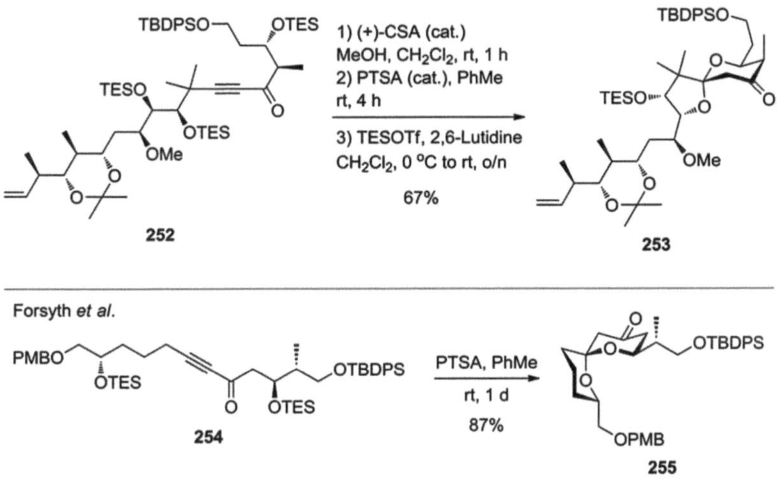

Scheme 62 DIHMA approaches toward calyculin C and okadaic acid [132–135]

Similarly, Sharma et al. [136] have utilized a DIHMA in their synthesis of the antimicrobial fungal metabolite dinemasone A **258** (Scheme 63). Unfortunately DIHMA of the ynone **256** afforded the singly anomerically stabilized spiroacetal **257a**, albeit in good yield. Epimerization to the desired doubly anomerically stabilized spiroacetal **257b** under the best conditions gave a 1:1 mixture of the anomers. Subsequent hydrogenolysis then gave dinemasone A **258** in high yield.

2.10 Ring Opening of Epoxides

Another of the well-established methods for the synthesis of α-hydroxy spiroacetals is through the intramolecular ring opening of an epoxide (Scheme 64).

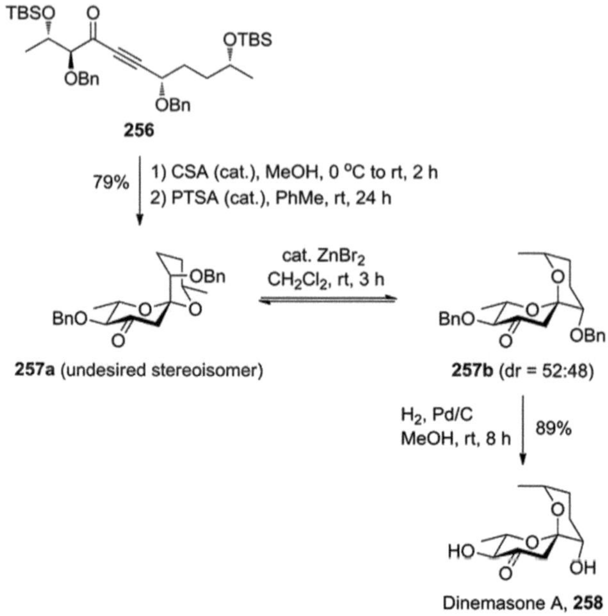

Scheme 63 Sharma et al.'s DIHMA approach towards dinemasone A [136]

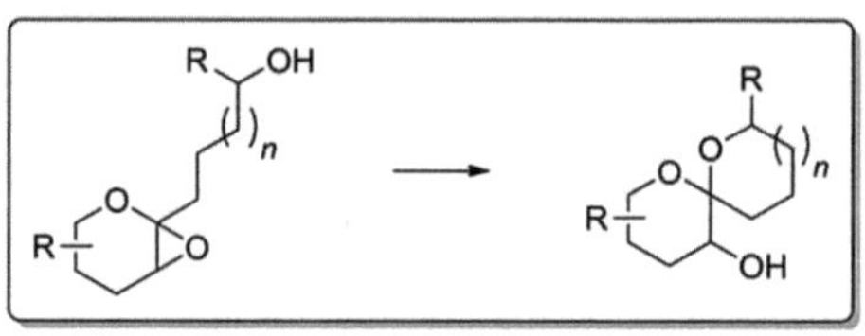

Scheme 64 Ring opening of epoxides in the synthesis of spiroacetals

Tan and coworkers [137, 138] have described several methods for the selective formation of either the kinetic or thermodynamic spiroacetal resulting from intramolecular ring opening of carbohydrate-derived epoxides **259–260** (Scheme 65).

The epoxides themselves were obtained by treatment of the parent glycal with DMDO. The selectivity of the *threo* glycal-derived epoxides **259** for either the "retention" spiroacetal **261a** (i.e., retention of stereochemistry at C1) or "inversion" spiroacetal **261b** could be controlled by the choice of reagent, with the thermodynamic preference for **261a** able to be reversed through the use of hydrogen bond-driven methanol catalysis. In the case of the *erythro* glycal-derived epoxides **260**, the inversion spiroacetal **262a** is favored under each set of conditions, though the selectivity is diminished somewhat when the spirocyclization is performed in the presence of methanol.

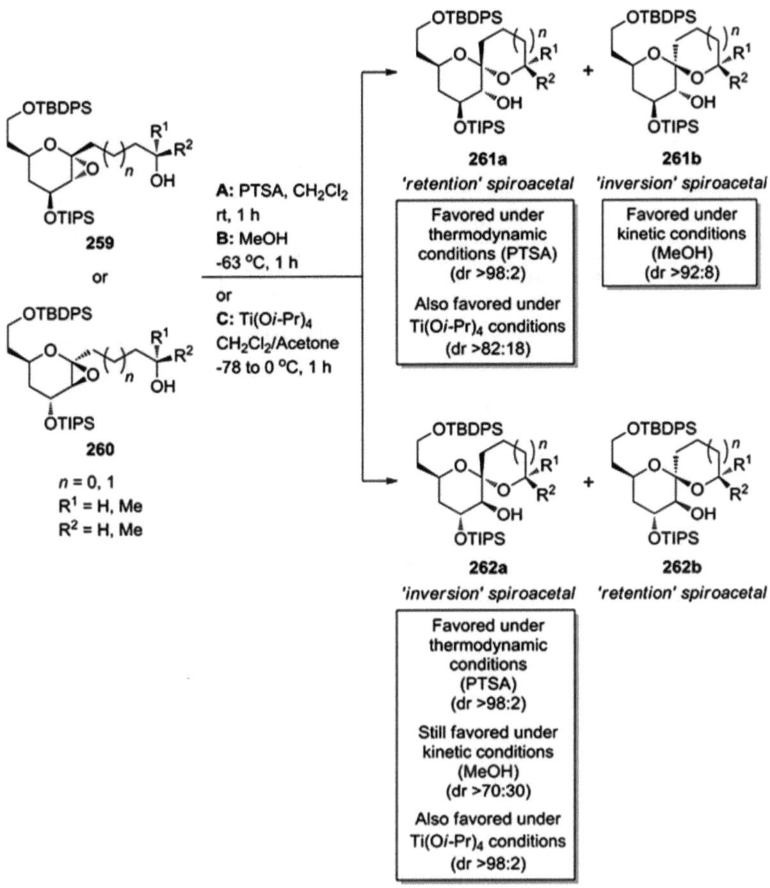

Scheme 65 Tan et al.'s selective syntheses of spiroacetals via intramolecular epoxide ring opening [137, 138]

This method has also been extended to the synthesis of mono(benzannulated) spiroacetals [139, 140] (Scheme 66). For both the *threo* and *erythro* glycal-derived epoxides, the retention spiroacetals **265a** and **266b** could be obtained selectively using the Ti(O*i*-Pr)$_4$ protocol. The inversion spiroacetals **265b** and **266a** could be accessed upon addition of MeOH or AcOH to the reaction mixture; however, the selectivity was much lower.

The effect of substitution on the aryl ring in this system was also studied by the same authors [140] (Scheme 67). The retention spiroacetals **268** were obtained under spontaneous cyclization conditions with good selectivity across most ring sizes, except for the electron-poor NO_2-substituted analogue. The selectivity could be completely reversed, in some cases, using the previously developed methanol-induced spirocyclization conditions to give the inversion spiroacetals **269** predominantly.

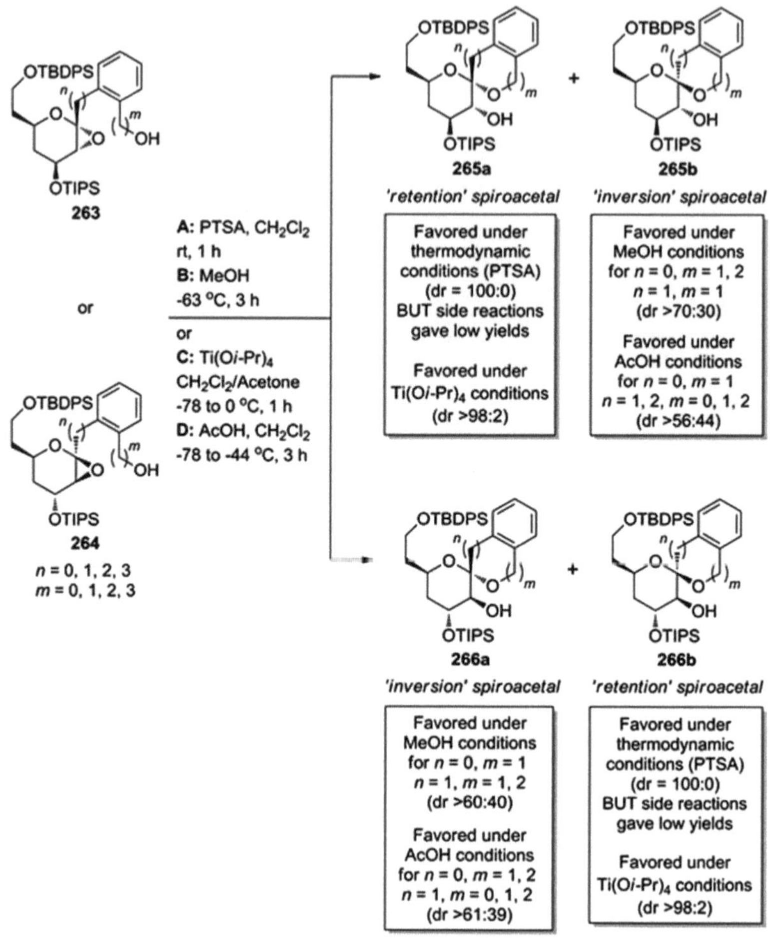

Scheme 66 Tan et al.'s selective syntheses of mono(benzannulated) spiroacetals via intramolecular epoxide ring opening [139, 140]

Robertson et al. [141] have used an intramolecular epoxide ring-opening in their studies toward the sawaranospirolides (Scheme 68). Treatment of the intermediate acid **271** with *m*-CPBA gave, after spontaneous spirocyclization, spirolactone **272** in low yield over four steps. Removal of the TIPS- and benzyl protecting groups then afforded the enantiomer of the natural product sawaranospirolide C, **273**.

A similar epoxidation–spirocyclization method has been used recently by Denmark et al. [142, 143] in their synthesis of papulacandin D, as well as in an independent study [144] toward papulacandin analogues (Scheme 69).

The epoxidation had to be carried out under basic conditions to prevent premature acid-catalyzed spirocyclization of the glycal **274**. The spiroacetals were obtained as mixtures of anomers, but treatment with catalytic acid induced complete epimerization to the desired anomeric spiroacetal **275**.

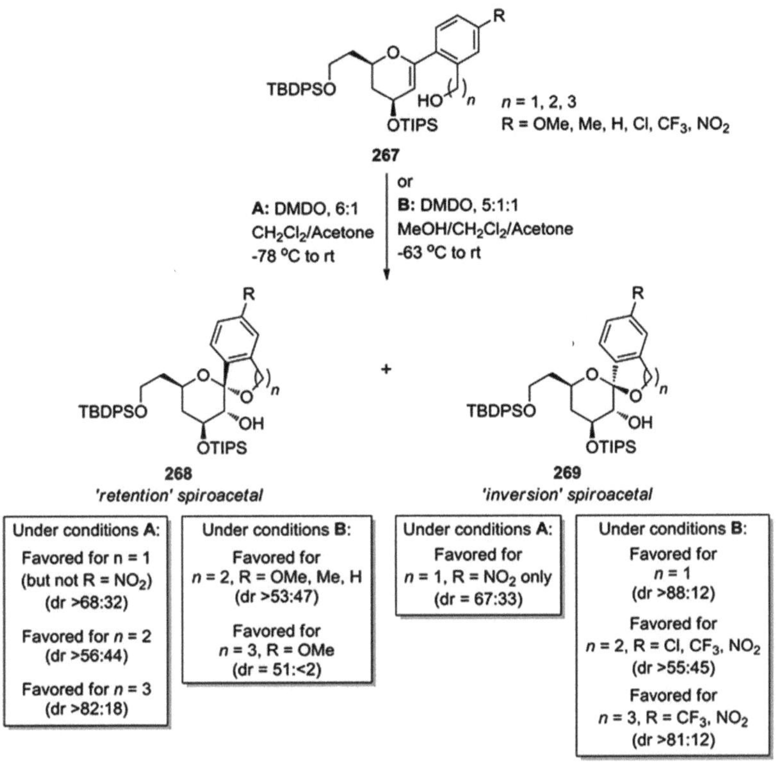

Scheme 67 Effect of aryl substitution on the spirocyclization of glycal epoxides **267** [140]

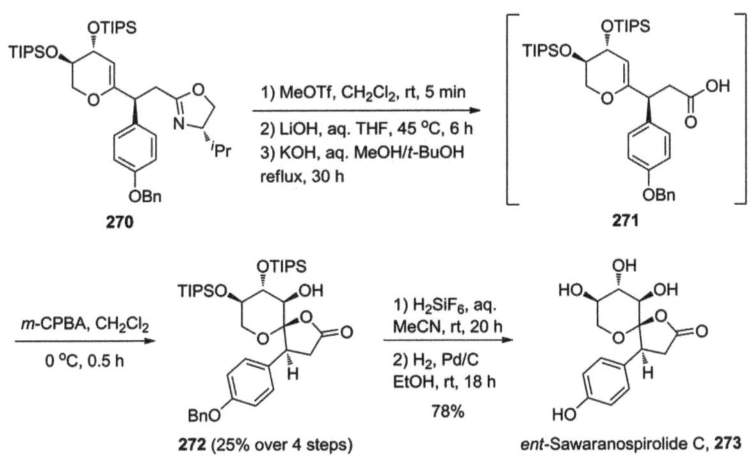

Scheme 68 Robertson et al.'s epoxide ring-opening approach toward the sawaranospirolides [141]

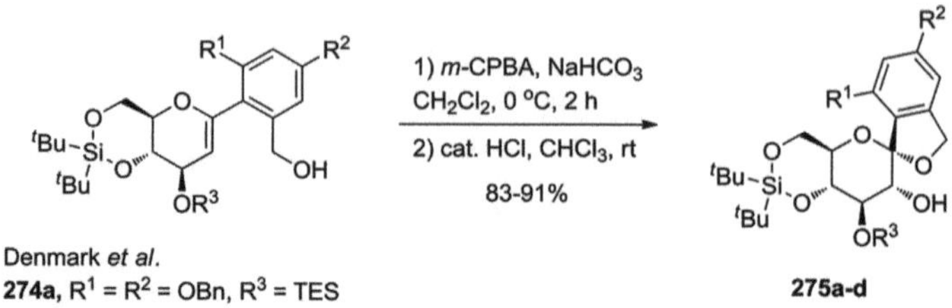

Denmark *et al.*
274a, $R^1 = R^2 = OBn$, $R^3 = TES$
Pieters *et al.*
274b, $R^1 = R^2 = OBn$, $R^3 = TIPS$
274c, $R^1 = OBn$, $R^2 = H$, $R^3 = TIPS$
274d, $R^1 = R^2 = H$, $R^3 = TIPS$

Scheme 69 Intramolecular epoxide ring-opening approaches toward papulacandin-like spiroacetals [142–144]

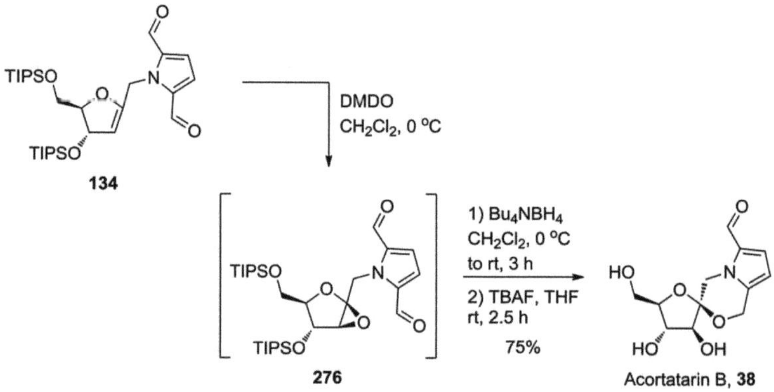

Scheme 70 Tan et al.'s intramolecular epoxide ring-opening approach toward acortatarin B [78]

Tan et al. [78] have also applied an intramolecular epoxide ring-opening in their synthesis of acortatarin B (Scheme 70).

Epoxidation of glycal **134** with DMDO followed by in situ reduction of the pyrrole aldehyde and subsequent attack on the epoxide afforded the spiroacetal core of acortatarin B in high yield as a single diastereomer. Cleavage of the TIPS-ethers with TBAF then afforded the natural product **38** in excellent yield.

2.11 Cycloaddition Approaches Toward Spiroacetals

Several different cycloaddition strategies have been employed in the synthesis of spiroacetals, including [4+2]-, [3+2]-, [2+2+2]- [145–147]-, and [2+2+1]-cycloadditions [148, 149] (Scheme 71).

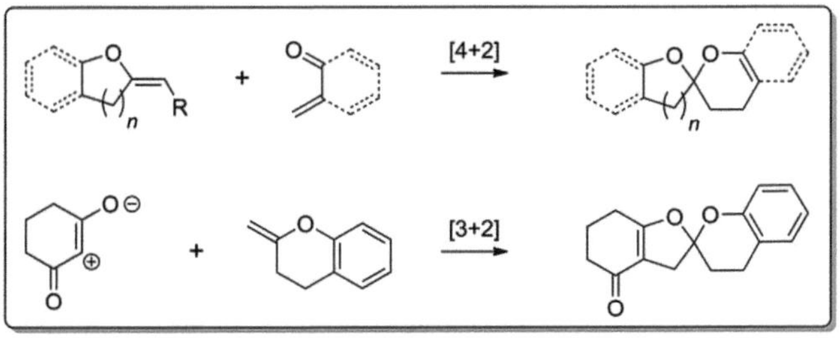

Scheme 71 Cycloaddition approaches for the synthesis of spiroacetals

2.11.1 [4+2]-Cycloadditions Toward Spiroacetals

The hetero-Diels–Alder reaction (HDA) has been used extensively for the construction of spiroacetals [4], notably in the synthesis of the reveromycins [150–153]. In particular, the recent development of mild methods for the generation of *ortho*-quinone methides (*o*-QM) in the presence of sensitive five-membered exocyclic enol ethers has revealed an alternative route to benzannulated 5,6-spiroacetals.

The first of these methods, reported by Pettus et al. [154], generates an *o*-QM from **277** at low temperature via a *t*-BuMgCl initiated cascasde [155], which can then undergo a reverse electron demand-HDA (Scheme 72). A contemporaneous report from Bray [156] made use of a similar strategy, where the *o*-QM is generated from precursor **281** using *i*-PrMgCl. Alternatively, Bray [157] demonstrated that the use of a more stable dienophile, γ-methylene-γ-butyrolactone **287**, allowed generation of the *o*-QM from **288** under thermal conditions. The corresponding spiroacetals were obtained in good yield.

Accordingly, Pettus and Huang have utilized their method in model work toward berkelic acid **96** [158]; however, completion of a diastereoselective formal synthesis of **96** required some modification of their strategy to incorporate the requisite subsituents [159] (Scheme 73).

Earlier model studies by Pettus and Huang [158] had established that the chiral methyl group on the enol ether directed selective formation of the *exo*-spiroacetal possessing the correct spiroacetal configuration. However, this was prior to the reassignment of berkelic acid relative stereochemistry and the configuration of the substituents on the five-membered ring turned out to be incorrect. Enol ether **290**, possessing the correct side chain stereochemistry, was selective for the *endo*-spiroacetal **292** in the HDA. Equilibration with TFA improved the diastereomeric ratio to 10:1 in favor of the desired configuration at the spirocenter. Subsequent benzylic oxidation then afforded the tetracyclic core **293** of berkelic acid.

Cascade processes involving a HDA have also been used for the synthesis of berkelic acid, where both the enol ether and *o*-QM are generated and reacted in situ. Fañanás et al. [160] first reported this type of cascade procedure for the synthesis of 5,6-benzannulated spiroacetals in 2009 (Scheme 74). The three-component, Pd-

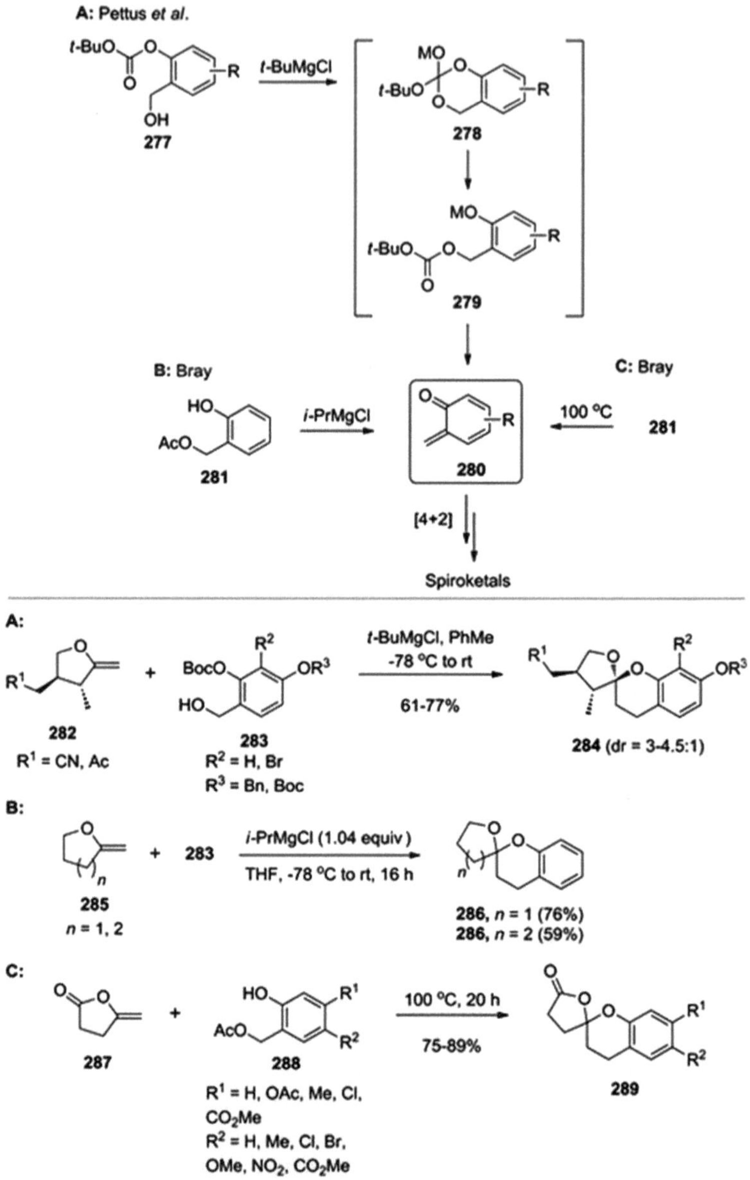

Scheme 72 Recently disclosed mild methods for the formation of *o*-quinone methides

catalyzed condensation/cycloisomerization/cycloaddition cascade afforded spiroacetals **299** in good to excellent yield as single diastereomers.

Two groups have since taken advantage of this cascade-type approach for the synthesis of berkelic acid. Fañanás et al. [161] modified their method to incorporate the final ring of the tetracyclic scaffold and adapted the sequence to the large-scale synthesis of **96** (Scheme 75).

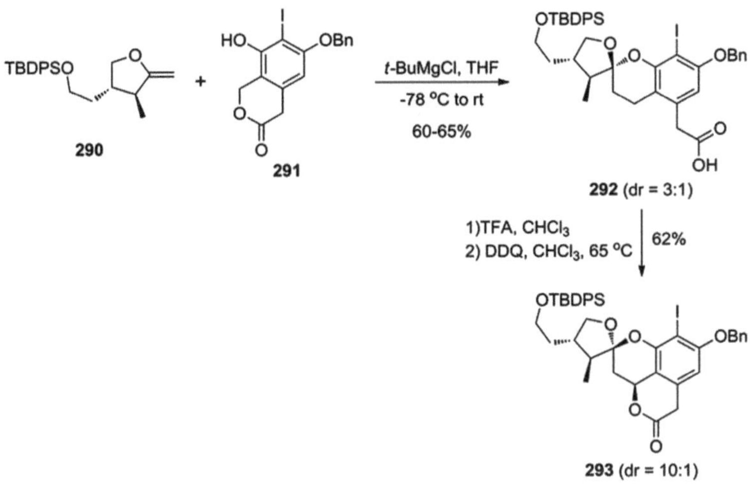

Scheme 73 Pettus et al.'s HDA approach toward berkelic acid [159]

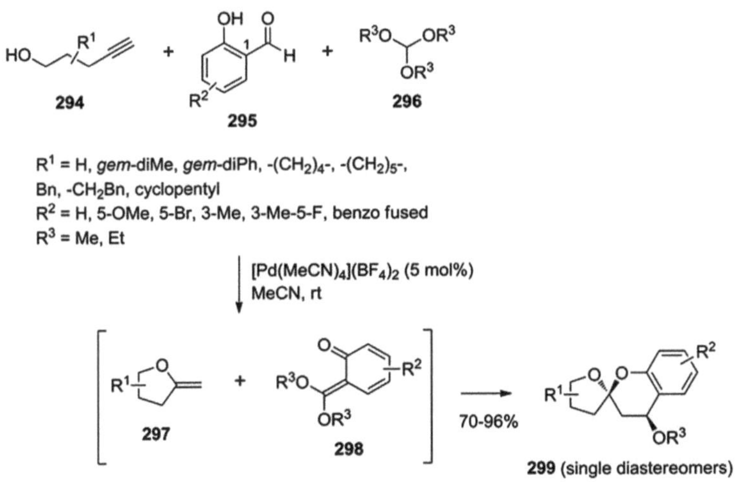

Scheme 74 Fañanás et al.'s Pd-catalyzed HDA cascade approach toward spiroacetals [160]

Thus, AgOTf-catalyzed cycloisomerization of alkyne diol **300** gives the dienophile, while cycloisomerization of aldehyde **301** generates the *o*-QM. Cycloaddition of **302** and **303** then affords the 5,6-spiroacetal core **304**, which is isolated after hydrogenation of the double bond in 83 % yield as a 2:1 mixture of diastereomers.

Similarly, based upon the proposed union of two naturally occurring components, De Brabander et al. [162] have also used a Ag-initiated cycloisomerization/dearomatization/cycloaddition cascade for the synthesis of **96** (Scheme 76).

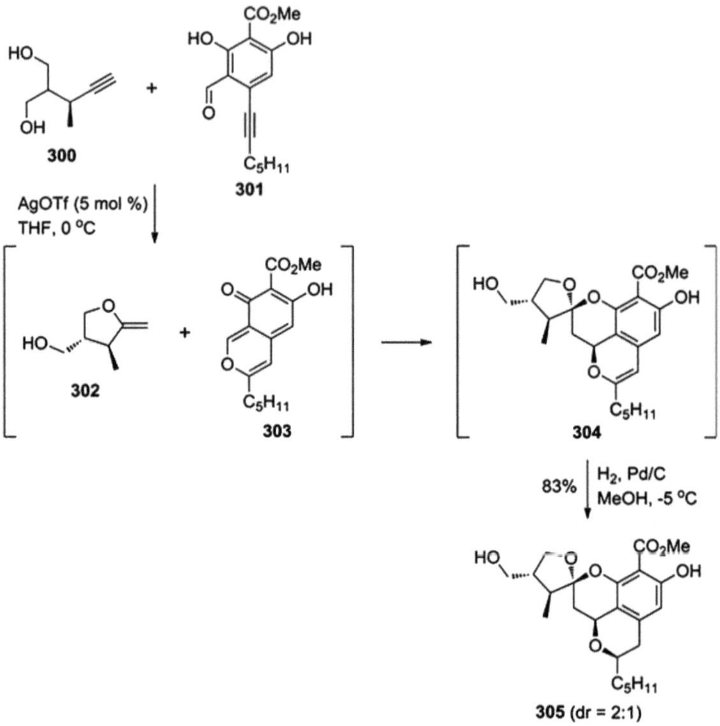

Scheme 75 Fañanás et al.'s Ag(I)-catalyzed HDA cascade approach toward berkelic acid [161]

Treatment of spicifernin-inspired alkynol **307** and pulvilloric acid-inspired fragment **306** with $AgSbF_6$ resulted in cycloisomerization of **307** to enol ether **309**; formation of *o*-QM **308** from **306**; and subsequent HDA of **308** and **309** to give an inseparable mixture of spiroacetal **311** as the major component, along with a mixture of four other diastereomers in an approximate ratio of 6:4. Careful regioselective demethylation to the free acid allowed separation of the diastereomers, and berkelic acid **96** was therefore obtained in 46 % after one recycle.

Alternatively, the HDA approach toward mono(benzannulated) spiroacetals can make use of aliphatic dienes, rather than *o*-QM, as the 4π reaction partner. This strategy has been demonstrated most recently by Belmont et al. [163], where HDA between a variety of furoquinolines **313** and acrolein affords the spiroacetals **314** in good yield and selectivity (Scheme 77).

The application of a HDA strategy for the synthesis of bis(benzannulated) spiroacetals as exemplified by the rubromycin family of natural products has also attracted attention. An initial report [164] of the HDA between an *o*-QM and chroman-derived enol ethers **315** required high temperature and pressure to afford 6,6-bis(benzannulated) spiroacetals (Scheme 78). Unfortunately, the corresponding 5,6-spiroacetals could not be accessed using this method due to the difficulty in obtaining the requisite enol ether **318**, which under methylenation conditions

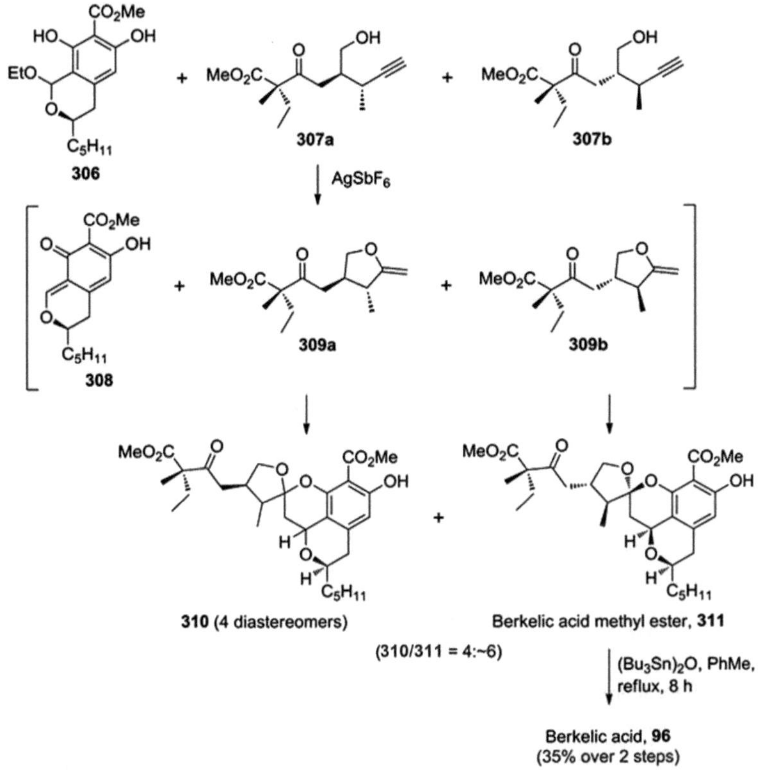

Scheme 76 De Brabander et al.'s Ag(I)-initiated bio-inspired synthesis of berkelic acid [162]

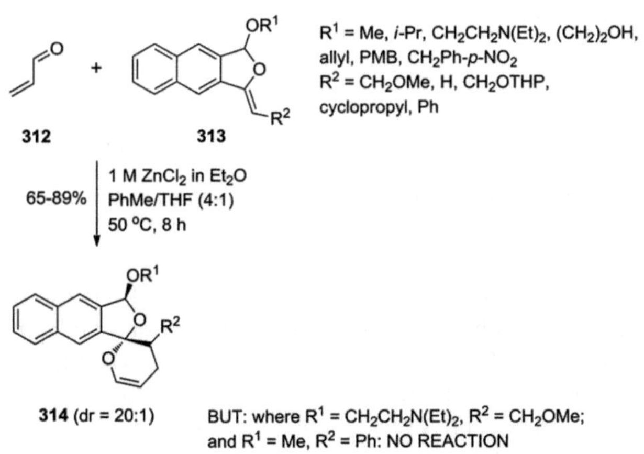

Scheme 77 Belmont et al.'s HDA approach to mono(benzannulated) spiroacetals [163]

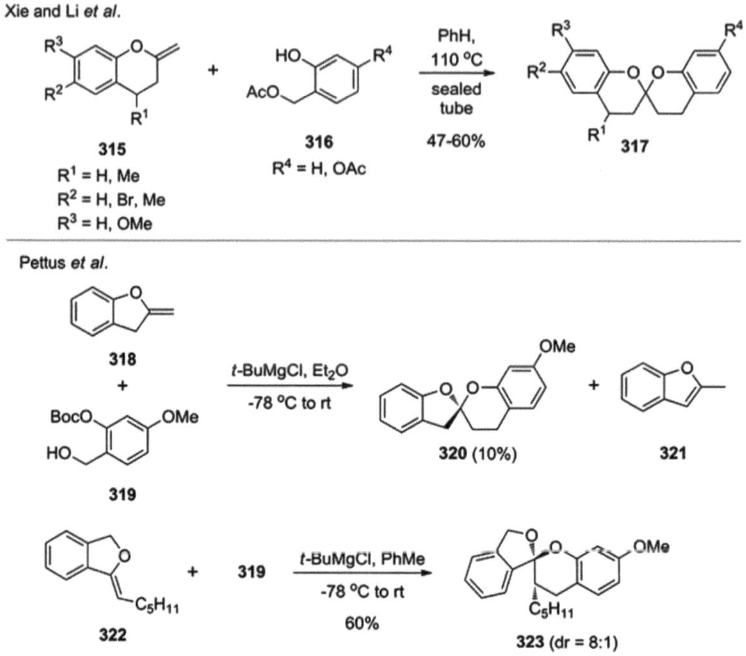

Scheme 78 HDA approaches toward bis(benzannulated) spiroacetals [154, 164]

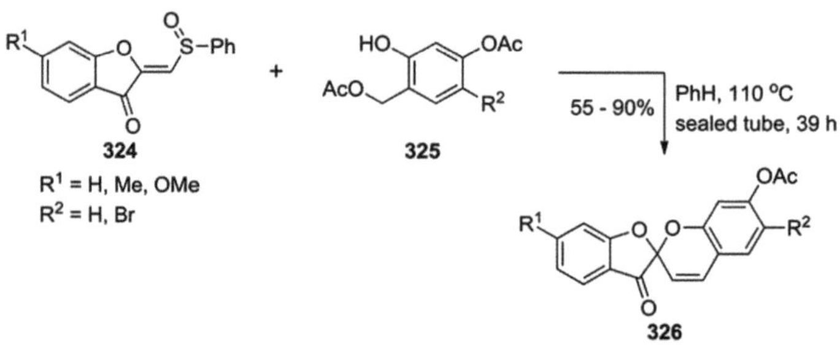

Scheme 79 Xie et al.'s vinyl sulfoxide-based HDA approach toward bis(benzannulated) spiroacetals [165]

isomerized to the endocyclic isomer exclusively. Similarly, Pettus and coworkers [154] were only able to obtain bis(benzannulated) spiroacetal **320** in poor yield when using enol ether **318** as the dienophile. However, when the enol ether **322** (which is less prone to isomerization) is submitted to the same conditions, paecilospirone-like spiroacetal **323** is obtained in moderate yield.

Thus, access to the rubromycin 5,6-bis(benzannulated) spiroacetal framework would necessitate a different approach. One solution [165] has been to use more stable benzofuranone-derived vinyl sulfoxides (e.g., **324**) as the dienophile (Scheme 79).

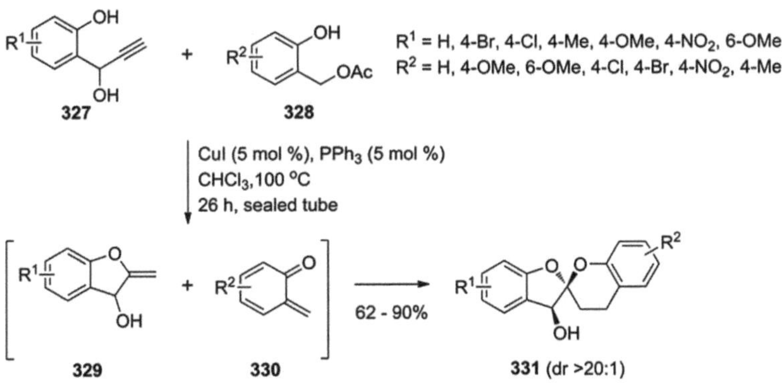

Scheme 80 Xue et al.'s Cu(I)-catalyzed cycloisomerization/HDA cascade approach towards spiroacetals [166]

Following the HDA, elimination of the sulfoxide occurs under the reaction conditions to afford the 5,6-spiroacetals **326** in good yield.

The cycloisomerization/cycloaddition cascade approach has been extended to bis (benzannulated) 5,6-spiroacetals recently by Xue et al. [166] (Scheme 80). Thus, Cu (I)-catalyzed cycloisomerization of alkynol **327** gives the enol ether **329**, which then undergoes HDA with the *o*-QM **330**, generated in situ from **328** under the thermal reaction conditions. The bis(benzannulated) spiroacetals **331** were obtained in good yield and selectivity (dr >20:1), favoring the doubly anomeric configuration.

2.11.2 [3+2]-Cycloadditions Toward Spiroacetals

An alternative cycloaddition approach to bis(benzannulated) spiroacetals has been explored by Pettus and coworkers in their synthesis of γ-rubromycin **58**. Based on the [3+2]-cycloaddition of an enol ether with a β-diketone-derived zwitterion [167], early studies on simple substrates afforded the 5,6-spiroacetal in moderate yield [168, 169].

In their successful synthesis of racemic γ-rubromycin **58** [170], highly functionalized chroman **333** and naphthoquinone **332** underwent oxidative [3+2]-cycloaddition (Scheme 81). The *o*- and *p*-quinone spiroacetals **334** and **335** were obtained as a 1:2 mixture in 58 % yield. Fortuitously, the demethylation conditions also catalyzed isomerization of **334**, converting the mixture of quinones to (±)-γ-rubromycin **58** in moderate yield.

2.12 *Furan Oxidation Strategies Toward Spiroacetals*

Oxidative transformation of a furan (Scheme 82) has been applied in the synthesis of many natural product scaffolds; spiroacetals have not escaped attention [171].

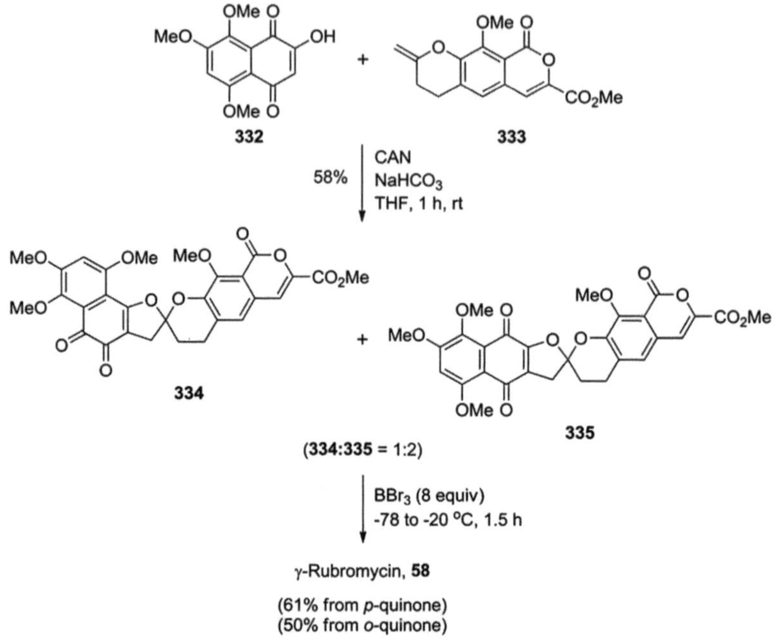

Scheme 81 Pettus et al.'s [3+2]-cycloaddition-based synthesis of γ-rubromycin [170]

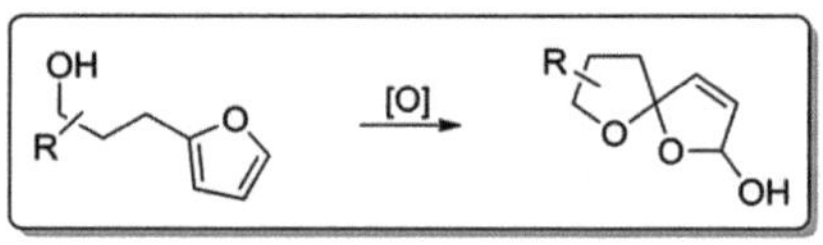

Scheme 82 Furan oxidation approaches for the synthesis of spiroacetals

A recent example of this approach has been reported by Yang and coworkers [172] in their synthesis of crassalactone D **338**, using *m*-CPBA as the oxidant (Scheme 83). Mohapatra et al. [173] used the same conditions in their synthesis of pyrenolide D (**341a**), while Robertson et al. [141] have used a furan oxidative spirocyclization in studies toward the sawaranospirolides (Scheme 84).

An alternative approach to 5,5-spirolactones has been developed by Vassilikogiannakis et al., where HDA of a furan with singlet oxygen followed by intramolecular ring opening and dehydration gives the dihydrofuran spiroacetals **347** or **349** (Scheme 85). This method has been used for the synthesis of crassalactone D **338**, as well as two epimeric spiroacetals in a sequence toward pyrenolide D, **354a** and **354b** [174].

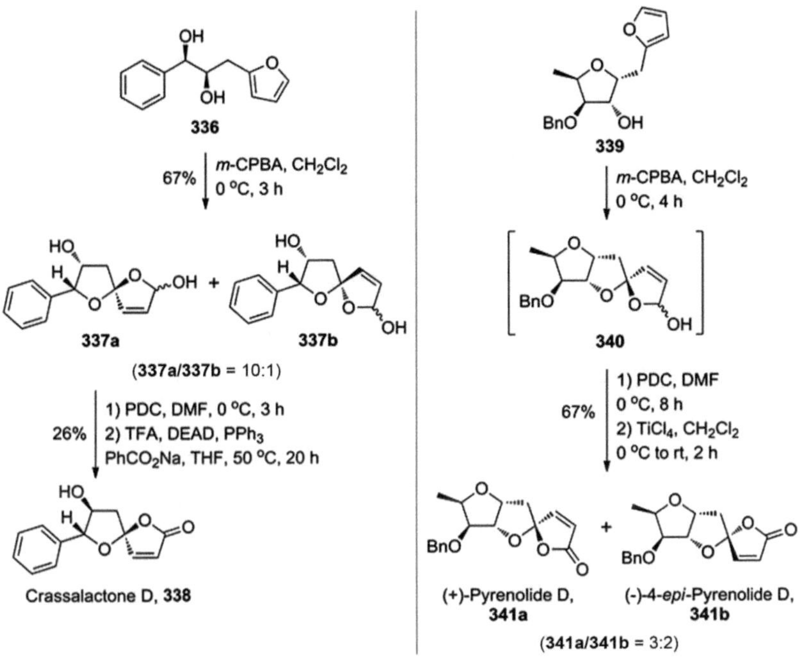

Scheme 83 Furan oxidation approaches toward crassalactone D and pyrenolide D [172, 173]

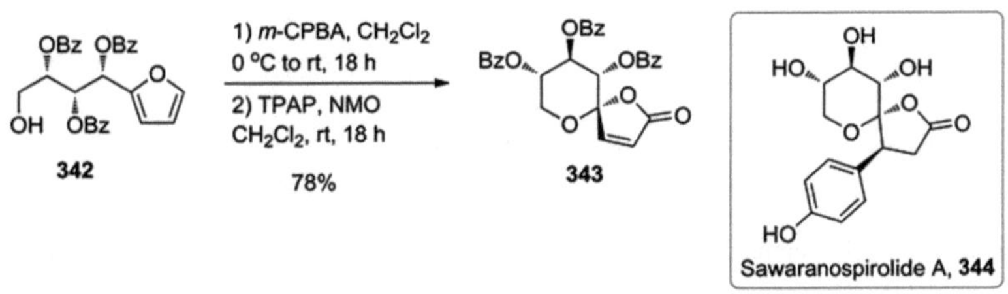

Scheme 84 Robertson et al.'s furan oxidation approach toward the sawaranospirolides [141]

2.13 *Intramolecular Hydrogen Abstraction (IHA) for the Synthesis of Spiroacetals*

The application of this method (Scheme 86) in the syntheses of a variety of spiroacetal natural products has been reviewed recently [175]. The mild reaction conditions enable construction of spiroacetals from very sensitive substrates that would not withstand more classical spirocyclization strategies (e.g., acid-catalyzed dehydrative spirocyclization).

Formation of five-membered rings using this method is favored over six-membered rings due to the relative stabilities of the six- and seven-membered

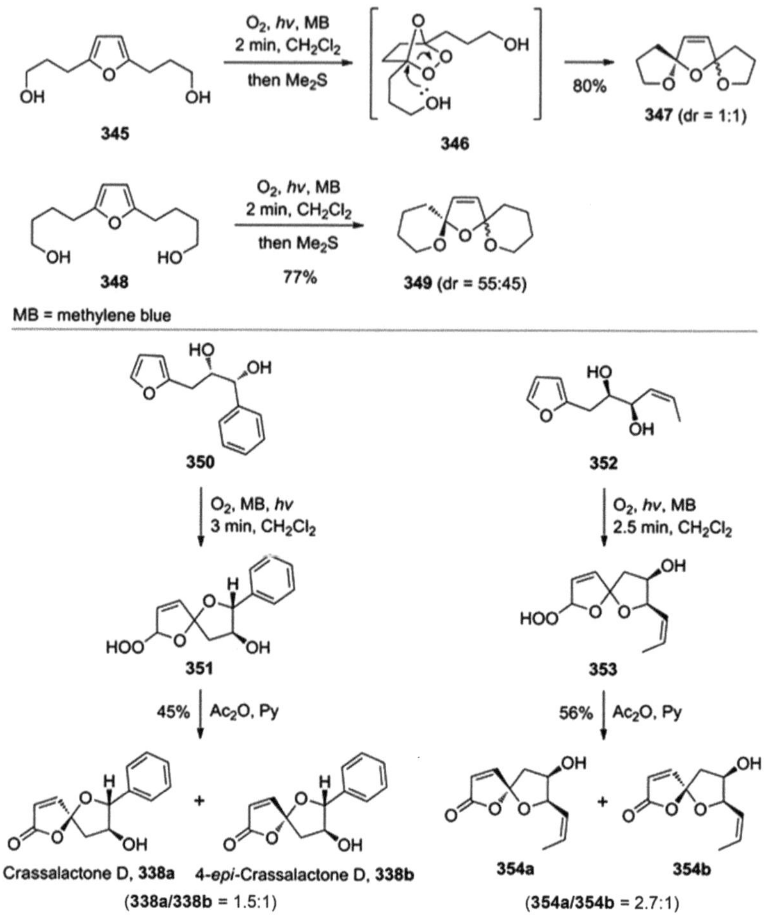

Scheme 85 Vassilikogiannakis et al.'s oxidative approach toward 5,5-spirolactones [174]

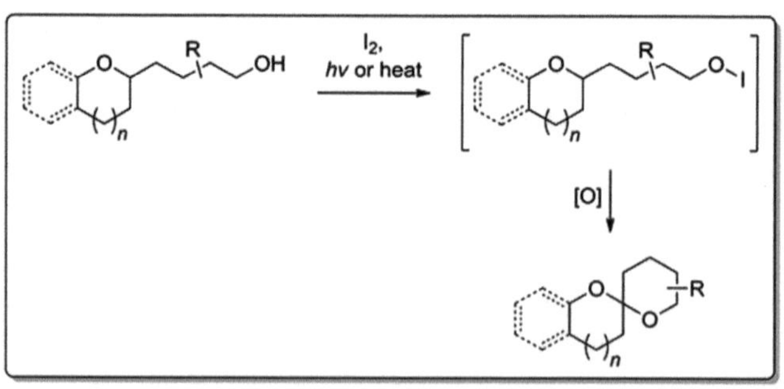

Scheme 86 Intramolecular hydrogen abstraction approaches for the synthesis of spiroacetals

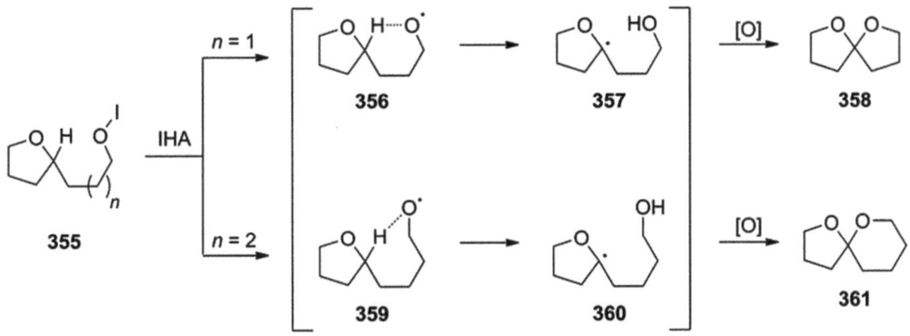

Fig. 5 Transition states in the IHA approach toward spiroacetals

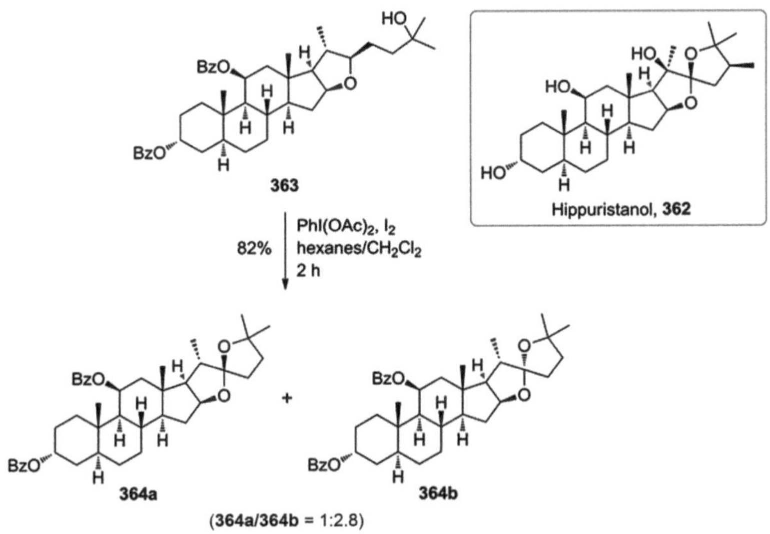

Scheme 87 Deslongchamps et al.'s application of an IHA in the synthesis of hippuristanol analogues [176]

intramolecular hydrogen abstraction (IHA) transition states **356** and **359**, respectively (Fig. 5).

Deslongchamps and coworkers [176] have used this method for the construction of the 5,5-spiroacetal in analogues of hippuristanol **362**, in an analogous manner to Suárez et al. and Shair et al.'s [177, 178] earlier syntheses of the cephalostatin and ritterazine scaffolds (Scheme 87). The 5,5-spiroacetals **364a** and **364b** were obtained as a 1:2.8 mixture in high yield.

The use of intramolecular hydrogen abstraction for the formation of both mono- and bis(benzannulated) spiroacetals has been explored recently by Brimble and coworkers [179–181] (Scheme 88). The observed mixtures of 5,5- and 6,5-spiroacetals were attributed to the weaker influence of the anomeric effect in five-membered rings, while the 5,6- and 6,6-spiroacetals could undergo

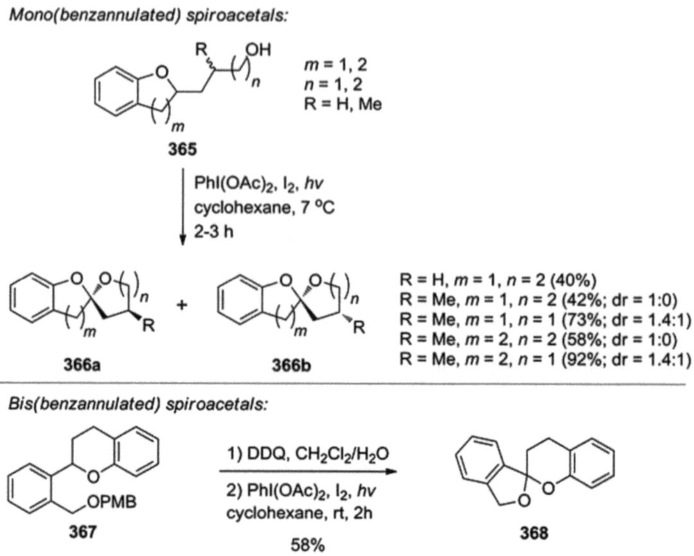

Scheme 88 Brimble et al.'s use of IHA in the synthesis of bis(benzannulated) spiroacetals [179–181]

equilibration under the reaction conditions to afford the doubly anomeric structures. For the bis(benzannulated) spiroacetal **368**, cleavage of the PMB-ether **367**, followed by the IHA smoothly provided the paecilospirone-like spiroacetal **368** in moderate yield.

This method has been exploited in the syntheses of the bioactive *Salvia miltiorrhiza* metabolites cryptoacetalide **371a** [182, 183] and danshenspiroketallactone **375a** [184] (Scheme 89).

The tetralin-lactone framework of cryptoacetalide was assembled via cyclotrimerization of triyne intermediate **369**, while the naphthalene-fused lactone core of danshenspiroketallactone, **374**, was constructed via directed metallation and subsequent lactonization of the diethylamide **372**. In each case, irradiation of the free alcohol in the presence of iodobenzene diacetate and iodine (i.e., Suárez oxidation conditions [185, 186]) afforded the spirocyclic lactone as an inseparable mixture of the natural product and its epimer.

Alternatively, Xue and coworkers [187] reported a system using hypoiodite generated by oxidation of an organoiodide salt (e.g., TBAI/*m*-CPBA) in the presence of fluoride ion (as a promoter) to assemble bis(benzannulated) spiroacetals (Scheme 90).

Formation of 5,5-spiroacetals was favored over the 5,6- and 5,7-spiroacetals. The 5,7-spiroacetals in particular required longer reaction times and were obtained in lower yields than their 5,5- and 5,6-counterparts. Electron-donating groups on the benzofuranone decrease the yield slightly; the same substituents on the pendant phenol exerted a slightly larger negative effect on the yield of the spiroacetal product. Attempted synthesis of the mono(benzannulated) spiroacetals **379** using this method failed.

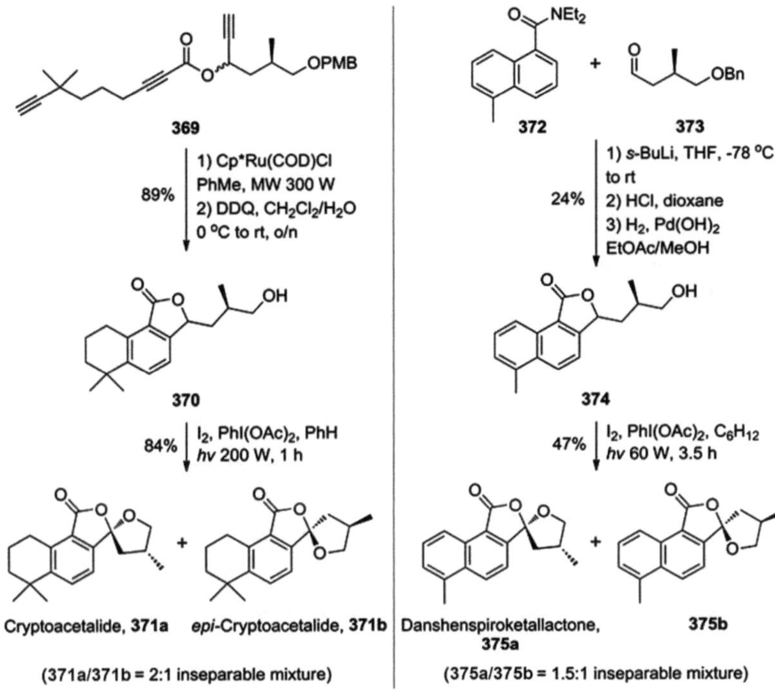

Scheme 89 IHA-based syntheses of cryptoacetalide and danshenspiroketallactone [183, 184]

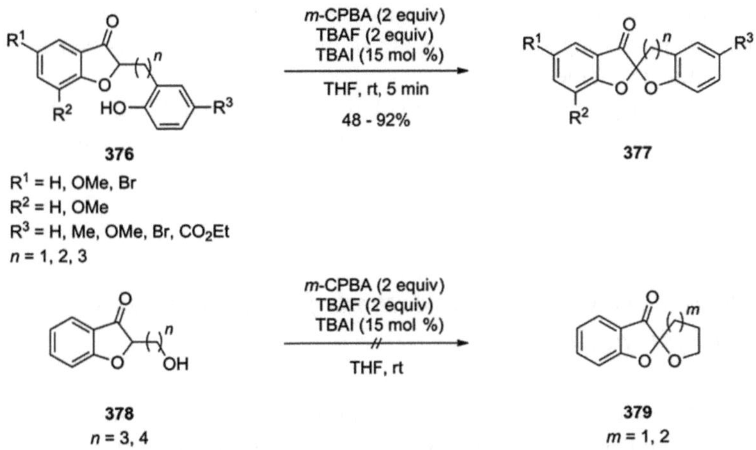

Scheme 90 Xue et al.'s hypoiodite approach toward bis(benzannulated) spiroacetals [187]

The same research group has recently applied this method to a formal synthesis of γ-rubromycin **58** [188] (Scheme 91). Treatment of phenol **380** with hypoiodite in the presence of TBAF as a fluoride ion source afforded the 5,6-bis(benzannulated)

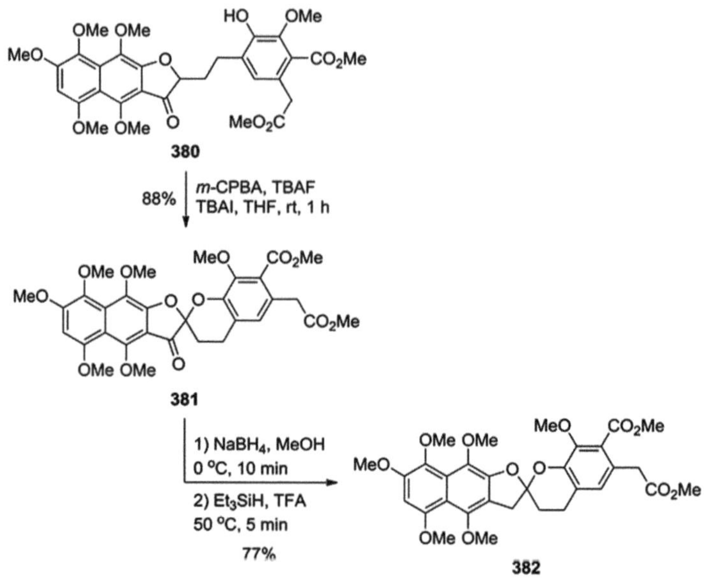

Scheme 91 Xue et al.'s hypoiodite-based synthesis of γ-rubromycin [188]

spiroacetal **381** in high yield. The formal synthesis of **58** was completed by reduction of the benzylic ketone via the alcohol to give the advanced γ-rubromycin intermediate **382** in good yield.

This method has also been incorporated into an inverse electron demand-HDA/ring-opening/spirocyclization cascade [189] for the synthesis of bis(benzannulated) 5,5-spiroacetals (Scheme 92).

Hydrolysis of the intermediate HDA-derived *N,O*-acetal **385** gives phenol **376**, which then feeds into the fluoride ion-promoted hypoiodite cycle, affording the bis (benzannulated) spiroacetals **377** in low to moderate yield.

2.14 Reductive Cyclization Methods for the Synthesis of Spiroacetals

Recently a Ni-catalyzed reductive cyclization/intermolecular cross-coupling has been reported by Peng et al. [190] for the synthesis of aryl-substituted spiroacetals (Scheme 93).

Spiroacetals **390** and **392** were obtained in low to moderate yield from the precursors **388** and **391** via the proposed reaction cascade illustrated below (Scheme 94):

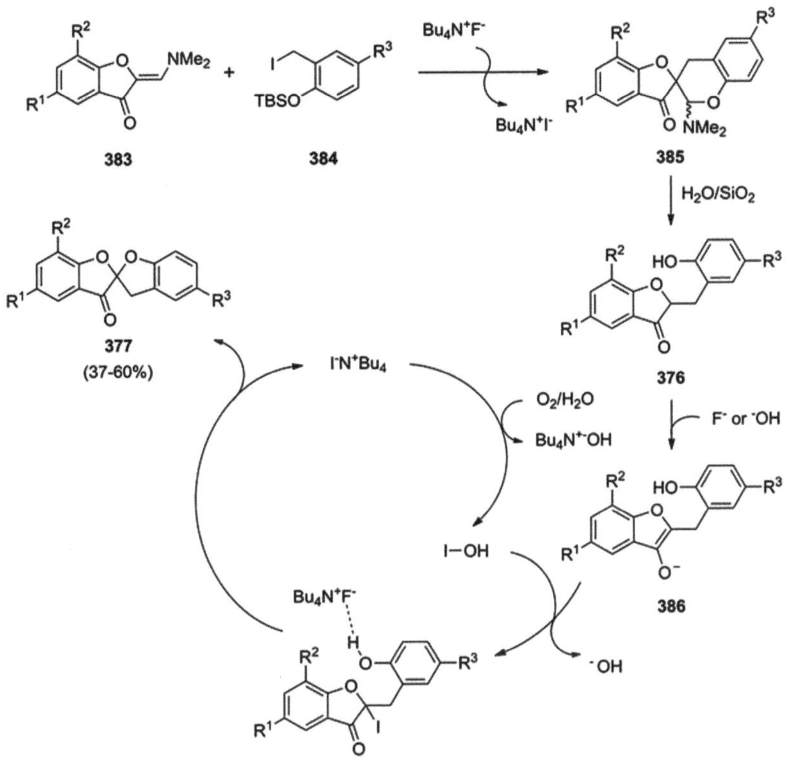

Scheme 92 Xue et al.'s HDA/IHA cascade sequence for the synthesis of bis(benzannulated) spiroacetals [189]

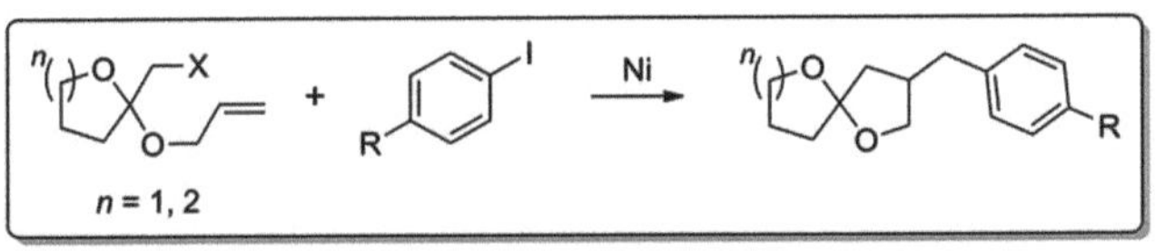

Scheme 93 Reductive cyclization methods for the synthesis of spiroacetals

2.15 *Ring-Closing Metathesis (RCM) for the Synthesis of Spiroacetals*

Unsaturated spiroacetals may be accessed via RCM of diene-subsituted tetrahydrofurans and tetrahydropyrans (Scheme 95).

Hsung and coworkers [191] have used simple γ- and δ-lactone-derived acetal-tethered dienes (e.g. **398**) in the RCM reaction (Scheme 96). This approach has most recently been exploited for the synthesis of (+)-aigialospirol **400**.

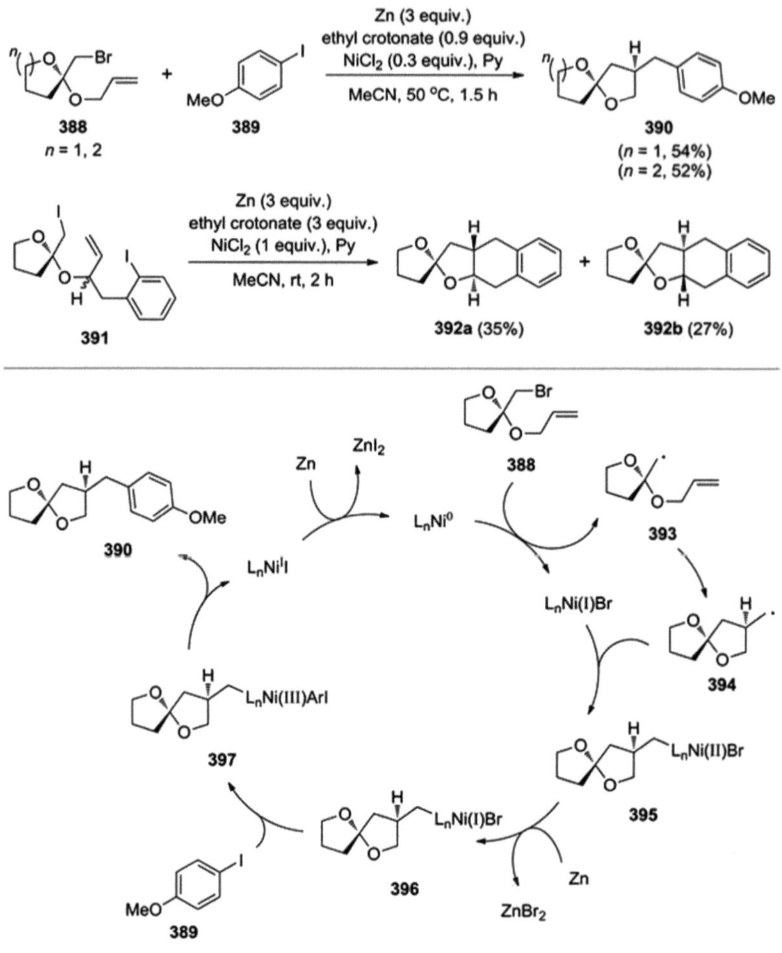

Scheme 94 Peng et al.'s Ni-catalyzed reductive cyclization approach toward spiroacetals [190]

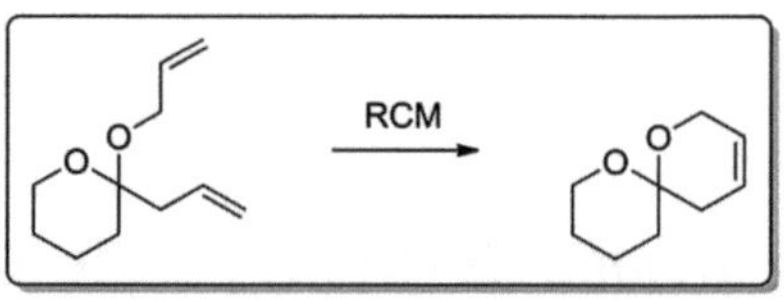

Scheme 95 Ring-closing metathesis for the synthesis of spiroacetals

Most recently, papulacandin-like spiroacetals have been accessed using this method [192] to provide a spiroacetal-diene **402**, which can then be converted to the benzannulated spiroacetal **404** via cycloaddition with electron-deficient alkynes, e.g., **404** (Scheme 97).

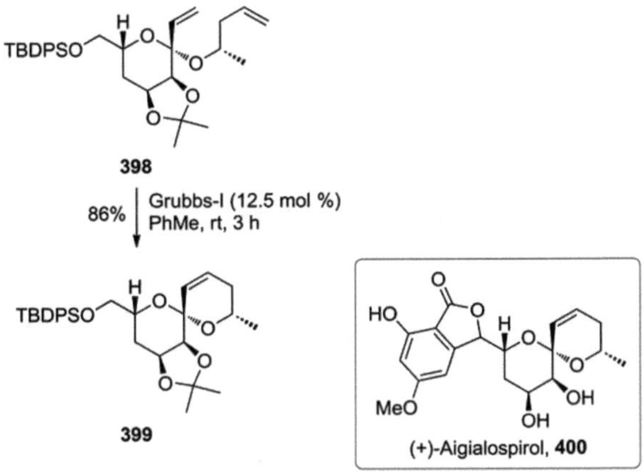

Scheme 96 RCM strategy in the synthesis of aigialospirol [191]

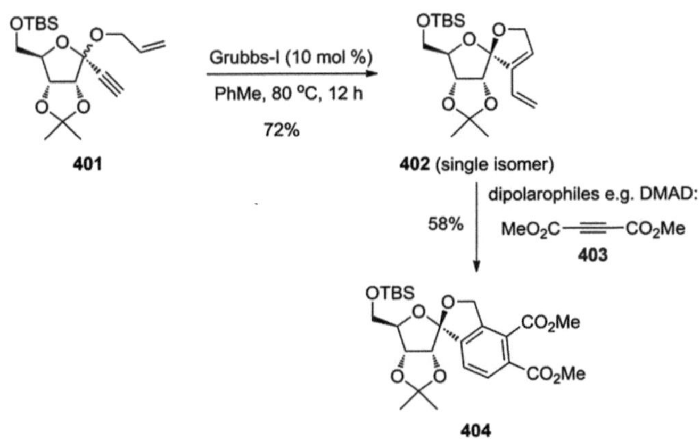

Scheme 97 RCM strategy in the synthesis of papulacandin-like spiroacetals [192]

A related, ring-rearrangement metathesis (RRM) strategy has been reported by Blanchard and coworkers [193] for the synthesis of a variety of spiroacetals (Scheme 98). Treatment of the oxabicycles **405** with second-generation Grubbs catalyst initiated the RRM cascade, affording the unsaturated 5,6-spiroacetals **409** in excellent yields. The 6,6-spiroacetal **411** could also be accessed using the oxabicyclo[3.2.1]octenone **410** as the substrate for the RRM.

The 6,5,6- and 6,6,6-bis-spiroacetals **413** and **415** could also be accessed by substituting the TBS-protected alcohol with a second allyl substituent, allowing sequential RRM–RRM reactions (Scheme 99). The 6,5,6-bis(spiroacetal) **413** was obtained in moderate yield, while the yield of the 6,6,6-bis(spiroacetal) **415** was much higher.

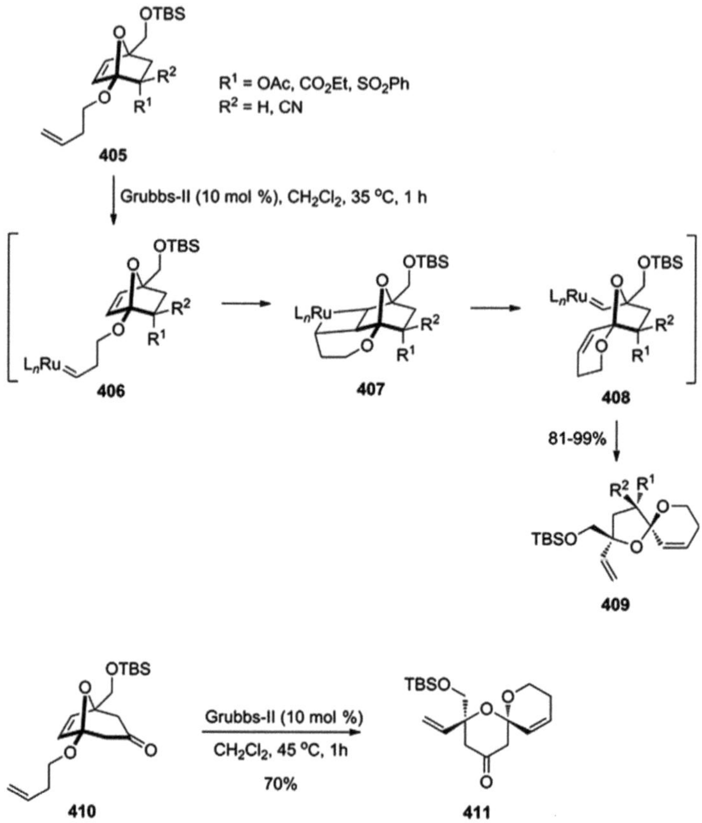

Scheme 98 Blanchard et al.'s ring-rearrangement metathesis strategy for the synthesis of spiroacetals [193]

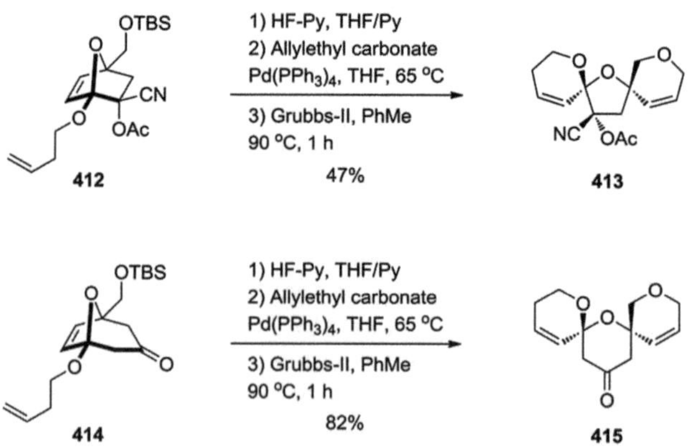

Scheme 99 Application of RRM in the synthesis of bis(spiroacetals) [193]

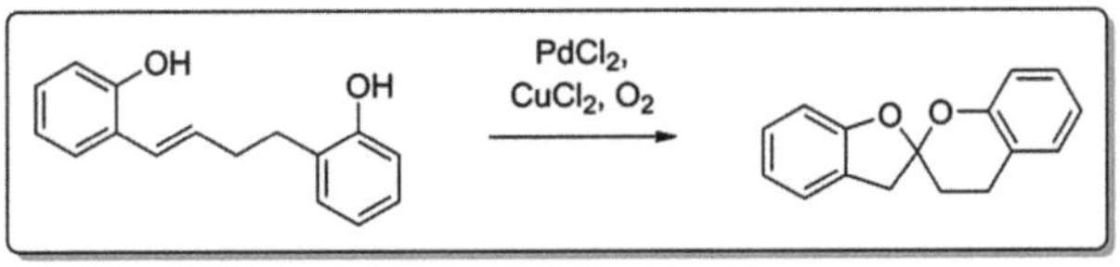

Scheme 100 Wacker-type cyclization for the synthesis of spiroacetals

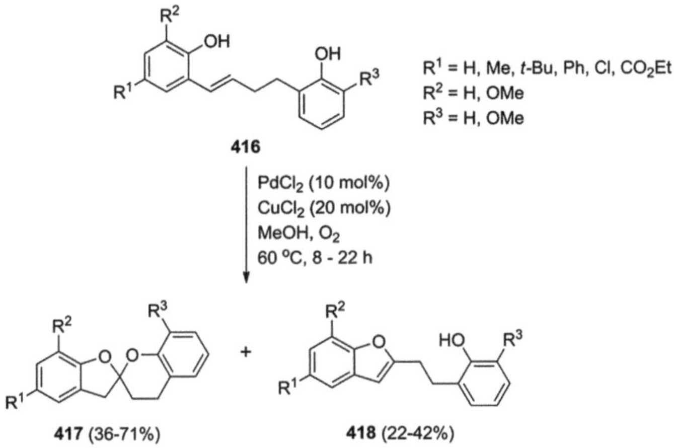

Scheme 101 Wacker-type cyclization for the synthesis of bis(benzannulated)spiroacetals [194]

2.16 Wacker-Type Cyclization

A novel Wacker-type approach toward bis(benzannulated) spiroacetals has been reported recently by Xue and coworkers [194] (Scheme 100).

In their study, a small range of simply substituted olefins **416** were subjected to Wacker conditions (cat. $PdCl_2/CuCl_2$, O_2, MeOH, 60 °C), giving mixtures of the desired bis(benzannulated) spiroacetals **417** in moderate yield, together with partially cyclized benzofuran **418** (Scheme 101).

Electron-withdrawing substituents at R^1 (i.e., Ph, Cl, CO_2Et) required longer reaction times and resulted in a higher proportion of the unwanted benzofuran byproduct. Electron-donating substituents (Me, *t*-Bu) at the same position had a beneficial effect on the spirocyclization, presumed to be due to increased nucleophilicity of the Pd–benzofuran complex.

2.17 Cyclopropane Ring-Opening Approaches Toward Spiroacetals

A recent addition to the range of ring-expansion strategies for the synthesis of spiroacetals includes ring opening of spirocyclopropanes (Scheme 102).

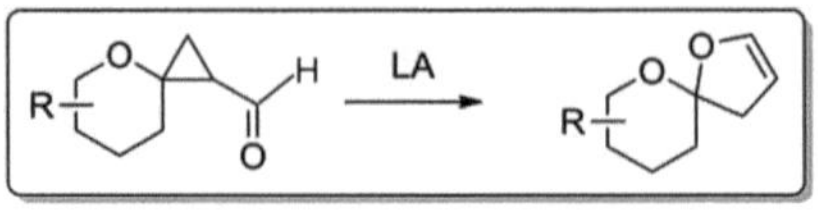

Scheme 102 Cyclopropane ring opening for the synthesis of spiroacetals

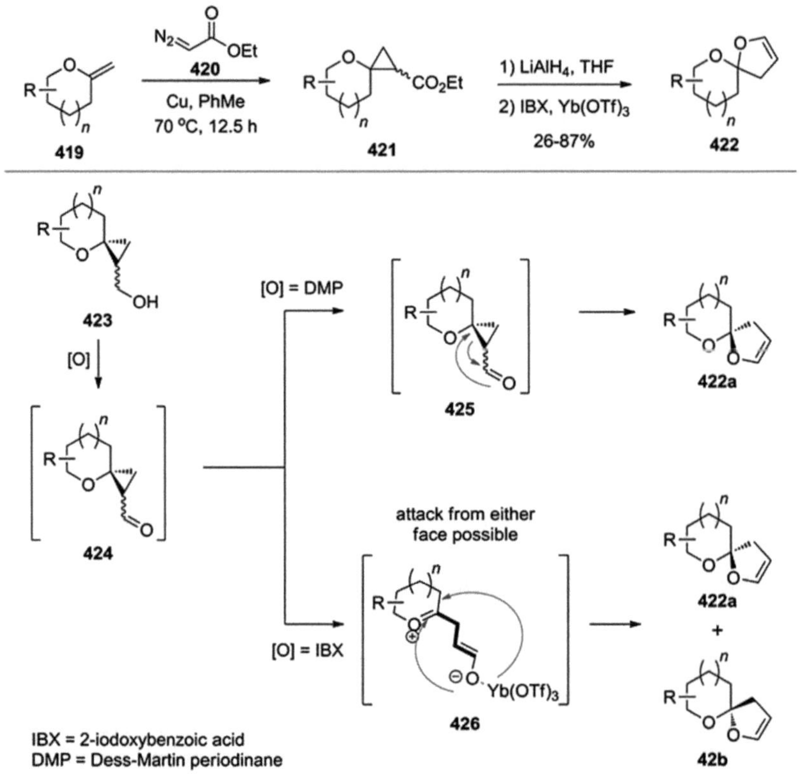

Scheme 103 Intramolecular ring opening of a cyclopropane for the synthesis of spiroacetals [195]

Werz et al. [195] have reported the synthesis of unsaturated spiroacetals **422** via intramolecular ring opening of spirocyclopropanes **421** (Scheme 103). Oxidation of the alcohol and subsequent acid-catalyzed ring opening affords the spiroacetals **422** in low to good yield.

Shair and coworkers [123] have reported a NBS-promoted intramolecular cyclopropane ring opening for the synthesis of the eastern half of cephalostatin 1 (Scheme 104).

Treatment of the cyclopropane **427** with NBS induced attack of the primary alcohol onto the cyclopropane, giving the bromomethylene spiroacetal **428** as the product. Subsequent reductive removal of the bromide and protection of the

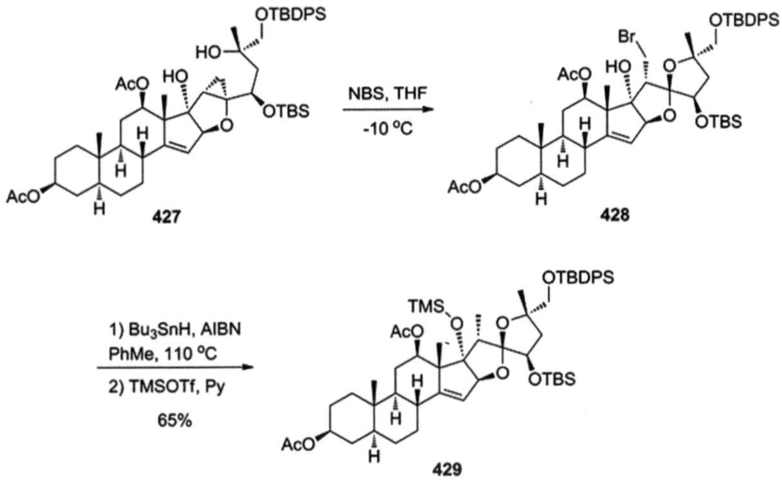

Scheme 104 Intramolecular ring opening of a cyclopropane for the synthesis of the eastern half of cephalostatin 1 [123]

remaining alcohol as a TMS-ether afforded the spiroacetal **429** in good yield over 3 steps.

2.18 Rearrangements

A variety of rearrangements have been used for the synthesis of spiroacetals. For example, Kita et al. used a double Pummerer-like rearrangement in the first total synthesis of γ-rubromycin [196], while acid-catalyzed [153, 197–200] and oxidative rearrangements [201], have also been reported. These methods can, however, be highly substrate dependent. Recent reports of rearrangements applied in the synthesis of spiroacetals are detailed below.

2.18.1 Anionic Homo-Fries Rearrangement/Lactonization

Schmalz et al. [202] recently reported a tandem anionic homo-Fries rearrangement/lactonization [203] for the synthesis of cyclo-mumbaistatin analogue **436** (Scheme 105).

The requisite anthracene **431** was obtained by reduction and methylation of the anthraquinone **430**. Halogen-metal exchange at low temperature generated the arylmagnesium species **432**, which underwent the anionic homo-Fries rearrangement and lactonization upon heating to 50 °C. The spirolactone **435** was thus obtained in good yield. Reoxidation of the anthracene with silver oxide and

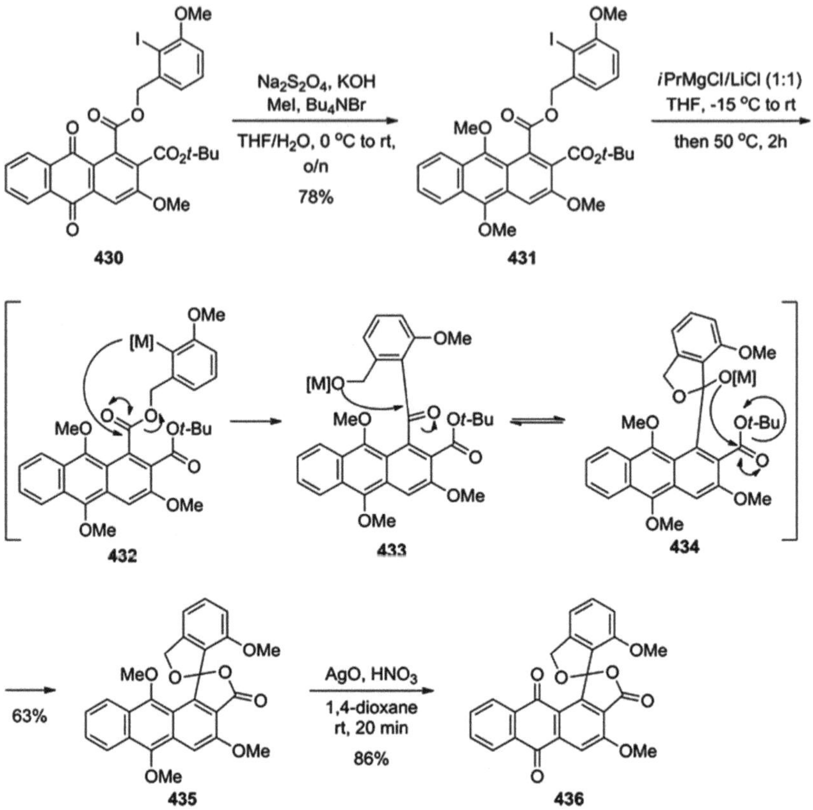

Scheme 105 Anionic homo-Fries rearrangement in the synthesis of a spiroacetal [202]

nitric acid then afforded the anthraquinone spirolactone core of cyclo-mumbaistatin **436** in high yield.

2.18.2 Ferrier Rearrangement

Smith et al. [28] have made use of an initially unanticipated Ferrier rearrangement for the synthesis of the southern hemisphere of spirastrellolide B (Scheme 106).

The spiroacetal **438** was initially targeted through a dehydrative spirocyclization strategy. However, following antiselective reduction of ketone **437**, treatment of the mixture with aq. $HClO_4$ afforded the spiroacetal **438** in moderate yield in addition to the Ferrier rearrangement byproducts **439** and **440**.

Ultimately, downstream modification of **438** toward the target molecule was unsuccessful; nevertheless the authors recognized the potential to make use of unsaturated spiroacetal **439** instead. Thus, the reaction sequence was optimized to maximize the yield of **439**, proposed to arise from dehydration of the intermediate hemiacetal **441**, followed by Ferrier rearrangement (Scheme 107). By inclusion of

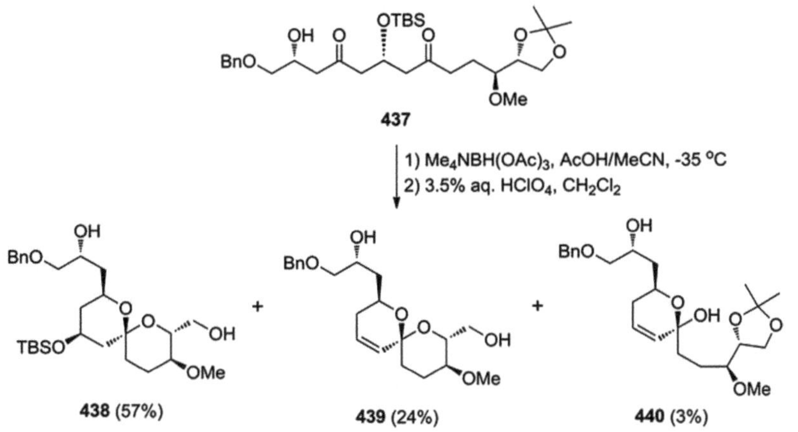

Scheme 106 Unexpected Ferrier rearrangement in the synthesis of the southern hemisphere of spirastrellolide B [28]

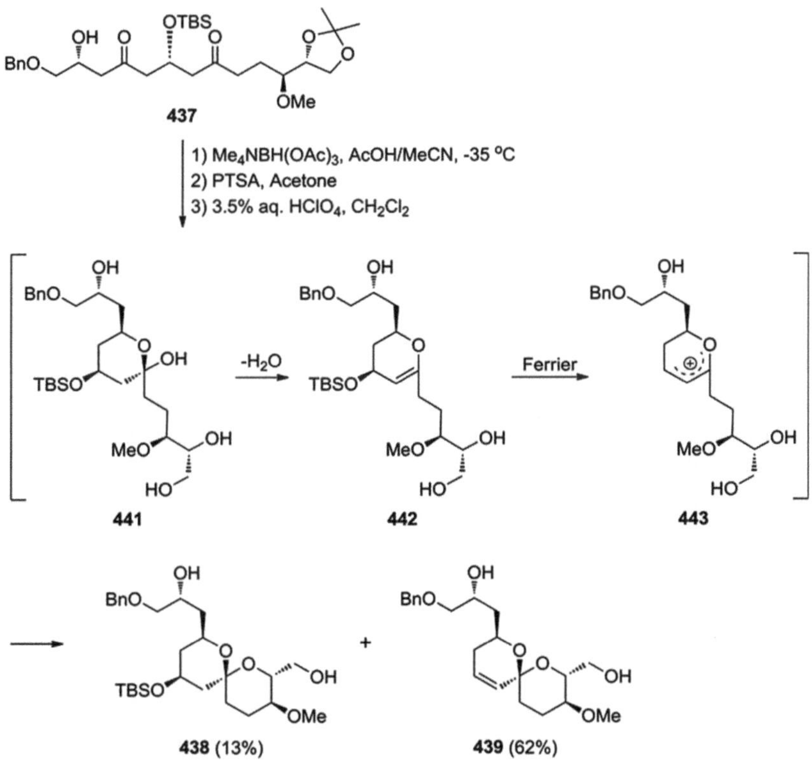

Scheme 107 Optimization of the Ferrier rearrangement in the synthesis of the southern hemisphere of spirastrellolide B [28]

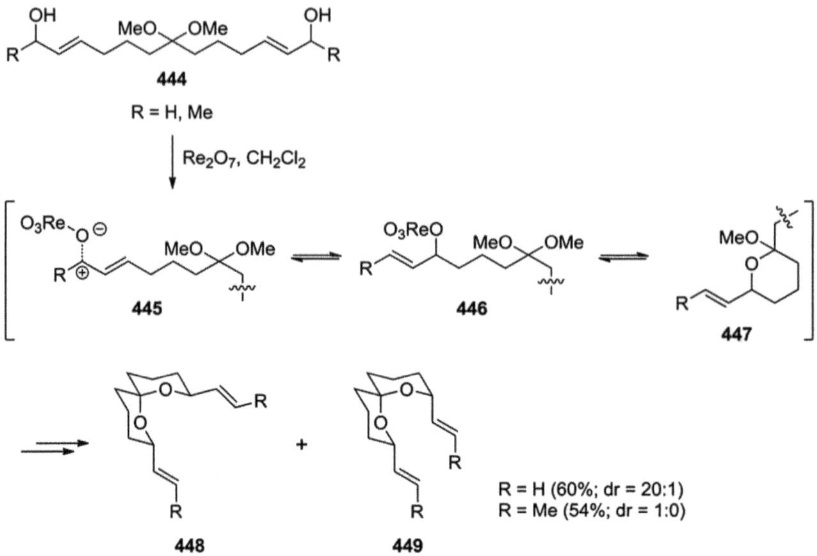

Scheme 108 Xie and Floreancig's Re(VII)-catalyzed allylic alcohol rearrangement strategy for the synthesis of spiroacetals [204]

PTSA in the reaction sequence, the yield of the unsaturated spiroacetal **443** could be improved from 24 % to 62 %.

2.18.3 Rhenium-Catalyzed Allylic Alcohol Rearrangement

Xie and Floreancig [204] have reported a tandem rhenium(VII)-catalyzed allylic alcohol rearrangement/nucleophilic addition sequence for the synthesis of spiroacetals (Scheme 108). Treatment of allylic alcohols **444** with Re_2O_7 in dichloromethane gives rise to the spiroacetals **448** and **449** in moderate yield.

Secondary alcohols were found to react more rapidly than their primary counterparts due to greater stabilization of the intermediate ion pair **445**. As each step is in equilibrium, the thermodynamically most stable spiroacetal predominates.

3 Conclusion

Spiroacetals have been found to be a common motif in complex natural products and as such are a popular synthetic target. The need to synthesize these scaffolds in a selective and efficient manner has driven efforts to develop new methods for their synthesis. A wide variety of techniques have emerged, including dehydrative methods, cycloadditions, oxidative methods, reductive couplings, ring-opening methods, radical-based processes, and multicomponent cascades. Interestingly, new methods for the selective synthesis of non-anomeric spiroacetals have included the use of chiral phosphoric acids and their derivatives in dehydrative cyclizations as well as hydrogen bonding-driven kinetic spirocyclizations.

It is expected that as the structures of new spiroacetal-containing natural products come to light, so too will new developments in the methods for their synthesis.

References

1. Mead KT, Brewer BN (2003) Curr Org Chem 7(3):227–256
2. Aho JE, Pihko PM, Rissa TK (2005) Chem Rev 105(12):4406–4440
3. Raju BR, Saikia AK (2008) Molecules 13(8):1942–2038
4. Rizzacasa MA, Pollex A (2009) Org Biomol Chem 7(6):1053–1059
5. Sperry J, Wilson ZE, Rathwell DCK et al (2010) Nat Prod Rep 27(8):1117–1137
6. Carter RG, Kuiper DL (2011) Sci Synth Stereosel Synth 2:863–914
7. Wilsdorf M, Reissig HU (2012) Angew Chem Int Ed 51(38):9486–9488
8. McLeod MC, Brimble MA, Rathwell DCK et al (2012) Pure Appl Chem 84(6):1379–1390
9. Green JC, Burnett GLIV, Pettus TRR (2012) Pure Appl Chem 84(7):1621–1631
10. Deslongchamps P, Rowan DD, Pothier N et al (1981) Can J Chem 59(7):1105–1121
11. Deslongchamps P (1983) Stereoelectronic effects in organic chemistry. Pergamon Press, New York
12. Kirby AJ (1983) The anomeric effect and related stereoelectronic effects at oxygen. Springer-Verlag Berlin, New York
13. Perron F, Albizati KF (1989) Chem Rev 89(7):1617–1661
14. Pothier N, Goldstein S, Deslongchamps P (1992) Helv Chim Acta 75(2):604–620
15. Palmes JA, Aponick A (2012) Synthesis. doi:10.1055/s-0032-1317489
16. Favre S, Vogel P, Gerber-Lemaire S (2008) Molecules 13(10):2570–2600
17. Capecchi T, de Koning CB, Michael JP (2000) J Chem Soc Perkin Trans 1(16):2681–2688
18. Ballini R, Petrini M, Rosini G (2008) Molecules 13(2):319–330
19. Doubsky J, Streinz L, Saman D et al (2004) Org Lett 6(26):4909–4911
20. Gaunt MJ, Jessiman AS, Orsini P et al (2003) Org Lett 5(25):4819–4822
21. Waters SP, Fennie MW, Kozlowski MC (2006) Tetrahedron Lett 47(30):5409–5413
22. Smith AB, Condon SM, McCauley JA (1998) Acc Chem Res 31(1):35–46
23. Enders D, Lenzen A (2003) Synlett (14):2185–2187
24. Smith AB, Adams CM (2004) Acc Chem Res 37(6):365–377
25. Smith AB, Wuest WM (2008) Chem Commun (45):5883–5895
26. Smith AB, Sfouggatakis C, Risatti CA et al (2009) Tetrahedron 65(33):6489–6509
27. Smith AB, Lin Q, Doughty VA et al (2009) Tetrahedron 65(33):6470–6488
28. Smith AB, Smits H, Kim DS (2010) Tetrahedron 66(33):6597–6605
29. Ollivier A, Sinibaldi ME, Toupet L et al (2010) Tetrahedron Lett 51(31):4147–4149
30. Goubert M, Canet I, Sinibaldi ME (2009) Tetrahedron 65(21):4182–4189
31. Yadav JS, Chetia L (2007) Org Lett 9(22):4587–4589
32. Tursun A, Canet I, Aboab B et al (2005) Tetrahedron Lett 46(13):2291–2294
33. Ardes-Guisot N, Ouled-Lahoucine B, Canet I et al (2007) Tetrahedron Lett 48 (48):8511–8513
34. Melendez J, Alonso F, Yus M (2006) Tetrahedron Lett 47(7):1187–1191
35. Ramu E, Rao BV (2008) Tetrahedron Asymmetry 19(15):1820–1823
36. De Greef M, Zard SZ (2007) Org Lett 9(9):1773–1776
37. Favre S, Gerber-Lemaire S, Vogel P (2007) Org Lett 9(24):5107–5110

38. Dias LC, Correia VG, Finelli FG (2007) Tetrahedron Lett 48(43):7683–7686
39. Crimmins MT, Azman AM (2012) Synlett 23(10):1489–1492
40. Allais F, Cossy J (2006) Org Lett 8(17):3655–3657
41. Canterbury DP, Micalizio GC (2010) J Am Chem Soc 132(22):7602–7604
42. Chen JLY, Brimble MA (2011) J Org Chem 76(22):9417–9428
43. Marjanovic J, Kozmin SA (2007) Angew Chem Int Ed 46(46):8854–8857
44. Crimmins MT, O'Bryan EA (2010) Org Lett 12(19):4416–4419
45. Lynch JE, Zanatta SD, White JM et al (2011) Chem Eur J 17(1):297–304
46. Fernandes RA, Ingle AB (2010) Synlett (1):158–160
47. Brimble MA, Finch OC, Heapy AM et al (2011) Tetrahedron 67(5):995–1001
48. Paterson I, Findlay AD, Noti C (2009) Chem Asian J 4(4):594–611
49. Guérinot A, Lepesqueux G, Sable S et al (2010) J Org Chem 75(15):5151–5163
50. Dias LC, Salles AG (2009) J Org Chem 74(15):5584–5589
51. Paley RS, Laupheimer MC, Erskine NAK et al (2011) Org Lett 13(1):58–61
52. Haym I, Brimble MA (2009) Synlett (14):2315–2319
53. Wilson ZE, Hubert JG, Brimble MA (2011) Eur J Org Chem 2011(20–21):3938–3945
54. Sudhakar G, Kadam VD, Bayya S et al (2011) Org Lett 13(20):5452–5455
55. Yuen TY, Yang SH, Brimble MA (2011) Angew Chem Int Ed 50(36):8350–8353
56. Wang X, Han Z, Wang Z et al (2012) Angew Chem Int Ed 51(4):936–940
57. Wang F, Qu M, Lu X et al (2012) Chem Commun 48(50):6259–6261
58. Waters SP, Fennie MW, Kozlowski MC (2006) Org Lett 8(15):3243–3246
59. Brasholz M, Sörgel S, Azap C et al (2007) Eur J Org Chem (23):3801–3814
60. Venkatesh C, Reißig HU (2008) Synthesis (22):3605–3614
61. Rathwell DCK, Yang SH, Tsang KY et al (2009) Angew Chem Int Ed 48(43):7996–8000
62. Lowell AN, Fennie MW, Kozlowski MC (2011) J Org Chem 76(16):6488–6502
63. Awasaguchi KI, Miyazawa M, Uoya I et al (2010) Synlett 2010(16):2392–2396
64. Borrero NV, Aponick A (2012) J Org Chem 77(19):8410–8416
65. Guérinot A, Serra-Muns A, Gnamm C et al (2010) Org Lett 12(8):1808–1811
66. Commandeur M, Commandeur C, Cossy J (2011) Org Lett 13(22):6018–6021
67. Henryon V, Férézou JP (2009) Synthesis (20):3477–3487
68. Jung ME, Salehi-Rad R (2009) Angew Chem Int Ed 48(46):8766–8769
69. Jung ME, Chaumontet M, Salehi-Rad R (2010) Org Lett 12(12):2872–2875
70. Yadav JS, Rajender V, Rao YG (2010) Org Lett 12(2):348–350
71. McLeod MC, Wilson ZE, Brimble MA (2012) J Org Chem 77(1):400–416
72. Fuwa H, Sasaki M (2008) Org Lett 10(12):2549–2552
73. Fuwa H, Noji S, Sasaki S (2010) Org Lett 12(22):5354–5357
74. Butkevich AN, Corbu A, Meerpoel L et al (2012) Org Lett 14(19):4998–5001
75. Sun Z, Winschel GA, Borovika A et al (2012) J Am Chem Soc 134(19):8074–8077
76. Čorić I, List B (2012) Nature 483(7389):315–319
77. Danishefsky SJ, Pearson WH (1983) J Org Chem 48(21):3865–3866
78. Wurst JM, Verano AL, Tan DS (2012) Org Lett 14(17):4442–4445
79. Tomas L, Boije af Gennaes G, Hiebel MA et al (2012) Chem Eur J 18(24):7452–7466
80. Gueyrard D, Haddoub R, Salem A et al (2005) Synlett 2005(03):520–522
81. Sridharan V, Vologdin N, Virolleaud MA et al (2011) Synthesis (13):2085–2090
82. Li X, Li J, Mootoo DR (2007) Org Lett 9(21):4303–4306
83. Tony KA, Li X, Dabideen D et al (2008) Org Biomol Chem 6(7):1165–1169
84. Tony KA, Dabideen D, Li J et al (2009) J Org Chem 74(20):7774–7780
85. Utimoto K (1983) Pure Appl Chem 55(11):1845–1852
86. Fukuchi I, Sanda F, Endo T (2001) Macromolecules 34(13):4296–4298
87. Trost BM, Horne DB, Woltering MJ (2003) Angew Chem Int Ed 42(48):5987–5990
88. Trost BM, Weiss AH (2007) Angew Chem Int Ed 46(40):7664–7666
89. Ramana CV, Suryawanshi SB, Gonnade RG (2009) J Org Chem 74(7):2842–2845
90. Alcaide B, Almendros P, Alonso JM (2011) Org Biomol Chem 9(12):4405–4416

91. Brenzovich WE (2012) Angew Chem Int Ed 51(36):8933–8935
92. Antoniotti S, Genin E, Michelet V et al (2005) J Am Chem Soc 127(28):9976–9977
93. Liu B, De Brabander JK (2006) Org Lett 8(21):4907–4910
94. Sherry BD, Maus L, Laforteza BN et al (2006) J Am Chem Soc 128(25):8132–8133
95. Tlais SF, Dudley GB (2012) Beilstein J Org Chem 8:1287–1292
96. Fang C, Pang Y, Forsyth CJ (2010) Org Lett 12(20):4528–4531
97. Trost BM, O'Boyle BM, Hund D (2009) J Am Chem Soc 131(41):15061–15074
98. Genin E, Antoniotti S, Michelet V et al (2005) Angew Chem Int Ed 44(31):4949–4953
99. Messerle BA, Vuong KQ (2006) Pure Appl Chem 78(2):385–390
100. Messerle BA, Vuong KQ (2007) Organometallics 26(12):3031–3040
101. Selvaratnam S, Ho JHH, Huleatt PB et al (2009) Tetrahedron Lett 50(10):1125–1127
102. Ho JHH, Hodgson R, Wagler J et al (2010) Dalton Trans 39(17):4062–4069
103. Ho JHH, Choy SWS, MacGregor SA et al (2011) Organometallics 30(21):5978–5984
104. Wang X, Paxton TJ, Li N et al (2012) Org Lett 14(15):3998–4001
105. Ravindar K, Reddy MS, Lindqvist L et al (2010) Org Lett 12(19):4420–4423
106. Aponick A, Li CY, Palmes JA (2009) Org Lett 11(1):121–124
107. Tlais SF, Dudley GB (2011) Beilstein J Org Chem 7:570–577
108. Nishizawa M, Imagawa H, Yamamoto H (2010) Org Biomol Chem 8(3):511–521
109. Ravindar K, Reddy MS, Deslongchamps P (2011) Org Lett 13(12):3178–3181
110. Volchkov I, Sharma K, Cho EJ et al (2011) Chem Asian J 6(8):1961–1966
111. Zhang Y, Xue J, Xin Z et al (2008) Synlett (6):940–944
112. Li X, Chianese AR, Vogel T et al (2005) Org Lett 7(24):5437–5440
113. Bartlett PA, Mori I, Bose JA (1989) J Org Chem 54(13):3236–3239
114. Kometani T, Takeuchi Y, Yoshii E (1983) J Org Chem 48(14):2311–2314
115. Guo J, Duffy KJ, Stevens KL et al (1998) Angew Chem Int Ed 37(1–2):187–190
116. Qin D, Ren RX, Siu T et al (2001) Angew Chem Int Ed 40(24):4709–4713
117. Wei WG, Zhang YX, Yao ZJ (2005) Tetrahedron 61(50):11882–11886
118. Elsley DA, MacLeod D, Miller JA et al (1992) Tetrahedron Lett 33(3):409–412
119. Uchiyama M, Oka M, Harai S et al (2001) Tetrahedron Lett 42(10):1931–1934
120. Conway JC, Quayle P, Regan AC et al (2006) Heterocycles 67(1):85–88
121. Conway JC, Urch CJ, Quayle P et al (2006) Synlett (5):776–780
122. Tiecco M, Testaferri L, Bagnoli L et al (2006) Tetrahedron Asymmetry 17(19):2768–2774
123. Fortner KC, Kato D, Tanaka Y et al (2010) J Am Chem Soc 132(1):275–280
124. Sun H, Abbott JR, Roush WR (2011) Org Lett 13(10):2734–2737
125. Crimmins MT, O'Mahony R (1990) J Org Chem 55(23):5894–5900
126. Jackson KL, Henderson JA, Motoyoshi H et al (2009) Angew Chem Int Ed 48 (13):2346–2350
127. Choi PJ, Rathwell DCK, Brimble MA (2009) Tetrahedron Lett 50(26):3245–3248
128. Snaddon TN, Buchgraber P, Schulthoff S et al (2010) Chem Eur J 16(40):12133–12140
129. Wilson ZE, Brimble MA (2010) Org Biomol Chem 8(6):1284–1286
130. McLeod MC, Wilson ZE, Brimble MA (2011) Org Lett 13(19):5382–5385
131. Davy JA, Moreau B, Wulff JE (2012) Synthesis 44(12):1854–1862
132. Rauhala V, Nättinen K, Rissanen K et al (2005) Eur J Org Chem 2005(19):4119–4126
133. Habrant D, Stewart AJW, Koskinen AMP (2009) Tetrahedron 65(38):7927–7934
134. Habrant D, Koskinen AMP (2010) Org Biomol Chem 8(19):4364–4373
135. Trygstad TM, Pang Y, Forsyth CJ (2009) J Org Chem 74(2):910–913
136. Sharma GVM, Srikanth G, Reddy PP (2012) Org Biomol Chem 10(40):8119–8124
137. Potuzak JS, Moilanen SB, Tan DS (2005) J Am Chem Soc 127(40):13796–13797
138. Moilanen SB, Potuzak JS, Tan DS (2006) J Am Chem Soc 128(6):1792–1793
139. Liu G, Wurst JM, Tan DS (2009) Org Lett 11(16):3670–3673
140. Wurst JM, Liu G, Tan DS (2011) J Am Chem Soc 133(20):7916–7925
141. Robertson J, Chovatia PT, Fowler TG et al (2010) Org Biomol Chem 8(1):226–233
142. Denmark SE, Regens CS, Kobayashi T (2007) J Am Chem Soc 129(10):2774–2776

143. Denmark SE, Kobayashi T, Regens CS (2010) Tetrahedron 66(26):4745–4759
144. van der Kaaden M, Breukink E, Pieters RJ (2012) Beilstein J Org Chem 8:732–737
145. McDonald FE, Zhu HYH, Holmquist CR (1995) J Am Chem Soc 117(24):6605–6606
146. Yamamoto Y, Hashimoto T, Hattori K et al (2006) Org Lett 8(16):3565–3568
147. Yamamoto Y, Yamashita K, Hotta T et al (2007) Chem Asian J 2(11):1388–1399
148. Leeuwenburgh MA, Appeldoorn CM, van Hooft PAV et al (2000) Eur J Org Chem 2000 (6):873–877
149. Ghosh SK, Hsung RP, Liu J (2005) J Am Chem Soc 127(23):8260–8261
150. El Sous M, Ganame D, Tregloan P et al (2010) Synthesis (23):3954–3966
151. El Sous M, Ganame D, Tregloan P et al (2004) Org Lett 6(17):3001–3004
152. Cuzzupe AN, Hutton CA, Lilly MJ et al (2001) J Org Chem 66(7):2382–2393
153. McRae KJ, Rizzacasa MA (1997) J Org Chem 62(5):1196–1197
154. Marsini MA, Huang Y, Lindsey CC et al (2008) Org Lett 10(7):1477–1480
155. Van De Water RW, Magdziak DJ, Chau JN et al (2000) J Am Chem Soc 122(27):6502–6503
156. Bray CD (2008) Org Biomol Chem 6(15):2815–2819
157. Bray CD (2008) Synlett (16):2500–2502
158. Huang Y, Pettus TRR (2008) Synlett (9):1353–1356
159. Wenderski TA, Marsini MA, Pettus TRR (2011) Org Lett 13(1):118–121
160. Barluenga J, Mendoza A, Rodríguez F et al (2009) Angew Chem Int Ed 48(9):1644–1647
161. Fañanás FJ, Mendoza A, Arto T et al (2012) Angew Chem Int Ed 51(20):4930–4933
162. Bender CF, Yoshimoto FK, Paradise CL et al (2009) J Am Chem Soc 131(32):11350–11352
163. Bantreil X, Vaxelaire C, Godet T et al (2011) Org Biomol Chem 9(13):4831–4841
164. Zhou G, Zheng D, Da S et al (2006) Tetrahedron Lett 47(20):3349–3352
165. Zhou G, Zhu J, Xie Z et al (2008) Org Lett 10(5):721–724
166. Li X, Xue J, Huang C et al (2012) Chem Asian J 7(5):903–906
167. Roy SC, Mandal PK (1996) Tetrahedron 52(6):2193–2198
168. Lindsey CC, Wu KL, Pettus TRR (2006) Org Lett 8(11):2365–2367
169. Wu KL, Wilkinson S, Reich NO et al (2007) Org Lett 9(26):5537–5540
170. Wu KL, Mercado EV, Pettus TRR (2011) J Am Chem Soc 133(16):6114–6117
171. Merino P, Tejero T, Delso JI et al (2007) Curr Org Chem 11(12):1076–1091
172. Yang Z, Tang P, Gauuan JF et al (2009) J Org Chem 74(24):9546–9549
173. Thirupathi B, Reddy PP, Mohapatra DK (2011) J Org Chem 76(23):9835–9840
174. Pavlakos E, Georgiou T, Tofi M et al (2009) Org Lett 11(20):4556–4559
175. Sperry J, Liu YC, Brimble MA (2010) Org Biomol Chem 8(1):29–38
176. Ravindar K, Reddy MS, Lindqvist L et al (2011) J Org Chem 76(5):1269–1284
177. Betancor C, Freire R, Pérez-Martin I et al (2005) Tetrahedron 61(11):2803–2814
178. Phillips ST, Shair MD (2007) J Am Chem Soc 129(20):6589–6598
179. Liu YC, Sperry J, Rathwell DCK et al (2009) Synlett (5):793–797
180. Jay-Smith M, Furkert DP, Sperry J et al (2011) Synlett (10):1395–1398
181. Sperry J, Liu YC, Wilson ZE et al (2011) Synthesis (9):1383–1398
182. Asari F, Kusumi T, Zheng GZ et al (1990) Chem Lett 19(10):1885–1888
183. Zou Y, Deiters A (2010) J Org Chem 75(15):5355–5358
184. Chorley DF, Chen JLY, Furkert DP et al (2012) Synlett 23(1):128–130
185. Martín A, Salazar JA, Suárez E (1996) J Org Chem 61(12):3999–4006
186. Dorta RL, Martín A, Salazar JA et al (1998) J Org Chem 63(7):2251–2261
187. Wei W, Li L, Lin X et al (2012) Org Biomol Chem 10(17):3494–3499
188. Wei L, Xue J, Liu H et al (2012) Org Lett 14(20):5302–5305
189. Wei W, Wang Y, Yin J et al (2012) Org Lett 14(4):1158–1161
190. Yan CS, Peng Y, Xu XB et al (2012) Chem Eur J 18(19):6039–6048
191. Yang JH, Liu J, Hsung RP (2008) Org Lett 10(12):2525–2528
192. Subrahmanyam AV, Palanichamy K, Kaliappan KP (2010) Chem Eur J 16(28):8545–8556
193. Mandel J, Dubois N, Neuburger M et al (2011) Chem Commun 47(37):10284–10286
194. Xin Z, Zhang Y, Tao H et al (2011) Synlett (11):1579–1584

195. Brand C, Rauch G, Zanoni M et al (2009) J Org Chem 74(22):8779–8786
196. Akai S, Kakiguchi K, Nakamura Y et al (2007) Angew Chem Int Ed 46(39):7458–7461
197. van Hooft PAV, van Swieten PF, van der Marel GA et al (2001) Synlett (02):269–271
198. Keller VA, Martinelli JR, Strieter ER et al (2002) Org Lett 4(3):467–470
199. Buffet MF, Dixon DJ, Ley SV et al (2004) Org Biomol Chem 2(8):1145–1154
200. Winkler JD, Mikochik PJ (2004) Org Lett 6(21):3735–3737
201. Waller DL, Stephenson CRJ, Wipf P (2007) Org Biomol Chem 5(1):58–60
202. Neufeind S, Hülsken N, Neudörfl JM et al (2011) Chem Eur J 17(9):2633–2641
203. Miller JA (1987) J Org Chem 52(2):322–323
204. Xie Y, Floreancig PE (2011) Chem Sci 2(12):2423–2427

Index

J. Cossy (ed.), *Synthesis of Saturated Oxygenated Heterocycles I*, Topics in Heterocyclic Chemistry 35, DOI 10.1007/978-3-642-41473-2,

MIX
Papier aus verantwortungsvollen Quellen
Paper from responsible sources
FSC® C105338

If you have any concerns about our products,
you can contact us on
ProductSafety@springernature.com

In case Publisher is established outside the EU,
the EU authorized representative is:
Springer Nature Customer Service Center GmbH
Europaplatz 3, 69115 Heidelberg, Germany

Printed by Libri Plureos GmbH
in Hamburg, Germany